ISNM 71:
International Series of Numerical Mathematics
Internationale Schriftenreihe zur Numerischen Mathematik
Série internationale d'Analyse numérique
Vol. 71

Birkhäuser Verlag
Basel · Boston · Stuttgart

General Inequalities 4

In memoriam Edwin F. Beckenbach

4th International Conference on General Inequalities, Oberwolfach, May 8–14, 1983

Edited by
W. Walter

1984

Birkhäuser Verlag
Basel · Boston · Stuttgart

Editor

Prof. Dr. Wolfgang Walter
Universität Karlsruhe
Mathematisches Institut I
Kaiserstrasse 12
D–7500 Karlsruhe 1 (FRG)

CIP-Kurztitelaufnahme der Deutschen Bibliothek

General inequalities / . . . Internat. Conference
on General Inequalities. – Basel ; Boston ;
Stuttgart : Birkhäuser
1 mit d. Erscheinungsorten Basel, Stuttgart. –
1 mit Parallelt. : Allgemeine Ungleichungen
NE: Internationale Tagung über Allgemeine
Ungleichungen; PT
4. Oberwolfach, May 8 – 14, 1983 : in memoriam
Edwin F. Beckenbach. – 1984.
(International series of numerical mathematics ;
Vol. 71)
ISBN 3–7643–1644–6
NE: Beckenbach, Edwin F.: Festschrift; GT

Printed in Germany
ISBN 3–7643–1644–6

FOREWORD

The Fourth International Conference on General Inequalities was held from May 8 to May 14, 1983, at the Mathematisches Forschungsinstitut Oberwolfach (Black Forest, Germany). The organizational committee consisted of L. Losonczi (Lagos and Debrecen) and W. Walter (Karlsruhe). Dr. A. Kovačec served extremely well as a secretary of the conference.

The meeting was attended by 42 participants from 17 countries. In the opening address, W. Walter had to report on the unexpected death of E.F. Beckenbach. He died of a stroke in September 1982; only a few days earlier he had received the award for Distinguished Service to Mathematics from the Mathematical Association of America.

Beckenbach was one of the founding fathers of the General Inequalities conferences. He served with energy and devotion as an organizer and as editor of the proceedings of those conferences. He was also engaged in the preparations for the present conference, which the participants decided should be held in memoriam Edwin F. Beckenbach. In a brief memorial lecture, M. Goldberg gave a survey of Beckenbach's mathematical activities and his services to the mathematical community.

Inequalities play a significant role in many branches of mathematics. Correspondingly, the participants represented many different fields among which classical inequalities still provided a steady source of new developments. Lectures also included differential and functional inequalities, bounds for eigenvalues, inequalities in functional analysis, convexity and its generalizations, inequalities in number theory and probability theory, as well as mathematical programming and economics.

As in earlier conferences, the problems and remarks sessions produced a vivid exchange of results, methods and hypotheses.

The participants experienced anew the creative, congenial and stimulating atmosphere at the Institute.

The conference was closed by L. Losonczi, who expressed the thanks of the participants for the excellent working conditions in the Institute and for the hospitality of its leaders and staff.

L. Losonczi

W. Walter

PARTICIPANTS

R.P. AGARWAL, National University of Singapore, Kent Ridge, Singapore

C. ALSINA, Universitat Politècnica de Barcelona, Barcelona, Spain

D. BRYDAK, Wyższa Szkoła Pedagogiczna, Kraków, Poland

B. CHOCZEWSKI, University of Mining and Metallurgy, Kraków, Poland

A. CLAUSING, Westfälische Wilhelms-Universität, Münster, West Germany

W. EICHHORN, Universität Karlsruhe, Karlsruhe, West Germany

W.N. EVERITT, University of Birmingham, Birmingham, England

F. FEHÉR, Rheinisch-Westfälische Technische Hochschule Aachen, Aachen, West Germany

I. FENYÖ, Polytechnical University of Budapest, Budapest, Hungary

C.H. FITZGERALD, University of California at San Diego, San Diego, California, USA

M. GOLDBERG, Technion-Israel Institute of Technology, Haifa, Israel

W. HAUSSMANN, Gesamthochschule Duisburg, Duisburg, West Germany

F. HUCKEMANN, Technische Universität Berlin, Berlin, West Germany

H.-H. KAIRIES, Technische Universität Clausthal, Clausthal-Zellerfeld, West Germany

H. KÖNIG, Universität des Saarlandes, Saarbrücken, West Germany

H. KÖNIG, Universität Kiel, Kiel, West Germany

A. KOVAČEC, An der Niederhaid 21, Wien, Austria

M. KWAPISZ, ul. 23 Marca 91c, 41, Sopot, Poland

V. LAKSHMIKANTHAM, University of Texas at Arlington, Arlington, Texas, USA

L. LOSONCZI, L. Kossuth University, Debrecen, Hungary

E.R. LOVE, University of Melbourne, Parkville, Victoria 3052, Australia

R.N. MOHAPATRA, American University of Beirut, Beirut, Lebanon

R.J. NESSEL, Rheinisch-Westfälische Technische Hochschule Aachen, Aachen, West Germany

Z. PÁLES, Kossuth Lajos University, Debrecen, Hungary

J. RÄTZ, Universität Bern, Bern, Switzerland

D.K. ROSS, La Trobe University, Bundoora, Victoria 3083, Australia

D.C. RUSSELL, York University, Downsview, Ontario, Canada

PARTICIPANTS (Continued)

B. SAFFARI, Université de Paris-Orsay, Orsay, France

S. SCHAIBLE, University of Alberta, Edmonton, Canada

W. SCHEMPP, Universität Siegen, Siegen, West Germany

A. SKLAR, Illinois Institute of Technology, Chicago, Illinois, USA

B. SMITH, California Institute of Technology, Pasadena, California, USA

R.P. SPERB, Eidgenössische Technische Hochschule Zürich, Zürich, Switzerland

R. STENS, Rheinisch-Westfälische Technische Hochschule Aachen, Aachen, West Germany

G. TALENTI, Istituto Matematico dell'Università, Firenze, Italy

E. TURDZA, Koniewa 59/70, Kraków, Poland

P.M. VASIĆ, Faculty of Electrotechnics, Beograd, Yugoslavia

P. VOLKMANN, Universität Karlsruhe, Karlsruhe, West Germany

B.J. WALLACE, La Trobe University, Bundoora, Victoria 3083, Australia

W. WALTER, Universität Karlsruhe, Karlsruhe, West Germany

CH.-L. WANG, University of Regina, Regina, Saskatchewan, Canada

K. ZELLER, Universität Tübingen, Tübingen, West Germany

SCIENTIFIC PROGRAM OF THE CONFERENCE

Monday, May 9

Opening of the conference — W. WALTER

Early morning session — Chairman: I. FENYÖ

W.N. EVERITT: Hardy-Littlewood integral inequalities

C.H. FITZGERALD: Opial-type inequalities that involve higher order derivatives or differences

Late morning session — Chairman: I. FENYÖ

E.R. LOVE: Links between some generilizations of Hardy's integral inequality

Problems and remarks

Early afternoon session — Chairman: HEINZ KÖNIG

F. FEHÉR: A weak-type inequality and convergence

R.J. NESSEL: Marchaud-type inequalities

Late afternoon session — Chairman: R.J. NESSEL

R.N. MOHAPATRA: Inequalities related to sequence-spaces [p,q]

B. SMITH: Strong pointwise convergence

Problems and remarks

* * * * *

Tuesday, May 10

Early morning session — Chairman: E.R. LOVE

V. LAKSHMIKANTHAM: Differential inequalities at resonance

P. VOLKMANN: Konvergenz der sukzessiven Approximation für Systeme gewöhnlicher Differentialgleichungen

R.P. AGARWAL: Difference calculus with applications to difference equations

Late morning session — Chairman: C. ALSINA

D. BRYDAK: Generalized convex functions and differential inequalities

S. SCHAIBLE	An application of Farkas' lemma to non-convex optimization theory

Early afternoon session Chairman: B. SAFFARI

D.K. ROSS:	Inequalities for ratios of integrals
Zs. PÁLES:	Inequalities for comparison means

Late afternoon session Chairman: D.C. RUSSELL

B. SAFFARI:	On the best constant in a remarkable inequality of Delange
P.M. VASIĆ:	The Jensen-Steffensen inequality as a function of the index set

Problems and remarks

* * * * *

Wednesday, May 11

M. GOLDBERG:	In memoriam Edwin F. Beckenbach

Early morning session Chairman: B. CHOCZEWSKI

A. SKLAR:	Extension of functions satisfying certain systems of inequalities
C. ALSINA:	Schur-concave t-norms and triangle functions
J. RÄTZ:	On unilaterally bounded orthogonally additive mappings

Late morning session Chairman: J. RÄTZ

B. CHOCZEWSKI:	Stability of some iterative functional equations
E. TURDZA:	Stability of an iterative linear equation
M. KWAPISZ:	Functional inequalities and existence results

Afternoon excursion and discussion

* * * * *

Thursday, May 12

Early morning session Chairman: W.N. EVERITT

G. TALENTI:	Estimates of eigenvalues of Sturm-Liouville problems

HERMANN KÖNIG:	Some inequalities for the eigenvalues of a compact operator
R. SPERB:	Inequalities in elliptic problems derived from a maximum principle

Late morning session — Chairman: A. SKLAR

HEINZ KÖNIG:	A Dini-type theorem in superconvex analysis
K. ZELLER:	Positivity in summability

Early afternoon session — Chairman: D.K. ROSS

I. FENYÖ:	On an integral inequality
L. LOSONCZI:	Inequalities of Young type

Late afternoon session — Chairman: P.M. VASIĆ

D.C. RUSSELL:	Remark on an inequality of N. Ozeki
A. KOVAĆEC:	On aspects of the matrix method for algebraic inequalities

Problems and remarks

* * * * *

Friday, May 13

Early morning session — Chairman: K. ZELLER

W. EICHHORN:	Inequalities in the theory of economic inequality
C. WANG:	Inequalities and mathematical programming II
M. GOLDBERG:	New inequalities for ℓ_p norms

Late morning session — Chairman: W. EICHHORN

R.J. WALLACE:	Optimal strategies for locating zeroes of derivatives
R. STENS:	Error estimates for sampling approximation

Afternoon session — Chairman: V. LAKSHMIKANTHAN

W. SCHEMPP:	Über eine Ungleichung der Radarortung
A. CLAUSING:	A t-entropy inequality
H.-H. KAIRIES:	An inequality for Krull solutions

Problems and remarks

Closing of the conference — L. LOSONCZI

PREFACE

The conferences on General Inequalities are starting to develop a tradition. The fourth conference, henceforth referred to as "GI 4", was held at the Mathematisches Forschungsinstitut Oberwolfach/Germany from May 9 to May 14, 1983. These Proceedings contain most of the research articles which were presented at GI 4. There are also a few contributions by authors who were unable to attend the conference. Inequalities play a significant role in many branches of mathematics, and the articles of this volume reflect the broad range of results, methods and applications, which comprise the field of inequalities. For this reason, the editor has divided the material in several chapters. In keeping with tradition, the morning and afternoon sessions were closed with a problems and remarks section. The last chapter gives an account of those activities.

The drawings in the Proceedings are also part of the tradition. In the earlier volumes the artist was Mrs. Irmgard Süss. This time they come from the hand of Mrs. Joy Russell. The editor is grateful for her beautiful drawings from Oberwolfach and its surroundings which greatly enhance the appearance of this book.

Professor Edwin F. Beckenbach, one of the inaugurators of the GI conferences, died unexpectedly of a stroke in September 1982. These conferences and their participants owe him a great debt. He was chairman of all the previous conferences, and he edited their Proceedings with unusual care, expertise and dedication. He was also engaged in the early preparation of GI 4. The conference was held

in memoriam Edwin F. Beckenbach .

Professor Goldberg, Ed's colleague at UCLA for many years, gave a memorial lecture and wrote the memorial article in this volume. He gives a vivid picture of Ed Beckenbach, the man and the mathematician, which reflects his generous and open-minded personality, his scientific achievements and his many services to the mathematical community.

The editor is pleased to express his appreciation to all those who have devoted their work and knowledge to make this volume a worthy successor of the previous Proceedings. Several colleagues have participated in the refereeing process. Dr. Reinhard Redlinger, Dr. Norbert Steinmetz and Dipl.-Math. Christian U. Huy did most of the proofreading, and Mr. Huy also compiled the index. Ms. Irene Jendrasik has typed nearly half of the articles and performed all technical preparations of the final typescript with extraordinary care and expertise. The editor is deeply grateful to them all and - last not least - to Birkhäuser Verlag for kind expression of interest and friendly collaboration.

Karlsruhe, July 1984

Wolfgang Walter, Editor
Universität Karlsruhe

CONTENTS

INEQUALITIES IN ANALYSIS AND APPROXIMATION

INEQUALITIES OF FUNCTIONAL ANALYSIS

FUNCTIONAL INEQUALITIES

INEQUALITIES FOR DIFFERENTIAL OPERATORS

INEQUALITIES IN ECONOMICS, OPTIMIZATION AND APPLICATIONS

PROBLEMS AND REMARKS

SKETCHES

by

Joy Russell

In memoriam
Edwin F. Beckenbach

IN MEMORIAM EDWIN F. BECKENBACH

Edwin Ford Beckenbach was born in Dallas, Texas, in 1906. He obtained his Bachelor and Master degrees at the Rice Institute (now Rice University) in Houston, Texas, where he received his Ph.D. in 1931. From 1931 to 1933 he was a Fellow of the National Research Council. In this capacity he spent one year at Princeton. The second year was divided between Ohio State University at Columbus, and the University of Chicago. "A National Research Fellowship at that time was the highest honor that could be bestowed on a young Ph.D.", recalls Magnus Hestenes, who first met Edwin Beckenbach in 1932.

In 1933, Ed Beckenbach went back to Houston, Texas, to join the Rice faculty. He stayed there until 1940, then went for two years, as an Assistant Professor, to the University of Michigan at Ann Arbor. In 1942 he moved to the University of Texas at Austin, where he served as Associate Professor for three years. In 1945 he finally moved to the University of California at Los Angeles (UCLA), where he became Professor of Mathematics in 1945. Ed retired in 1974 after a long and distinguished career, in which he had great impact until his very last days, not only within the UCLA Mathematics Department, but also on the mathematical community as a whole.

Ed's Ph.D. thesis, carried out under the supervision of Lester Ford, was on minimal surfaces. His interest in this subject grew deeper through his papers with Tibor Rado (published in 1933), where it was shown that the logarithms of the distance-function and the area-deformation function for minimal surfaces are both subharmonic. This led to a series of articles by Ed on subharmonic functions, harmonic functions, convex functions, and certain generalizations of these classes. Ed's involvement in complex function theory also led to several 3-dimensional analogues of function-theoretical results such as the Cauchy and Morera theorems, and to numerous inequalities for subharmonic and other types of functions, which triggered a life-long interest in general inequalities.

In the mid 60's Ed wrote intensively on inequalities. In his papers he treated classical topics such as the Hölder and Minkowski inequalities, as well as inequalities related to conformal maps and differential geometry of surfaces. At the same time he continued his work on minimal surfaces, and maintained his enthusiasm for geometrical interpretation of analytic theorems. In particular, Ed's work was strongly influenced by Nevanlinna's two fundamental theorems for meromorphic functions, which he extended, partially in collaboration with Gerald Hutchinson and Thomas Cootz (both Ed's Ph.D. students), to meromorphic minimal surfaces. These extensions imply, among other things, the following remarkable geometrical interpretation of Picard's theorem: If S is a non-constant meromorphic minimal surface, then there can be at most two points in space that can neither "feel" nor "see" the surface; and if there are two such points, then S is a plane surface and the points are on the plane. This type of "intuitive" result demonstrates Ed's deep geometrical insight. Another example of Ed's analytic-geometric attitude is given in a later paper with Fook Eng and Edward Tafel (again, Ph.D. students of his), where it was shown that the fundamental theorem of algebra, suitably formulated, holds for rational minimal surfaces. In his prolific career, Ed published some 75 research papers, most of which appeared in leading American journals.

Prior to Ed's arrival at UCLA in 1945, the Mathematics Department there was not very large, and produced no Ph.D.s. The first UCLA Ph.D. in Mathematics, Dr. William Gustin, received his doctorate in 1947 from Ed. All told, Ed had ten Ph.D. students during the twenty-four year period 1947-1970.

Ed is well remembered by his students. Dr. Thomas Cootz, a 1966 Ph.D. of Ed's, had the following things to say in a gathering of the UCLA Mathematics Department in memory of Ed Beckenbach last October: "He was a very easy man to get along with for a student. He was soft-spoken and never angry or sharp... The word that best describes his manner towards me, as well as other students, was 'friend'... He always seemed to be on my side... He was a very good thesis advisor, in the sense that he gave me advice when I needed it, and left me alone when I just needed to go off and explore things."

Ed was also administratively active at UCLA, where he was Acting Chairman of the Mathematics Department between 1954 and 1955.

Another major achievement of Ed's, soon after he came to the West Coast, was the establishment of the Institute for Numerical Analysis (INA) at UCLA. In 1946, in view of the development of large computers, Edward Condon, the Director of the National Bureau of Standards (NBS), had decided to promote research in numerical analysis, by setting up two groups, equipped with the largest and most modern computers available at that time, one in Washington, D.C., and the other on the West Coast. This led to the creation of the INA at UCLA, through a chain of events, best described in Ed's own words written in 1980 in a letter (a copy of which was kindly provided by Kirby Baker of UCLA) to Mina Rees, who headed in 1946 the Mathematical Division of the Office of Naval Research: "... I was excited at the thought of what a stimulus it would be, to West Coast mathematics and science in general, to have such a computer and such a flow of outstanding mathematicians at any of our institutions... Surprisingly, no one else seemed to share the vision, possibly in part because it seemed inevitable, ...that UC Berkeley or Stanford would get the installation... I can report with certainty only on my own enthusiasm and on my impression of my colleagues' apparent relative lack of it... A few days later I asked Chairman Daus if... the Mathematics Department, or UCLA was doing anything about making a bid for the Institute. He replied that as far as he knew they were not. I said I thought that was unfortunate and asked if he would object if I made an effort. He shrugged an indifferent assent. I gained no active support from any of my colleagues except Sokolnikoff, who at first only said it was a shame that 'they' were doing nothing about it, but who soon thereafter actively joined forces with me. We first got a quick go-ahead from Provost Dykstra... Then we sought the support of a few other departments and institutions, and received casual 'Sure, go ahead' responses... I agreed to write the proposal, and spent a couple of weeks carefully preparing an extensive... document, neatly divided into separate sections concerning climate, aircraft industry, RAND, educational institutions, libraries, and the region's wealth of mathematicians, scientists, and engineers... The proposal was submitted to the NBS... The affirmative response to our proposal came rather sooner than I had anticipated... As I walked into the Mathematics Department office one morning, someone -- Paul Hoel, I think -- was grinning and asked, 'Have you heard? we're getting the Institute! 'Oh, boy!' I exclaimed. 'That will really be great for our Department!'..." With these kinds of actions Ed was instrumental in putting UCLA on the international mathematical map.

Naturally, Ed was the first member of the UCLA Mathematics Department to join the INA. Already in 1949, Ed organized through INA a symposium on conformal maps. He also edited the proceedings of the symposium (*Construction and Applications of Conformal Maps*, NBS Applied Mathematics Series 1952), which was the beginning of Ed's long career as an editor.

The first director of the INA was John Curtiss, who took a year off from his duties as Chief of the NBS Applied Mathematics Laboratories in Washington, D.C. After being with the Institute for a while, Curtiss had the idea that a special journal for numerical analysis and computational mathematics was called for. While this idea failed, it led, with Ed's crucial help, to the establishing of what has become one of the most distinghished general mathematical journals, the *Pacific Journal of Mathematics (PJM)*. The creation of *PJM* is described in the above quoted letter Ed wrote to Mina Rees: "... John very much wanted to establish a journal of numerical analysis and computer mathematics...

The problem was that, while the Institute had ample and legitimate funds for manuscript preparation, our notions of a journal... ran afoul of Governmental publication restrictions. A proper publisher had to be found. We decided to call a formal meeting of representatives of such West Coast universities as we could assemble at the Institute, to discuss the matter... None of the visitors were interested in backing a journal in such a restricted mathematical field... Early in the fray, I knew that John's cause was lost, and I could see that he knew it, too. It was soon apparent, also, that with their rather contentious and myopic approach (as it seemed to me) the others would not come to agreement on anything else. Adopting a unity-through-exhaustion strategy, I kept quiet for a long while. At last I stood and spoke at length, saying that it was clear that we could agree only on a totally unrestricted mathematical journal, that we could succeed only through the unreserved cooperation of all, and that the Pacific area owed it to the mathematical community to carry its fair share of the mathematical-research publication burden -- which was quite heavy at the time. A practical person, John yielded on content; the others, having ventilated themselves, were quiescent and compliant; I was named chairman of an exploratory committee on which Frank [Franticek] Wolf of UC Berkeley and Arthur Erdelyi of Cal Tech also served; and thus the *Pacific Journal of Mathematics* was born ..."

Ed was the first managing editor of *PJM*. He was succeeded by Ernst Straus, another distinguished member of the UCLA Mathematics Department, who managed *PJM* during 1955-1958. Straus, who passed away last July, said in his recent recollections of the first days of *PJM*: "... These universities that founded and still run the PJM had essentially no money to contribute, but he [Ed] managed to get the money anyway because of the saving that they would gain from exchanging the *PJM* for journals from other institutions... I should stress a point, something I think I've not seen another editor, except perhaps Piranian of the *Michigan Journal*, engage in. Ed was determined that every sentence that got printed in the *PJM* should actually be a good English sentence. So, he spent a great deal of time rearranging phrases, putting in verbs, putting in periods and semicolons..., so that afterwards, it really read as if it were written by somebody who had an appreciation for the English language..." This attitude was so very typical of Ed's outstanding work as editor and author.

Before the creation of *PJM*, during the five year period 1946-1951, Ed was the editor of the "Mathematical Notes Department" (called the "Discussion and Notes Department" in 1946) of the *American Mathematical Monthly*, published by the Mathematical Association of America (MAA). In 1960 he was instrumental in having the MAA take over and expand the *Mathematics Magazine*. This was done in view of the failing health of its previous editor Glenn James, and it is now a successful magazine for elementary expository mathematics. Ed's responsiblity for establishing in 1981 the MAA newsletter *Focus*, should also be mentioned.

During his long professional life Ed was unusually active in serving the various mathematical societies in the USA. He was for years a member of the Council of the American Mathematical Society (AMS). He was also on the AMS Printing and Publishing Committee, and on several committees of the Mathematical Reviews during difficult and precarious times when the continued existence of this important publication was in doubt. Ed rendered many services to the National Council of Teachers of Mathematics. He served on the Conference Board of the Mathematical Sciences. And he was on the Membership Committee of the Society of Industrial and Applied Mathematics (SIAM).

Ed had a very special role with the MAA. As mentioned above, he was an Associate Editor of the *Monthly* and the founder of *Focus*, both published by the MAA. He was for many years a Visiting Lecturer for the MAA; and during 1971-1982 Chairman of the MAA Publication Committee, where he guided an unprecedented growth in journals, books and monographs. For these and many other services, Ed received the Distinguished Service Award of the MAA. It was awarded, posthumously, at the Joint Meeting of the Societies at Denver, Colorado, in January of 1983. Fortunately, the award was announced by Ivan Niven, President of the MAA, on August 1982, at the Annual Meeting of the MAA in Toronto, Canada, only days before Ed's death. No doubt, this announcement gave Ed much pleasure; it was a beautiful way to crown his unusual career. In an article describing the award in the February 1983 issue of the *Monthly*, writes Niven: "... In all his activities, Ed Beckenbach enlisted the cooperation of his colleagues by his skill at negotiation, his unfailing courtesy and consideration towards others, and his common sense and good humor..."

One of the most outstanding aspects of Ed's work concerns his prolific editing and writing. Among the texts he edited one finds the above mentioned *Construction and Applications of Conformal Maps* (NBS, 1952), *Modern Mathematics for the Engineer* (two volumes; McGraw-Hill, 1956, *Applied Combinatorial Mathematics* (Wiley, 1964), and *Concepts of Communication* (co-edited with Charles Tompkins; Wiley, 1971). These and other titles demonstrate a wide knowledge and interest, in a large variety of mathematical areas. Since 1964 Ed co-authored a very long list of elementary, intermediate, and collegiate-level texts. This includes, *College Algebra* (with Irvin Drooyan and William Wooton, 1964), *Modern Trigonometry* (with William Wooton and Mary Dolciani, 1965), *Analysis of Elementary Functions* (with Robert Sorgenfrey, 1970), *Modern Analytic Geometry* (with William Wooton et al., 1972), and over a dozen other titles. Ed also co-authored with Robert James the third edition of the *James and James Mathematics Dictionary* (Van Nostrand, 1968). All told, Ed co-authored and edited almost thirty volumes. His great success as an author is reflected by the fact that quite a few of his books were translated into Russian, French, Japanese, Spanish, Italian, and other foreign languages.

Ed's perhaps most distinguished contribution as an author sprang from his association with Richard Bellman (then at the RAND Corporation), which gave birth to two well known texts: *An Introduction to Inequalities* (Yale University Press, and Random House), and *Inequalities* (Springer Verlag), both published in 1961. The first book was, and probably still is, the most popular elementary text in the field, and it has been reprinted time and again in the past twenty years. The second book was an advanced text which has immediately become a standard reference, proudly standing next to the immortal *Inequalities* by Hardy, Littlewood, and Pólya. This is truly so, despite the modest dedication in the book: "To G.H. Hardy, J.E. Littlewood, and G. Pólya from two followers afar".

Soon after the appearance of his *Inequalities* Ed was recognized as a leader in this field, and became a major participant in every conference on inequalities. This holds, of course, for the Inequalities Symposia organized by Oved Shisha at the Wright-Patterson Air Force Base, Ohio, in 1965, and at the United States Air Force Academy, Colorado, in 1967. The third symposium in this series, at UCLA in 1969, was organized by Shisha together with Ed Beckenbach, Theodore Motzkin, and Robert Pohrer.

After the Shisha symposia were discontinued, a vacuum was created. This gap was filled by Ed Beckenbach, Janos Aczél, and the late Georg Aumann, who organized in 1976 the First International Conference on General Inequalities in Oberwolfach. In 1978 the Second Oberwolfach Inequalities Conference followed under the direction of Beckenbach, Aczél, and Wolfgang Walter; and in 1981 Beckenbach, Walter, and Marek Kuczma organized the third conference in this series. During 1982 Beckenbach, Walter, and Lászlő Losonczi were busy organizing the Fourth Oberwolfach Conference which was to be held in May of 1983. Unfortunately, Ed never lived to see this highly successful meeting, which was excellently managed by Walter and Losonczi, and dedicated to Ed.

Today, the Oberwolfach Inequalities Conferences have become a tradition, and it is very much regretted that Ed has ceased to be such a vital and important part of it. In organizing these conferences, he demonstrated a deep feeling of devotion, and a strong drive for scientific success. His contribution to these conferences, however, did not stop there. He took it upon himself to edit and prepare for print the proceedings of these conferences, *General Inequalities 1*, *General Inequalities 2*, and *General Inequalities 3* (Birkhäuser, 1978, 1980, and 1983). In this he was meticulously and devotedly assisted by Elaine Barth, whose long association with Ed began when he was elected the first Managing Editor of *PJM*, and she became the Technical Editor of this journal. The editing of *General Inequalities 3* was completed in 1983 by Walter.

Ed was truly a warm and generous person, loved by all who knew him. This is exemplified in the following story, told last October at the UCLA Memorial Gathering for Ed, by Ray Redheffer: "... On the last night at the last [1981] meeting of Oberwolfach, suddenly as if by magic, there appeared some really excellent German wine on all of those many, many tables. Now, I've been to at least twelve meetings there, perhaps more, and nothing of the sort ever happened before. Everybody else was just as bewildered as I was, and they were whispering to their neighbors, 'where did it come from?' Finally, the rumor got started that this was the gift of an oil millionaire from Texas. The Texas part was right! The few people who were 'in the know', finally insisted that Ed should stand up and accept our thanks. As you might guess, he did stand up, but instead of accepting any thanks, he thanked us for coming". Such was Ed, and so too is his beloved wife Alice who shared with him his generosity and warmth.

Ed lived a full life to the very last days: On one hand, driven by his life-long passion for tennis and travel, and on the other, consistently carrying out the many professional tasks and obligations he took upon himself. Last summer, only days before his death on September 5, 1982, he used to come in every day to UCLA, in order to complete editing the proceedings of the Third Oberwolfach Inequalities Conference, *General Inequalities 3*. He showed up in a tennis outfit, a tennis racket under one arm, and a heavy pile of mathematics under the other. This is the way we would all like to remember the mathematician and man, Edwin Beckenbach.

Moshe Goldberg
Los Angeles, California
August 1983

Inequalities for Sums and Integrals

Lorenzenhof

International Series of
Numerical Mathematics, Vol. 71

SOME EXAMPLES OF HARDY-LITTLEWOOD TYPE INTEGRAL INEQUALITIES

William N. Everitt

Abstract. The Hardy-Littlewood integral inequality, which is the starting point of this paper, is

$$\left(\int_0^\infty f'^2\right)^2 \le 4 \int_0^\infty f^2 \int_0^\infty f''^2$$

for which the number 4 is best possible and all cases of equality are known. In recent years this inequality has been extended to the general case

$$\left(\int_a^b \{pf'^2+qf^2\}\right)^2 \le K \int_a^b wf^2 \int_a^b w^{-1}((pf')'-qf)^2 ,$$

where p, q are real-valued functions, and w is a positive weight function, on an interval [a,b). Three examples of this general inequality are considered in this paper; these examples illustrate many of the general properties of this extended Hardy-Littlewood inequality.

1. INTRODUCTION

These notes report on a lecture given to the General Inequalities 4 Conference held at the Mathematical Research Institute, Oberwolfach in May 1983. Three examples of the Hardy-Littlewood type integral inequality are considered, for which best possible constants can be determined together with all the associated cases of equality. The methods required for the analysis of these inequalities depend on the properties of the Titchmarsh-Weyl m-coefficient for second-order linear quasi-differential equations; these methods are not discussed in these notes but reference is made to Evans and Everitt [1].

For the third example the analysis is not complete and a conjecture is made concerning the inequality concerned.

ACKNOWLEDGEMENT. The author is grateful to W.D. Evans (University College, Cardiff) and W.K. Hayman (Imperial College, London) for many discussions and for the collaboration which led to the results quoted in this paper from the papers referenced as [1], [2] and [3].

2. THE HARDY-LITTLEWOOD INEQUALITY

This inequality was first considered in 1932; see [7; section 7]. Details are also given in the book *Inequalities* [8; theorem 260]. The result may be stated as follows: Define the linear manifold $\Delta \subset L^2(0,\infty)$

(2.1) $\Delta = \{f\colon [0,\infty) \to R \mid f, f' \in AC_{loc}[0,\infty)$ <u>and</u> $f, f'' \in L^2(0,\infty)\}$,

then

(2.2) $$f' \in L^2(0,\infty) \quad \text{for all } f \in \Delta$$

and the following inequality holds

(2.3) $$\left(\int_0^\infty f'^2\right)^2 \le 4 \int_0^\infty f^2 \int_0^\infty f''^2 \quad \text{for all } f \in \Delta$$

with equality if and only if for some $\alpha \in R$ and some $\rho > 0$

(2.4) $$f(x) = \alpha Y(\rho x) \quad \text{for all } x \in [0,\infty)$$

where the extremal Y is given by, on $[0,\infty)$,

(2.5) $$Y(x) = e^{-x/2} \sin\left(\tfrac{1}{2} x\sqrt{3} - \tfrac{1}{3}\pi\right) .$$

True to the spirit of inequalities three proofs of this result are given in Hardy, Littlewood and Pólya, see [8; section 7.8].

3. THE GENERALIZED HARDY-LITTLEWOOD INEQUALITY

The following extensions to the result in section 2 are made:

(i) the interval $[0,\infty)$ is replaced by $[a,b)$ with $-\infty < a < b \le \infty$,

(ii) the space $L^2(0,\infty)$ is replaced by $L^2_w(a,b)$ where the weight function w satisfies

(3.1) $$w \in L_{loc}[a,b) \quad \text{and} \quad w(x) > 0 \quad (\text{almost all } x \in [a,b))$$

and $f\colon [a,b) \to R$ is in $L^2_w(a,b)$ if $\int_a^b wf^2 = \int_a^b w(x)f(x)^2dx < \infty$,

(iii) the second derivative f" is replaced by the symmetric quasi-derivative $(pf')'-qf$ where

(3.2) $$p,q: [a,b) \to R \quad \text{and} \quad p^{-1}, q \in L_{loc}[a,b),$$

(iv) the linear manifold Δ is replaced by

(3.3) $$\Delta = \{f: [a,b) \to R \mid f, pf' \in AC_{loc}[a,b); \; f, w^{-1}((pf')'-qf) \in L_w^2(a,b)\} .$$

(Note. The real, symmetric quasi-differential expression

(3.4) $$(pf')'-qf$$

can be extended to the consideration of complex-valued differential expressions which are also symmetric; however, no essential generalization is thereby obtained. For details see [1; section 11]).

With the extensions given above we then ask the following questions, which should be compared with the properties (2.2) and (2.3) of the Hardy-Littlewood inequality:

(i) Does the Dirichlet integral

(3.5) $$\int_a^b \{pf'^2+qf^2\}$$

exist in some sense for all $f \in \Delta$, given by (3.3) ?

(ii) Is there an inequality of the form, valid for all $f \in \Delta$,

(3.6) $$\left(\int_a^b \{pf'^2+qf^2\}\right)^2 \le K \int_a^b wf^2 \int_a^b w^{-1}((pf')'-qf)^2$$

where K is a number, independent of f, satisfying $0 < K < \infty$?

Note that if, in (3.6), we take $a = 0$, $b = \infty$ and $p(x) = 1$, $q(x) = 0$, $w(x) = 1$ for all $x \in [0,\infty)$ then the general case reduces to the Hardy-Littlewood inequality (2.3).

If (3.6) holds for some number K then the inequality is said to be valid; the smallest such value of K is then said to be the best possible number for the inequality; any $f \in \Delta$ which then gives equality is called a case of equality for the inequality.

Alternatively, even though the Dirichlet integral (3.5) is finite for all $f \in \Delta$, it may happen that the inequality (3.6) is not valid for any number K; in this case the inequality is said to be not valid and we write $K = \infty$.

The general theory of this inequality (3.6) is given in detail in the paper by Evans and Everitt [1]; earlier contributions are given in Everitt [5], Evans and Zettl [4]. This theory involves properties of the m-coefficient generated by the differential expression (3.4) in the space $L_w^2(a,b)$. The m-coefficient was introduced by Titchmarsh [9], essentially in 1940, but was heralded by the work of Weyl [10] in 1910. For a statement of the main results concerning the inequality (3.6) see [1; section 6]; details are not reproduced here since we are concerned in this note with considering examples of the general inequality.

However, four points do require a mention here:

(i) If in (3.6) we take $a = 0$, $b = \infty$, $p(x) = w(x) = 1$ and $q(x)=0$ $(x \in [0,\infty))$ then the general result reduces to the Hardy-Littlewood (2.3) together with all the cases of equality given by (2.4) and (2.5).

(ii) Conditions additional to (3.1) and (3.2) are required on the coefficients p, q and w to ensure that the Dirichlet integral (3.5) exists, in some sense, and is finite for all $f \in \Delta$; the most effective condition to adopt is to assume p, q and w chosen so that the differential expression (3.4) is in the strong limit-point condition at b in $L_w^2(a,b)$; for details see [1; section 3]; there is an extensive literature which gives explicit conditions on the coefficients p, q and w for the strong limit-point condition, i.e.

(3.7) $$\lim_{x\to b} (pf')(x)g(x) = 0 \qquad (f,g \in \Delta)$$

to hold, see [1, section 3]; this implies

(3.8) $$\int_a^{\to b} \{pf'^2+qf^2\} \equiv \lim_{x\to b} \int_a^x \{pf'^2+qf^2\}$$

exists and is finite for all $f \in \Delta$; an example is given in [1; section 3] to show that (3.4) may be strong limit-point and yet (3.8) only converges conditionally; this is also example 2 below.

(iii) The general theory of the inequality (3.6) shows that the number K satisfies

(3.9) $$1 < K \le \infty$$

and that this is a best possible result in the sense that the bounds are best possible; $K=1$ is impossible for any choice of coefficients p, q and w satisfying (3.1), (3.2) and (3.7); on the other hand, under these conditions, $K=\infty$ is possible (i.e. the inequality may be invalid), and every $K\in(1,\infty)$ can be realized as the best possible number of a valid inequality by suitable choice of the coefficients p, q and w; all these possibilities are illustrated in the examples given in the next section.

(iv) With the strong limit-point condition (3.7) satisfied the so-called Dirichlet formula is valid

$$(3.10)\qquad \int_a^{\to b}\{pf'^2+qf^2\} = -(pf')(a)f(a)+\int_a^b wfw^{-1}(-(pf')'+qf)$$

for all $f\in\Delta$; this result gives some idea of the requirement for $K>1$ in the general inequality (3.6); the boundary term $(pf')(a)f(a)$ in (3.10) has to be taken into account and does, in one sense, decide if the inequality is to be valid or not; if we reduce Δ to Δ_o where

$$\Delta_o = \{f\in\Delta:\ (pf')(a)=0 \text{ or } f(a)=0\}$$

then the Cauchy-Schwarz inequality gives, from (3.10),

$$\left(\int_a^{\to b}\{pf'^2+qf^2\}\right)^2 \le \int_a^b wf^2\int_a^b w^{-1}((pf')'-qf)^2 \qquad (f\in\Delta_o)\ ,$$

i.e. the inequality (3.6) with $K=1$ but valid only on Δ_o; in the Hardy-Littlewood case (2.3) it is the boundary term, i.e. $f'(0)f(0)$, which forces the number K in the inequality up to the best possible value 4; in some examples the boundary term forces $K=\infty$ and the inequality is not valid.

4. EXAMPLES

In this section we consider three examples which illustrate the properties of the general inequality given above; all these examples are also considered in [1; section 9].

1. Let $a=0$, $b=\infty$ and

$$p(x)=1\ ,\qquad q(x)=0\ ,\qquad w(x)=x^\alpha\qquad (x\in(0,\infty))\ ;$$

here α is a real parameter with $\alpha>-1$ in order that the weight w satisfies condition (3.1). This example is known to be strong limit-point at ∞ in $L_w^2(0,\infty)$ for all $\alpha > -1$.

The inequality in this case takes the form

$$\left(\int_0^\infty f'^2\right)^2 \le K(\alpha) \int_0^\infty x^\alpha f^2 \int_0^\infty x^{-\alpha} f''^2 \qquad (f \in \Delta(\alpha))$$

where the domain $\Delta(\alpha)$ is determined by

$$\Delta(\alpha) = \{f\colon [0,\infty) \to R \mid f, f' \in AC_{loc}[0,\infty) \ \underline{\text{and}} \ x^{\alpha/2} f, x^{-\alpha/2} f'' \in L^2(0,\infty)\} .$$

It is shown in Everitt and Zettl [6] that the best possible value of the number K in this case is given as follows

$$(4.1) \qquad K(\alpha) = \left\{\cos\left[\frac{\pi(\alpha+1)}{2\alpha+3}\right]\right\}^{-2} \qquad (\alpha \in (-1,\infty))$$

with all cases of equality determined in terms of the Hankel-Bessel function $H_\nu^{(1)}$ of order $\nu = \nu(\alpha) = (\alpha+2)^{-1}$ $(\alpha \in (-1,\infty))$, and type 1.

Note that $1 < K(\alpha) < \infty$ and is monotonic increasing for $\alpha \in (-1,\infty)$; also

$$\lim_{\alpha\to -1} K(\alpha) = 1 , \qquad \lim_{\alpha\to\infty} K(\alpha) = \infty .$$

For $\alpha = 0$ we have $K(0) = 4$ and the inequality reduces to the Hardy-Littlewood case (2.3) together with the cases of equality (2.4) and (2.5).

It is this example which demonstrates the general result that it is impossible for K to take the value 1 in the general inequality (3.6). We have $K(\alpha) > 1$ for all $\alpha \in (-1,\infty)$; clearly $K(-1) = 1$ but $\alpha = -1$ is not an admissable value of α for the general theory of (3.6) since then the weight function $w(x) = x^{-1}$ $(x \in (0,\infty))$ and this is not integrable down to the end-point 0, i.e. w does not then satisfy the condition (3.1).

This example also confirms the statement made in the previous section that every value for K in the range $(1,\infty)$ can be realised as the best possible constant by an appropriate choice of the coefficients p, q and w.

2. Let $a = 0$, $b = \infty$ and

$$p(x) = 1\,, \quad q(x) = -x\,, \quad w(x) = 1 \quad (x \in [0,\infty))\,;$$

this example is known to be strong limit-point at ∞ in $L^2(0,\infty)$, but the Dirichlet integral in this case, i.e.

$$\int_0^{\to\infty} \{f'^2 - xf^2\} \tag{4.2}$$

is known to be only contionally convergent, in general, on the set Δ given by

$$\Delta = \{f\colon [0,\infty) \to R \mid f, f' \in AC_{loc}[0,\infty) \ \underline{\text{and}}$$
$$f, f'' + xf \in L^2(0,\infty) \quad .$$

It is shown in [1; section 3] that there are elements of Δ for which the integral in (4.2) converges conditionally but for which

$$\int_0^\infty f'^2 = \int_0^\infty xf^2 = \int_0^\infty |f'^2 - xf^2| = \infty\,.$$

Notwithstanding this delicate situation this example yields a valid inequality in the form

$$\Big(\int_0^{\to\infty} \{f'^2 - xf^2\}\Big)^2 \le 4 \int_0^\infty f^2 \int_0^\infty (f'' + xf)^2 \quad (f \in \Delta)$$

where the number 4 is best possible, and equality holds if and only if, with $H^{(1)}_{1/3}$ the Hankel-Bessel function,

$$f(x) = \alpha \operatorname{re}[e^{-\pi i/3}(x+\lambda)^{1/2} H^{(1)}_{1/3}(2(x+\lambda)^{3/2}/3)] \quad (x \in [0,\infty))$$

where $\alpha \in R$, $\lambda = re^{\pi i/3}$ and r is a parameter, $r \in (0,\infty)$.

This example is considered in detail by Evans and Everitt [2].

3. Let $a = 0$, $b = \infty$ and

$$p(x) = 1\,, \quad q(x) = x^2 - \tau\,, \quad w(x) = 1 \quad (x \in [0,\infty))\,.$$

Here $\tau \in R$ is a real parameter; this case is strong limit-point at ∞ in $L^2(0,\infty)$ for all $\tau \in R$.

The analysis of this example is not complete but the known results are presented below.

The general inequality in this case takes the form

$$\Big(\int_0^\infty \{f'^2 + (x^2-\tau)f^2\}\Big)^2 \le K(\tau) \int_0^\infty f^2 \int_0^\infty \{f'' - (x^2-\tau)f\}^2 \quad (f \in \Delta)$$

where

$$\Delta = \{f: [0,\infty) \to R \mid f, f' \in AC_{loc}[0,\infty),$$
$$f \text{ and } f''-x^2 f \in L^2(0,\infty) \quad .$$

It is shown in Everitt [5], and Evans and Everitt [1] that the following results hold:

$$K(\tau) = \infty \qquad \text{for all } \tau \in R \text{ except } \tau \in (1,3,5,7,\ldots) ,$$
$$4 \le K(\tau) < \infty \quad \text{for all } \tau \in (1,3,5,7,\ldots) .$$

Note that this example illustrates the case when the general inequality (3.6) is not valid.

The case $\tau = 1$ has been analysed completely in Evans, Everitt and Hayman [3] and yields the result $K(1) = 4$ with

$$\left(\int_0^\infty \{f'^2+(x^2-1)f^2\}\right)^2 \le 4 \int_0^\infty f^2 \int_0^\infty \{f''-(x^2-1)f\}^2 \quad (f \in \Delta)$$

with equality if and only if for $\alpha \in R$

$$f(x) = \alpha \exp[-x^2/2] \quad (x \in [0,\infty)) .$$

The number 4 is best possible although this cannot be observed from the case of equality since this choice of $f \in \Delta$ renders both sides of the inequality equal to zero. There is no element of Δ which gives equality in the inequality <u>and</u> gives a positive value to the Dirichlet integral on the left-hand side. Nevertheless the number 4 is best possible; the proof of this is to be found in the general analysis in [1] and the special results given in [3].

The cases when $\tau \in (3,5,7,\ldots)$ have not yet been analysed in detail but it is conjectured that the best possible numbers are given by

$$K(\tau) = 4 \quad \text{for all } \tau \in (3,5,7,\ldots)$$

and that equality occurs if and only if

$$f(x) = \alpha H_n(x) \exp[-x^2/2] \quad (x \in [0,\infty))$$

where $n = \frac{1}{2}(\tau-1)$ and H_n is a Hermite polynomial.

This example is under further consideration.

5. Other examples. Additional examples of the general inequality (3.6) are considered in [1; section 9].

REFERENCES

1. W.D. Evans and W.N. Everitt, A return to the Hardy-Littlewood integral inequality. Proc. R. Soc. Lond. A 380 (1982), 447-486.

2. W.D. Evans and W.N. Everitt, On an inequality of Hardy-Littlewood type: I. UDDM Report DE 82:2 (1982) (Department of Mathematics, The University of Dundee).

3. W.D. Evans, W.N. Everitt and W.K. Hayman, On an inequality of Hardy-Littlewood type: II (to appear).

4. W.D. Evans and A. Zettl, Norm inequalities involving derivatives. Proc. R. Soc. Edinb. A 82 (1978), 51-70.

5. W.N. Everitt, On an extension to an integro-differential inequality of Hardy, Littlewood and Pólya. Proc. R. Soc. Edinb. A 69 (1971/72), 295-333.

6. W.N. Everitt and A. Zettl, On a class of integral inequalities. J. Lond. Math. Soc. (2) 17 (1978), 291-303.

7. G.H. Hardy and J.E. Littlewood, Some integral inequalities connected with the calculus of variations. Q. J. Math. (2) 3 (1932), 241-252.

8. G.H. Hardy, J.E. Littlewood and G. Pólya, Inequalities. Cambridge University Press, 1934.

9. E.C. Titchmarsh, Eigenfunction expansions I. Oxford University Press, 1962.

10. H. Weyl, Über gewöhnliche Differentialgleichungen mit Singularitäten und die zugehörigen Entwicklungen willkürlicher Funktionen. Math. Ann. 68 (1910), 220-269.

W.N. Everitt, Department of Mathematics, University of Brimingham, Birmingham B15 2TT, England

International Series of
Numerical Mathematics, Vol. 71

OPIAL-TYPE INEQUALITIES THAT INVOLVE HIGHER ORDER DERIVATIVES

Carl H. FitzGerald

Abstract. Estimates of integrals of the form $\int|y(x)y'(x)|dx$ are made in terms of integrals of the form $\int[y^{(n)}(x)]^2dx$ for functions y satisfying appropriate end point conditions. Certain extremal functions are shown to exist and have particular monotonicity properties. For each n, these extremals can be found explicitly by solving associated systems of linear equations. The best constants in the estimates can then be obtained.

I. INTRODUCTION

In 1960 Z. Opial proved an inequality which has been generalized in many directions [1,3]. The following version [3] was obtained by C. Olech.

THEOREM 1.1. *If* f *is absolutely continuous on* $[a,b]$ *and* $f(a) = f(b) = 0$, *then*

$$\int_a^b |f(x)f'(x)|dx \le \frac{b-a}{4}\int_a^b f'(x)^2dx. \tag{1.1}$$

Inequality (1.1) *is sharp exactly for functions* $f(x) = C(x-a)$ *for* $a \le x \le \frac{1}{2}(a+b)$ *and* $f(x) = C(b-x)$ *for* $\frac{1}{2}(a+b) \le x \le b$ *for* C *constant*.

The following inequality is known as Wirtinger's inequality [3].

THEOREM 1.2. *If* g *is periodic of period* 2π, $g'(x) \in L^2$, *and the integral of* g *over a period is zero, then*

(1.2) $$\int_0^{2\pi} g(x)^2dx \le \int_0^{2\pi} g'(x)^2dx$$

with equality if and only if $g(x) = A\cos x + B\sin x$ *for constants* A *and* B.

It is possible to use Theorem 1.2 to estimate the right side of inequality (1.1). In addition to the hypothesis of Theorem 1.1, suppose $f \in C^2[a,b]$ and $f'(a) = f'(b) = 0$. Consider a symmetric extension of f past b to the interval $[a, 2b-a]$. Let $G(x) = f'(x)$. Make a linear change of variable so that Theorem 1.2 can be applied to G. Combine the result with inequality (1.1) to obtain the following inequality.

(1.3) $$\int_a^b |f(x)f'(x)|dx \le \frac{(b-a)^3}{4\pi^2}\int_a^b f''(x)^2dx.$$

It is possible to extend inequality (1.3) to involve still higher derivatives on the right. However, already inequality (1.3) is not sharp since the extremal function for (1.1) is a piecewise polynomial and for (1.2) is a trigonometric function.

The goal of this paper is to obtain the best constants in inequality (1.3) and the inequalities involving higher order derivatives. The ideas of the Calculus of Variations are used to obtain the extremal functions.

2. PRELIMINARIES

The extremals for the inequalities proved in this paper are polynomial spline functions. The following lemmas about polynomials are used in the proof of the main theorem.

LEMMA 2.1. *If* $a < b$ *are real and* n *is an integer greater than one, then there exists a unique polynomial* $p(x) = \Sigma a_k x^k$ *of degree* $2n-1$ *or less which solves the interpolation problem:*

(2.1) $$p(a) = p'(a) = p^{(2)}(a) = \dots = p^{(n-1)}(a) = 0\,, \quad p(b) = 1$$

and

(2.2) $$p'(b) = p^{(3)}(b) = \dots = p^{(2n-3)}(b) = 0.$$

Proof. Let $a = x_1 < x_2 < \ldots < x_{2n} = b$. Consider the interpolation problem consisting of $p(x_m)$ given for $m = 1,2,\ldots,2n$. The resulting $2n$ equations can be regarded as linear equations in the unknown coefficients a_k of the polynomial solution. The determinant of this system of equations is the Vandermonde determinant:

$$\begin{vmatrix} 1 & x_1 & x_1^2 & \ldots & x_1^{2n-1} \\ 1 & x_2 & \cdot & \ldots & \\ 1 & x_3 & & & \\ \cdot & \cdot & & & \\ 1 & x_{2n} & \cdot & \ldots & x_{2n}^{2n-1} \end{vmatrix} = \prod_{i>j}(x_i - x_j). \tag{2.3}$$

Suppose the derivative at x_2 had been specified instead of the value at x_2. The corresponding determinant would be the derivative of the Vandermonde with respect to x_2 evaluated at x_2. There are many terms in that derivative. These terms may even differ in sign depending on whether the factor differentiated was (x_2-x_1) or a later factor of the form (x_i-x_2) with $i > 2$. If x_2 is moved to x_1, the latter terms drop out since (x_2-x_1) is a factor in each of them. Hence the remaining terms all have the same sign and add up; and the determinant is non-zero. In fact it equals

$$\prod_{i>2}(x_i-x_1)\prod_{i>j>1}(x_i-x_j). \tag{2.4}$$

The interpolation problem in which $p(x_1)$, $p'(x_1)$, $p(x_3)$, $\ldots$, $p(x_{2n})$ are specified has one and only one solution.

Suppose $p^{(2)}(x_3)$ is to be specified in place of $p(x_3)$. The second derivative of expression (2.4) with respect to x_3 results in many terms. As x_3 tends to $x_1 = x_2$, the remaining terms all have the same sign.

Since one can continue in this way the conditions listed in equation (2.1) can be easily handled.

For the right end point conditions (2.2), the argument is essentially the same. The only concern is that the order of derivative never be so high that the resulting polynomial is identically zero. There is no problem in this case. If x_q is the variable in one of the later differentiations, there are the factors involving x_q and points that have already been taken to b and the n-factors involving x_q and $x_1, x_2, \ldots, x_n$ that are at a. Thus the determinant is non-zero and the conditions on $p(x)$ are solved by one and only one polynomial.

It is possible to deduce a geometrical property of the polynomial of Lemma 2.1.

LEMMA 2.2. *If* $p(x)$ *is the polynomial of Lemma* 2.1, *then* $p(x)$ *is monotonically increasing from* a *to* b.

Proof. Consider the conditions satisfied by the polynomial $q(x) = p'(x)$. Clearly $q(a) = q'(a) = \ldots = q^{(n-2)}(a) = 0$ and $q(b) = q^{(2)}(b) = \ldots = q^{(2n-4)}(b) = 0$.

Suppose $q'(r) = 0$ at some point $a < r < b$. That would mean $2n-1$ conditions would be satisfied by this polynomial of degree $2n-2$ or less. By a similar discussion as given in the proof of Lemma 2.1, this set of conditions has a unique solution. In this case, the solution is $q(x) = 0$. But that would imply $p(x)$ equals a constant, which is not true. Thus $q'(r)$ is never zero for $a < r < b$. Since $p(a) = 0$ and $p(b) = 1$, $p(x)$ must be monotonically increasing on $[a,b]$.

3. THE MAIN THEOREM

A version of Opial's inequality using higher order derivatives follows.

THEOREM 3.1. *For each* $n \geq 2$, *there is a constant* C_n *such that for any real interval* $[a,b]$ *if* $y \in C^n[a,b]$ *and*

$$y(a) = y'(a) = \ldots = y^{(n-1)}(a) = 0 \tag{3.1}$$

and

$$y(b) = y'(b) = \ldots = y^{(n-1)}(b) = 0, \tag{3.2}$$

then

$$\int_a^b |y(x)y'(x)|dx \leq C_n(b-a)^{2n-1} \int_a^b y^{(n)}(x)^2 dx. \tag{3.3}$$

Furthermore, the constant C_n *is sharp for a polynomial spline of degree* $2n-1$ *which has a knot at the midpoint of* $[a,b]$, *is symmetric about the midpoint, and is in* $C^{2n-2}[a,b]$.

Proof. The first part of the proof shows it suffices to consider functions of a particular form. Inequality (3.3) indicates the ratio of the integral on the left side divided by the integral on the right is bounded from above. This ratio is not changed if the function y is multiplied by a non-zero constant. Also the ratio is unchanged if the domain $[a,b]$ of the function and the integrals is translated rigidly.

On the other hand changing the domain by a linear transformation $x = \lambda t$ for a positive constant λ does change the ratio. Note $dx = \lambda dt$. If $Y(t) = y(x)$, then $Y'(t) = y'(x)\lambda$. If $A = \lambda^{-1}a$ and $B = \lambda^{-1}b$, then $\int_A^B |Y(t)Y'(t)|dt = \int_a^b |y(x)y'(x)|\lambda\ \lambda^{-1}dx = \int_a^b |y(x)y'(x)|dx$. However the other integral does change. Since $Y^{(n)}(t) = y^{(n)}(x)\lambda^n$, $\int_A^B Y^{(n)}(t)^2 dt = \int_a^b y^{(n)}(x)^2\ \lambda^{2n}\ \lambda^{-1}dx = \int_a^b y^{(n)}(x)^2 dx \cdot \lambda^{2n-1}$.

Thus is suffices to prove inequality (3.3) for some choice of $[a,b]$, for example, $[-1,1]$.

Furthermore, it suffices to prove inequality (3.3) for polynomials of high degree. Suppose y is a polynomial that satisfies the hypothesis of Theorem 3.1 and which makes the ratio of the integrals close to the supremum. We will modify y without changing the ratio substantially. Let $\hat{y}$ be defined by

$$\hat{y}(x) = \begin{cases} \int_{-1}^{x} |y'(t)|dt & \text{for } -1 \le x \le c \\ \int_{x}^{1} |y'(t)|dt & \text{for } c \le x \le 1 \\ 0 & \text{otherwise} \end{cases} \tag{3.4}$$

where c is chosen in $(-1,1)$ so that $\hat{y}$ is continuous.

Clearly $\hat{y}(x) \ge |y(x)|$ and $|\hat{y}'(x)| = |y'(x)|$ for all but at most finitely many points of $[-1,1]$. The integral on the left side of inequality (3.3) is at least as large for $\hat{y}$ as for y. The integral on the right is the same for $\hat{y}$ as for y.

Let $m = 1+2\varepsilon\div(1-|c|)$ where ε is still to be chosen. Define $\tilde{y}$ as follows:

$$\tilde{y}(x) = \begin{cases} \hat{y}(c+m(x-c+\varepsilon)) & \text{for } x \le c-\varepsilon \\ \hat{y}(c) & \text{for } c-\varepsilon \le x \le c+\varepsilon \\ \hat{y}(c+m(x-c-\varepsilon)) & \text{for } c+\varepsilon \le x. \end{cases}$$

For ε small and positive, m is close to one; and the ratio of integrals is essentially the same for $\tilde{y}$ as for $\hat{y}$.

Note $\tilde{y}(x)$ is zero near -1 and $+1$. Let $\varphi(t) \in C^{\infty}(-\infty,\infty)$, have support in $[-\frac{1}{2}\varepsilon,\frac{1}{2}\varepsilon]$, be non-negative, and have $\int\varphi(t)dt = 1$. Complete the mollification of y as follows:

$$y^*(x) = \int_{-\infty}^{\infty}\tilde{y}(x+t)\varphi(t)dt.$$

For small ε, the ratio of integrals is not changed essentially. Also y^* satisfies the end point conditions (3.1) and (3.2), and y^* is monotonic on either side of c.

To examine the value of c, we define q as follows:

$$q(x) = \begin{cases} y^*(c+(1+c)x) & \text{for } x \le 0 \\ y^*(c+(1-c)x) & \text{for } 0 \le x. \end{cases} \tag{3.5}$$

Since y^* is non-negative and monotonically increasing from -1 to c and decreasing from c to 1, clearly

$$\begin{aligned}\int_{-1}^{c}|y^*(x)y^{*\prime}(x)|dx &= \int_{-1}^{c}y^*(x)y^{*\prime}(x)dx \\ &= \tfrac{1}{2}y^*(x)^2\Big|_{-1}^{c} = \tfrac{1}{2}y^*(c)^2 \\ &= \int_{-1}^{0}|q(x)q'(x)|dx \\ &= \int_{0}^{1}|q(x)q'(x)|dx \\ &= \int_{0}^{1}|y^*(x)y^{*\prime}(x)|dx\end{aligned}$$

Hence, the integral on the left of (3.3) is the same for q as for y^*.

Let R be the value of the integral on the right of (3.3) for y^*. Define

$$R_1 = \int_1^0 q^{(n)}(x)^2 dx \quad \text{and}$$

$$R_2 = \int_0^1 q^{(n)}(x)^2 dx.$$

If R_1 is less than $\frac{1}{2}R$, then extend q to $(0,1]$ symmetrically and mollify as before to make the ratio of integrals of (3.3) larger. Similarly for R_2. But, since y^* gives close to the supremum ratio, neither R_1 nor R_2 can be substantially less than $\frac{1}{2}R$.

If equation (3.5) is regarded as a definition of y^* in terms of q, the following equation holds.

$$\int_{-1}^1 y^{*(n)}(x)^2 dx = R_1(1+c)^{1-2n} + R_2(1-c)^{1-2n} \tag{3.6}$$

If c were substantially different from zero, the right side of (3.6) would be substantially more than R. Thus, since y^* makes the ratio close to its supremum value, c must be close to 0; in which case the transformation (3.5) does not change the value of the ratio significantly. It can be assumed that $c = 0$. Also y^* can be assumed to be symmetric about 0, and $y^*(0)$ can be made equal to one by dividing by $y^*(0)$.

Thus it suffices to prove the inequality (3.3) for functions that satisfy the hypothesis of Theorem 3.1 and that are monotonically increasing on $[-1,0]$ and decreasing on $[0,1]$, that are symmetric about 0, and that have maximum equal 1.

In the second part of the proof, the Calculus of Variations is used non-rigorously to suggest a candidate for an extremal function. Technically this part could be omitted since, in the third part, a direct proof will be given which proves the candidate does in fact maximize the ratio of integrals.

Suppose $y(x)$ maximizes the ratio of integrals, satisfies the hypothesis of Theorem 3.1 and is monotonically increasing on $[-1,0]$, is symmetric about $x = 0$, and is equal to one at $x = 0$. Clearly

$$\int_{-1}^{1}|y(x)y'(x)|dx = \int_{-1}^{0}y(x)y'(x)dx - \int_{0}^{1}y(x)y'(x)dx$$

$$= \frac{1}{2}y(x)^2\Big|_{x=-1}^{x=0} - \frac{1}{2}y(x)^2\Big|_{x=0}^{x=1} = 1.$$

Let $h(x)$ be a function in $C^\infty[-1,1]$ such that $y(x) + sh(x)$ has the monotonicity properties of y and satisfies the hypothesis of Theorem 3.1 for $|s|$ sufficiently small. Then

$$\int_{-1}^{1}|[y(x)+sh(x)][y(x)+sh(x)]'|dx = 1.$$

On the other hand,

$$\int_{-1}^{1}\left(\frac{d^n}{dx^n}[y(x)+sh(x)\right)^2dx$$

$$= \int_{-1}^{1}y^{(n)}(x)^2 + 2sy^{(n)}(x)h^{(n)}(x) + s^2h^{(n)}(x)^2dx.$$

Since y is suppose to be an extremal function, the variation of the integral is zero. Thus

$$\int_{-1}^{1}y^{(n)}(x)h^{(n)}(x)dx = 0. \tag{3.7}$$

Consider functions $h(x)$ with support in $[-1,0]$. Then $\int_{-1}^{0}y^{(n)}(x)h^{(n)}(x)dx = 0$. The Fundamental Lemma of the Calculus of Variation implies that $y^{(2n)}(x)$ exists and equals zero in $(-1,0)$. Thus $y(x)$ is a polynomial of degree $2n-1$ on $[-1,0]$. Symmetry implies y is a polynomial of degree $2n-1$ on $[0,1]$, but y need not be a polynomial on $[-1,1]$.

Consider $h(x)$ on $[-1,1]$ with the minimum assumptions on the behavior at the end points. At $x=-1$, the hypothesis of Theorem 3.1 is to be satisfied by $y+sh$; consequently, $h(-1) = h'(-1) = \ldots = h^{(n-1)}(-1) = 0$. Similarly for $x = 1$, $h(1) = h'(1) = \ldots = h^{(n-1)}(1) = 0$. At $x = 0$, $y(0)+sh(0) = 1$ implies $h(0) = 0$, otherwise there are no restrictions. In applying integration by parts to (3.7), we must remember that y may not have higher derivatives at $x = 0$. In fact we find that $y^{(n)}(0^-) = y^{(n)}(0^+)$, $y^{(n+1)}(0^-) = y^{(n+1)}(0^+)$, ...,

$y^{(2n-2)}(0^-) = y^{(2n-2)}(0^+)$. (The argument fails for $(2n-1)^{st}$ derivative because $h(0) = 0$.) By the symmetry of y, we concluded $y'(0) = 0$, $y^{(3)}(0) = 0, \ldots, y^{(2n-3)}(0) = 0$.

The facts about y on $[-1,0]$ are enough to determine the function. Indeed, each part of the hypothesis of Lemma 2.1 has been proved if the interval $[a,b]$ is interpreted as $[-1,0]$. Thus there is one and only one function that y can be on $[-1,0]$. By symmetry, y is also determined on $[0,1]$. Note there is a discontinuity of the $(2n-1)^{st}$ derivative at $x = 0$; the solution y is a polynomial spline with a knot at 0.

We now start the third part of the proof. We make a direct verification that the spline y found in second part of the proof does maximize the ratio of the integrals in inequality (3.3). Note that Lemma 2.2 implies that y is monotonically increasing on $[-1,0]$.

Suppose there is a function that makes the ratio larger. By the first part of the proof, there would have to be a function on $[-1,1]$ which is symmetric about zero, is monotonically increasing on $[-1,0]$, equals 1 at $x = 0$, and makes the ratio larger than the spline y does. Any such function could be expressed as $y(x) + h(x)$. Clearly $h(-1) = h'(-1) = \ldots = h^{(n-1)}(-1) = 0$ and $h(1) = h'(1) = \ldots = h^{(n-1)}(1) = 0$ and $h(0) = 0$. Also $\int_{-1}^{1}|(y+h)(y+h)'|dx = 1$, thus it should be that the integral on the right side of (3.3) is smaller for $y+h$ than for y.

$$\int_{-1}^{1}[y^{(n)}(x)+h^{(n)}(x)]^2dx$$

$$= \int_{-1}^{1}y^{(n)}(x)^2dx + 2\int_{-1}^{1}y^{(n)}(x)h^{(n)}(x)dx + \int_{-1}^{1}h^{(n)}(x)^2dx. \tag{3.8}$$

Consider the second integral on the right side of (3.8).

$$\int_{-1}^{1} y^{(n)}(x)h^{(n)}(x)dx = y^{(n)}(x)h^{(n-1)}(x)\Big|_{x=-1}^{x=0} + y^{(n)}(x)h^{(n-1)}(x)\Big|_{x=0}^{1}$$

$$- \int_{-1}^{1} y^{(n+1)}(x)h^{(n-1)}(x)dx$$

$$= - \int_{-1}^{1} y^{(n+1)}(x)h^{(n-1)}(x)dx$$

$$\vdots$$

The early pairs of evaluations at $x = 0$ cancel because $y \in C^{2n-2}[-1,1]$ and the final pair vanishes because $h(0) = 0$. The evaluations at -1 and $+1$ vanish because the factor involving h is zero in each case.

$$\int_{-1}^{1} y^{(n)}(x)h^{(n)}(x)dx = (-1)^n \int_{-1}^{1} y^{(2n)}(x)h(x)dx = 0,$$

because y is a polynomial of degree $2n-1$ on $[-1,0]$ and on $[0,1]$.

Since the third integral on the right side of (3.8) is positive, $y+h$ decreases the ratio of integrals. This contradiction to the choice of h shows the polynomial spline y maximizes the ratio, and Theorem 3.1 is proved.

Example. Let $n = 2$ and $[a,b] = [-1,1]$.

$$y(x) = \begin{cases} 1 - 3x^2 - 2x^3 & \text{for } -1 \le x \le 0 \\ 1 - 3x^2 + 2x^3 & \text{for } 0 \le x \le 1 \end{cases}$$

$$\int_{-1}^{1} |y(x)y'(x)|dx = 1$$

$$\int_{-1}^{1} y^{(2)}(x)^2 dx = \int_{-1}^{0} (-6-6x)^2 dx + \int_{0}^{1} (-6+6x)^2 dx$$

$$= 24.$$

If $y \in C^2[a,b]$ and $y(a) = y'(a) = 0$ and $y(b) = y'(b) = 0$, then

$$\int_a^b |y(x)y'(x)|dx \le \frac{1}{192}(b-a)^3 \int_a^b y^{(2)}(x)^2 dx.$$

This inequality has the correct constant. Note the value is

much smaller than was found in inequality (1.3).

4. VARIATIONS OF THE MAIN THEOREM

If more conditions of the form $y^{(k)}(a) = 0$ and $y^{(k)}(b) = 0$ are added to the hypothesis of Theorem 3.1, the constant in (3.3) cannot be decreased. The polynomial spline which was extremal can be approached by functions such that $y^{(k)}(a) = y^{(k)}(b) = 0$ for all $k = 0,1,2,..;$ and for these functions the values of the integrals in (3.3) tend to the values for the extremal for Theorem 3.1.

On the other hand conditions $y^{(n-1)}(a) = 0$ and $y^{(n-1)}(b) = 0$ can be deleted. Of course the constants in the inequality must be changed and the extremals change. The natural boundary conditions that replace $y^{(n-1)}(a) = 0$ and $y^{(n-1)}(b) = 0$ are $y^{(n)}(a) = 0$ and $y^{(n)}(b) = 0$. Lemmas 2.1 and 2.2 can be appropriately modified.

Example. For $n = 2$ and $[a,b] = [-1,1]$.

$$y(x) = \begin{cases} 1-\frac{3}{2}x^2 - \frac{1}{2}x^3 & \text{for } -1 \le x \le 0 \\ 1-\frac{3}{2}x^2 + \frac{1}{2}x^3 & \text{for } 0 \le x \le 1 \end{cases}$$

$$\int_{-1}^{1} |y(x)y'(x)|dx = 1$$

$$\int_{-1}^{1} y^{(2)}(x)^2 dx = \int_{-1}^{0}(-3-3x)^2 dx + \int_{0}^{1}(-3+3x)^2 dx$$

$$= 6.$$

Consequently if $y \in C^2[a,b]$ and $y(a) = y(b) = 0$, then

$$\int_a^b |y(x)y'(x)|dx \le \frac{(b-a)^3}{48} \int_a^b y^{(2)}(x)^2 dx.$$

The argument for inequality (1.3) could be extended to this hypothesis, but the constant in (1.3) is not best possible.

REMARKS. It is also possible to delete the condition $y^{(n-1)}(a) = 0$ and retain the condition $y^{(n-1)}(b) = 0$. The extremal is no longer symmetric.

It has been suggested that the first part of the proof of

Theorem 3.1 could be shortened by use of measure preserving rearrangements [2]. This approach does not appear to be helpful.

After the conference, Dr. A. Clausing pointed out that Lemma 2.1 had already been proved by Pólya in a different way. The reference is Satz II of "Bemerkungen zur Interpolation und zur Näherungstheorie der Balkenbiegung" by G. Pólya [Z. angew. Math. Mech. 11 (1931), 445-449].

REFERENCES

1. P.R. Beesack, Elementary proofs of some Opial-type integral inequalities. J. d'Analyse Math. 36 (1979), 1-14.

2. G.H. Hardy, J.E. Littlewood and G. Pólya, Inequalities. Cambridge University, Cambridge, 2nd Edition, 1952.

3. D.S. Mitrinović, Analytic Inequalities. Springer-Verlag, Berlin and New York, 1970.

This work was supported in part by NSF grant MCS81-03380.

Carl H. FitzGerald, Mathematics Department, University of California at San Diego, San Diego, California 92093, U.S.A.

International Series of
Numerical Mathematics, Vol. 71

TWO CONTRIBUTIONS TO INEQUALITIES

Alexander Kovačec

Abstract. We study a certain class of cyclic inequalities and - independently - give some new applications of majorization techniques.

1. INTRODUCTION

The two subjects of this paper, to be found in sections 3 and 4, can be read independently. In section 3 we study conditions for a function $s = s(x,y)$ of two variables in order that inequalities of the form $\sum_i s(x_i,x_{i+1}) \geq 0$ $(x_{n+1} = x_1)$ will hold. We give an application by proving an apparently new inequality.

Section 4 gives some unfamiliar applications of majorization techniques. We prove certain polynomial as well as integral inequalities.

2. NOTATION

The numeration of theorems, lemmas, formulas, etc. begins with 1 in each section. By independence of sections this will not lead to ambiguities. The symbols $\mathbb{N}$, $\mathbb{R}$ and $\mathbb{R}_+$ denote natural, real and nonnegative (we say positive) real numbers, respectively. Attributes, such as "positive", "increasing", etc. are to be understood in the weak sense ("increasing" = "nondecreasing", etc.).

3. A GENERAL CYCLIC INEQUALITY

In Theorem 2 (infra) we will give an interesting application of the following theorem.

THEOREM 1. *Let* $I \subseteq \mathbb{R}$ *be an interval and let* $s: I \times I \to \mathbb{R}$ *be a function such that*

(i) *the function* $x \mapsto s(x,x)$ $(x \in I)$ *is increasing*,

(ii) *the function* $x \mapsto s(a,x) + s(x,b)$ $(a,b,x \in I)$ *is increasing whenever* $x \geq \max\{a,b\}$.

Then, given $n \in \mathbb{N}$, *for each* $(x_1, x_2, \dots, x_n) \in I^n$ *there holds the inequality*

$$\sum_{i=1}^{n} s(x_i, x_{i+1}) \geq ns(a,a) \qquad (x_{n+1} = x_1,\ a = \min_{1 \leq i \leq n} x_i).$$

Proof. Let $\underline{x} = (x_1, x_2, \dots, x_n)$ and $\underline{e} = (1,1,\dots,1)$. For any subset $J \subseteq \{1,2,3,\dots,n\}$ let $\underline{e}(J)$ be the n-tuple with components $\underline{e}(J)_i = 1$ if $i \in J$ and $\underline{e}(J)_i = 0$ if $i \notin J$. We may assume $s(a,a) = 0$ for convenience. Clearly, for $\underline{x}$ there exists a decomposition

$$\underline{x} = a\underline{e} + d_1\underline{e}(J_1) + d_2\underline{e}(J_2) + \dots + d_k\underline{e}(J_k)$$

with $d_1, d_2, \dots, d_k > 0$ and $\{1,2,3,\dots,n\} \supset J_1 \supset J_2 \supset \dots \supset J_k \neq \emptyset$.

Let $G(\underline{x}) = \sum_{i=1}^{n} s(x_i, x_{i+1})$, and for $0 \leq m \leq k$ define $\underline{x}^{(m)} = a\underline{e} + \sum_{i=1}^{m} d_i e(J_i)$, in particular $\underline{x}^{(0)} = a\underline{e}$. Obviously, $G(\underline{x}^{(0)}) = 0$ since $s(a,a) = 0$. We assume that $G(\underline{x}^{(m)}) \geq 0$, where $m < k$ is fixed, and we shall prove that $G(\underline{x}^{(m+1)}) \geq 0$. Let $d = d_{m+1}$ and $J = J_{m+1}$. We can write

$$\underline{e}(J) = \underline{e}(I_1) + \underline{e}(I_2) + \dots + \underline{e}(I_s), \tag{1}$$

where each of the I's denotes an interval modulo n, that is, with some $i \leq j$ has either the form $\{i, i+1, \dots, j\}$ or $\{j, j+1, \dots, n, 1, \dots i\}$. Let us also assume that s is minimal in the representation (1) so that the union of no two intervals I can be replaced by a new interval. (For example, if $n = 10$ and $J = \{1,2,3,5,6,8,10\}$ then $I_1 = \{10,1,2,3\}$, $I_2 = \{5,6\}$, $I_3 = \{8\}$ (apart from renumbering the intervals) is the only admissible representation.) For $1 \leq p \leq s$ let $\underline{x}^{(m,p)} = \underline{x}^{(m)} + \sum_{i=1}^{p} d\underline{e}(I_i)$ and, by convention, $\underline{x}^{(m,0)} = \underline{x}^{(m)}$. Assume that $G(\underline{x}^{(m,p-1)}) \geq 0$ is true. Let $I_p = \{i, \dots, j\}$ $(i \leq j)$ and $\underline{u} = \underline{x}^{(m,p-1)}$, $\underline{v} = \underline{x}^{(m,p)}$. Then $u_q = v_q$ for $q \notin I_p$ and $u_q = D$, $v_q = D+d$ for $q \in I_p$, where $D = a + d_1 + d_2 + \dots + d_m$. Hence

$$G(\underline{x}^{(m,p)})-G(\underline{x}^{(m,p-1)}) = G(\underline{v})-G(\underline{u}) = \sum_{q=i-1}^{j} s(v_q,v_{q+1})-s(u_q,u_{q+1}) .$$

The middle terms are positive due to (i),

$$s(v_q,v_{q+1})-s(u_q,u_{q+1}) = s(D+d,D+d)-s(D,D) \geq 0 \quad \text{for} \quad i \leq q < j .$$

The remaining two terms are

$$s(v_{i-1},v_i)+s(v_j,v_{j+1})-s(u_{i-1},u_i)-s(u_j,u_{j+1})$$

$$= s(v_{i-1},D+d)+s(D+d,v_{j+1})-[s(v_{i-1},D)+s(D,v_{j+1})] ,$$

which is positive due to (ii). Hence $G(\underline{v})-G(\underline{u}) \geq 0$, and a similar reasoning yields the same conclusion in the case $I_p = \{j,\dots,n,1,\dots,i\}$. This readily implies $G(\underline{x}^{(m+1)})-G(\underline{x}^{(m)}) = G(\underline{x}^{(m,s)})-g(\underline{x}^{(m,0)}) \geq 0$, which by induction completes the proof of $G(\underline{x}) = G(\underline{x}^{(k)}) \geq 0$. □

A very simple implication is the following well known inequality.

COROLLARY 1. _If_ $a_1,a_2,\dots,a_n > 0$, _then_

$$n \leq \frac{a_1}{a_2} + \frac{a_2}{a_3} + \dots + \frac{a_n}{a_1} .$$

Equivalently,

$$\sqrt[n]{a_1 a_2 \cdots a_n} \leq \frac{1}{n} \sum_{i=1}^{n} a_i .$$

Proof. Put $I = (0,\infty)$ and $s(x,y) = \frac{x}{y}$ in Theorem 1 to obtain the first inequality. The inequality of the arithmetic and geometric means is a well known consequence obtained by putting $b_i = \frac{a_i}{a_{i+1}}$ (modulo n) and observing that $\prod_{i=1}^{n} b_i = 1$. □

THEOREM 2. _There holds the inequality_

$$n\left(\frac{3-2\sqrt{2}}{2\sqrt{2}}\right) \min_{1\leq i\leq n} \sqrt{x_i} + \left(1 - \frac{1}{2\sqrt{2}}\right) \sum_{i=1}^{n} \sqrt{x_i} \leq \sum_{i=1}^{n} \frac{x_i}{\sqrt{x_i+x_{i+1}}} \qquad (x_{n+1} = x_1)$$

for each $(x_1,x_2,\dots,x_n) \in (0,\infty)^n$. _If even_ $r(x_1,x_2,\dots,x_n) \in [1,2+\sqrt{13}]^n$ _for some_ $r > 0$, _then_

$$\frac{1}{\sqrt{2}} \sum_{i=1}^{n} \sqrt{x_i} \le \sum_{i=1}^{n} \frac{x_i}{\sqrt{x_i+x_{i+1}}} \qquad (x_{n+1} = x_1) .$$

<u>Proof.</u> Let us define

$$s(x,y) = \frac{x}{\sqrt{x+y}} - C\sqrt{x}$$

and let us look for restrictions on C in order that conditions (i) and (ii) of Theorem 1 are satisfied. (i) is satisfied if and only if $C \le 1/\sqrt{2}$. As to (ii), one sees that it is sufficient to require that

$$(2) \qquad \frac{d}{dx}\left(\frac{b}{\sqrt{b+x}} + \frac{x}{\sqrt{x+a}} - C\sqrt{x}\right) \ge 0 \quad \text{for } a,b,x \in I, \quad x \ge \max\{a,b\} .$$

This is equivalent with the inequality

$$b(b+x)^{-3/2}+x(x+a)^{-3/2}+Cx^{-1/2} \le 2(x+a)^{-1/2} \quad \text{for } a,b,x \in I, \ x \ge \max\{a,b\} .$$

In investigating this inequality we distinguish two cases. By homogeneity we may assume $r = 1$ if $I = [1,2+\sqrt{13}]$.

Case 1: $0 < a \le b \le x$. Upon homogeneity we may assume $a = 1$. One easily verifies for $1 \le b \le b' \le x$ that $b(b+x)^{-3/2} \le b'(b'+x)^{-3/2}$ so that we may assume the further simplification $b = x$. With these assumptions ($a = 1$, $b = x$) a simple calculation yields the condition

$$(3) \qquad 0 \le (1-D)x^3 + (4-3D)x^2+(4-3D)x-D , \quad D = \left(C + \frac{1}{2\sqrt{2}}\right)^2 , \quad x \ge 1 ,$$

which we require for the cases $\underline{x} \in (0,\infty)^n$ and $\underline{x} \in [1,2+\sqrt{13}]^n$. In the first case (3) has to hold for all $x \ge 1$. This is indeed guaranteed if we put $D = 1$, i.e. $C = 1 - 1/2\sqrt{2}$. In the second case originally $a,b,x \in [1,2+\sqrt{13}]$ and this remains true if we multiply these numbers by $1/a$, which leads to $a = 1$. For $C = 1/\sqrt{2}$, i.e. $D = 9/8$, (3) reduces to $P(x) \le 0$, where $P(x) = x^3-5x^2-5x+9 = (x-1)(x^2-4x-9)$. This inequality is easily shown to hold for $x \in [1,2+\sqrt{13}]$.

Case 2: $0 < b \le a \le x$. In this case we can assume $b = 1$. We have to show that

$$(4) \qquad (1+x)^{-3/2} + Cx^{-1/2} \le \min_{1\le a\le x} (x+2a)(x+a)^{-3/2} .$$

For fixed x, the function $a \to \phi(a) \equiv (x+2a)(x+a)^{-3/2}$ has a maximum for $x = 2a$ and no minimum in $(1,\infty)$. Hence, $\min_{1\le a\le x} \phi(a) = \min\{\phi(1),\phi(x)\} = \min\{(x+2)(x+1)^{-3/2}, \frac{3}{\sqrt{8x}}\}$. $\phi(1) \ge \phi(x)$ if and only if $P(x) \le 0$, i.e. certainly if $x \in [1,2+\sqrt{13}]$. One sees that (4) holds with $C = 1/\sqrt{2}$ for these x. In the other case $x \ge 2+\sqrt{13}$, where $C = 1 - 1/2\sqrt{2}$ is chosen, we have to show $(\sqrt{8}-1)^2(x+1)^3 \le 8(x^3+2x^2+x)$, which is easily shown to hold true.

Altogether, we have shown that we can choose $C = 1/\sqrt{2}$ if $x \in [1,2+\sqrt{13}]$ and $C = 1 - 1/2\sqrt{2}$ if $x \in (0,\infty)^n$ in order that the function $s(x,y)$ satisfies the hypotheses of Theorem 1, which then yields the conclusions. □

REMARKS. The numerical values of the constants are $1/2\sqrt{2} \approx 0.65$ and $1/\sqrt{2} \approx 0.71$. In spite of this small difference, we must not replace $1 - 2\sqrt{2}$ by $1/\sqrt{2}$ in the general case. An easily calculated counterexample emerges for $n = 9$. We put $\underline{x} = (1,2,2^2,\dots,2^8)$. For this $\underline{x}$ $\frac{1}{\sqrt{2}} \sum_{i=1}^{n} \sqrt{x_i} \approx 36.92$, but $\sum_{i=1}^{n} \frac{x_i}{\sqrt{x_i+x_{i+1}}} \approx 36.87$. Further results on this inequality are due to Professors FitzGerald, Talenti und König (private communication). V. Losert pointed out that the inequality remains true for $C = 1/\sqrt{2}$ whenever $n \le 6$. The author conjectures that $\frac{1}{\sqrt{2}} \sum_{i=1}^{\infty} \sqrt{x_i} \le \sum_{i=1}^{\infty} \frac{x_i}{\sqrt{x_i+x_{i+1}}}$ holds whenever the right hand side converges.

4. REMARKS ON MAJORIZATION

The starting point of this section is a well known inequality in majorization theory (associated with many names, such as Schur, Karamata, Tomić, Weyl and others), from which we derive some not too easily seen consequences (see [3] for the history of this inequality).

Given two n-tuples $\underline{u} = (u_1,u_2,\dots,u_n)$, $\underline{v} = (v_1,v_2,\dots,v_n)$ satisfying $u_1 \le u_2 \le \dots \le u_n$ and $v_1 \le v_2 \le \dots \le v_n$, we write

$$\underline{u} \prec \underline{v} \text{ iff for all } k \in \{1,2,\dots,n\} \text{ we have } \sum_{i=k}^{n} u_i \le \sum_{i=k}^{n} v_i ,$$

with equality for $k = n$.

The theorem in question (see [2, Theorem 108]) reads as follows.

THEOREM. *If* $F: \mathbb{R} \to \mathbb{R}$ *is a convex function and if* $(u_1, u_2, \dots, u_n) \prec (v_1, v_2, \dots, v_n)$, *then there holds the inequality* $\sum_{i=1}^{n} F(u_i) \le \sum_{i=1}^{n} F(v_i)$.

From this we derive a preliminary theorem.

THEOREM 1. *Assume* $\{f_i\}_{i=0}^{n}$ *and* $\{g_i\}_{i=0}^{n}$ *to be two sequences of positive real numbers, and assume* $\{x_i\}_{i=0}^{n}$ *to be an increasing such sequence. If*

(i) $$\sum_{i=0}^{n} f_i x_i = \sum_{i=0}^{n} g_i x_i ,$$

(ii) $$\sum_{i=0}^{n} f_i = \sum_{i=0}^{n} g_i ,$$

(iii) *for some* $k_o \in \{0, 1, 2, \dots, n\}$ *we have*

$$\sum_{i=0}^{k} (f_i - g_i) = \begin{cases} \le 0 & \text{for } k \le k_o , \\ \ge 0 & \text{for } k > k_o , \end{cases}$$

then for every convex function $F: \mathbb{R} \to \mathbb{R}$ *there holds the inequality*

$$\sum_{i=0}^{n} f_i F(x_i) \le \sum_{i=0}^{n} g_i F(x_i) .$$

Proof. Let us assume the f_i and g_i to be rational numbers first. In this case we may assume even that the f_i, g_i are natural numbers. Let $N = \sum f_i = \sum g_i$ and consider the following increasing N-tuples:

$$\underline{u} = (\underbrace{x_1, x_1, \dots, x_1}_{f_1}, \underbrace{x_2, x_2, \dots, x_2}_{f_2}, \dots, \underbrace{x_n, x_n, \dots, x_n}_{f_n}) ,$$

$$\underline{v} = (\underbrace{x_1, x_1, \dots, x_1}_{g_1}, \underbrace{x_2, x_2, \dots, x_2}_{g_2}, \dots, \underbrace{x_n, x_n, \dots, x_n}_{g_n}) .$$

Condition (i) guarantees that $\sum_{1}^{N} u_i = \sum_{1}^{N} v_i$ so that it is easily seen from (iii) that $\underline{u} \prec \underline{v}$. The conclusion follows immediately from the cited theorem above. For arbitrary real numbers one obtains the conclusion from the density of $\mathbb{Q}$ in $\mathbb{R}$. □

COROLLARY 1. *Let*

$$f(t) = \sum_{j=0}^{n} f_j t^j \quad \text{and} \quad g(t) = \sum_{j=0}^{n} g_j t^j$$

be two polynomials with positive coefficients f_j, g_j. *If* $f(1) = g(1)$, $f'(1) = g'(1)$ *and if for some* $k_o \in \{1,2,\dots,n\}$ *we have*

$$\sum_{j=0}^{k} (f_j - g_j) \begin{cases} \le 0 & \text{for } k \le k_o\,, \\ \ge 0 & \text{for } k > k_o\,, \end{cases}$$

then

$$f(t) \le g(t) \quad (t \ge 0)\ .$$

Proof. Choose $t \ge 0$ and define $F(x) = t^x$. Then $F\colon \mathbb{R} \to \mathbb{R}$ is convex. Put $x_i = i$ for $i = 0,1,2,\dots,n$. Then use Theorem 1. □

The following result was established by De Temple and Robertson [1]. Similar results can be found in the paper [5] of Menon.

COROLLARY 2. *For* $n \ge 1$, $a,b > 0$ *let*

$$h_n = h_n(a,b) = \frac{1}{n+1} \sum_{j=0}^{n} a^{n-j} b^j\ .$$

Then

$$h_{n-1} h_{n+1} - h_n^2 \ge 0\ .$$

Proof. For $0 \le m \le 2n$ define $D_m = |\{(k,j)\colon 0 \le k \le (n-1),\ 0 \le j \le (n+1),\ k+j = m\}|$ and $E_m = |\{(k,j)\colon 0 \le k \le n,\ 0 \le j \le n,\ k+j = m\}|$. One easily establishes the following table.

m	0	1	2	...	(n-1)	n	(n+1)	...	(2n-1)	2n
E_m	1	2	3	...	n	(n+1)	n	...	2	1
D_m	1	2	3	...	n	n	n	...	2	1

It follows from $\sum E_m = (n+1)^2$ and $\sum D_m = n(n+2)$ that the function

$$k \mapsto \sum_{\iota=0}^{k} \left(\frac{E_{2n-\iota}}{(n+1)^2} - \frac{D_{2n-\iota}}{n(n+2)} \right)$$

changes its sign from negative to positive as k increases from 0 to 2n. Putting $t = a/b$ it is (by homogeneity) sufficient to establish

$$h_{n-1}(t,1)h_{n+1}(t,1) \geq h_n^2(t,1) .$$

Now

$$h^2(t,1) = \frac{1}{(n+1)^2} \sum_{\iota=0}^{2n} E_{2n-\iota} t^\iota$$

and

$$h_{n-1}(t,1)h_{n+1}(t,1) = \frac{1}{n(n+2)} \sum_{\iota=0}^{2n} D_{2n-\iota} t^\iota .$$

It follows from the symmetry of the h_n's in a and b that their derivatives with respect to t for $t = 1$ are equal (to zero). Corollary 1 then gives the conclusion. □

REMARK. It follows from a well known trick of McLaurin ([2, p.52]) that

$$h_1^1 \leq h_2^{1/2} \leq h_3^{1/3} \leq \dots \leq h_n^{1/n} .$$

Finally, let us assume Theorem 1 to be stated in terms of the numbers $r_i = g_i - f_i$ so that conditions (i) and (ii) change to $\sum r_i x_i = 0$ and $\sum r_i = 0$. Similar remarks apply to (iii) and the conclusion. (Note that these formulations are equivalent, since by separating negative and positive r's in the new formulation, we arrive at the formulation of Theorem 1.) Let us call a function F defined on [0,1] a (+,-)-<u>wave</u> if it has the following properties:

(i) $F(0) = 0 = F(1)$;

(ii) There exists an $a \in [0,1]$ such that

$$F|_{[0,a)} \geq 0 \quad \text{and} \quad F|_{[a,1)} \leq 0 .$$

The integral analogue of the reformulated Theorem 1 is easily established:

THEOREM 2. *Let* $f,x\colon \mathbb{R}\to\mathbb{R}$ *be continuous functions*, x *increasing. Assume*

(i) $$\int_0^1 f(t)x(t)dt = 0 ;$$

(ii) *the function*

$$s \mapsto \int_0^s f(t)dt$$

is a $(+,-)$-*wave*.

Then for each convex continuous function $F\colon \mathbb{R}\to\mathbb{R}$

$$\int_0^1 f(t)F(x(t))dt \geq 0 .$$

Inequalities such as

$$\int_0^{2\pi} F(t)\cos t\, dt \geq 0$$

for each convex function F are immediate consequences of this theorem (see also [7]).

Acknowledgement. The author wants to thank the referee for streamlining the proof of Theorem 1 in Section 3.

REFERENCES

1. D.W. De Temple and J.M. Robertson, On generalized symmetric means of two variables. Univ. Beograd. Publ. Elektrotechn. Fak. Ser. Mat. Fiz. No. 634 - No. 677 (1979), 236-238.

2. G.H. Hardy, J.E. Littlewood and G. Pólya, Inequalities. Cambridge Univ. Press, London and New York, 2nd Edition, 1952.

3. A.W. Marshall and I. Olkin, Inequalities: Theory of Majorization and its Applications. Academic Press, New York, 1979.

4. A.W. Marshall and I. Olkin, Inequalities via Majorization - an introduction. In: E.F. Beckenbach and W. Walter, General Inequalities 3, pp.37-46, Birkhäuser Verlag, Basel, 1980.

5. K.V. Menon, Inequalities for symmetric functions. Duke Math. J. 35 (1968), 37-45.

6. D.S. Mitrinović, Analytic Inequalities. Springer Verlag, Berlin and New York, 1970.

7. P.M. Vasić, Notes on convex functions III: On the Fourier coefficients of convex functions. Univ. Beograd. Publ. Elektrotechn. Fak. Ser. Mat. Fiz. No. 602 - No. 633 (1978), 93-95.

Alexander Kovaćec, Institut für Mathematik, Universität Wien, Strudlhofgasse 4, A-1090 Wien, Austria.

International Series of
Numerical Mathematics, Vol. 71

LINKS BETWEEN SOME GENERALIZATIONS OF HARDY'S INTEGRAL INEQUALITY

E.R. Love

Abstract. Some generalizations of Hardy's Inequality were given by J. Kadlec and A. Kufner in 1967, and several more of a different kind by E.T. Copson in 1976. In this paper a generalization is given which contains typical inequalities of both kinds as quite special cases, and which estimates a norm of a fairly general integral transform.

1. INTRODUCTION

Hardy's Integral Inequality [2: Theorem 327] may be stated as follows. *If* $p > 1$, $\|\cdot\|_p$ *is the* $L^p(0,\infty)$ *norm, and* f *is measurable and non-negative on* $(0,\infty)$, *then*

$$\left\|\frac{1}{x}\int_0^x f(t)\,dt\right\|_p \le \frac{p}{p-1}\|f\|_p$$

and the constant on the right cannot be reduced.

In order to cope with certain singular situations regarding functions with zero traces, Kadlec and Kufner [3: Lemma 3] gave the following modifications of Hardy's Inequality, expressed for convenience in the notation of the present paper.

THEOREM K. *Let* F *be an infinitely differentiable complex-valued function with support in* $(0,1)$. *Let* $f(x) = F'(x)$ *and* $p \ge 1$.

(a) *If* λ, μ *are real,* $\lambda \neq 1$, *and* $R \ge e^{2|\mu/(\lambda-1)|}$, *then*

$$\left(\int_0^1 |F(x)|^p x^{-\lambda}\{\log(R/x)\}^{-\mu}\,dx\right)^{1/p}$$

$$\le \frac{2p}{|\lambda-1|}\left(\int_0^1 |f(x)|^p x^{p-\lambda}\{\log(R/x)\}^{-\mu}\,dx\right)^{1/p} .$$

b) *If instead* $\lambda = 1$, $\mu \neq 1$ *and* $R \geq 1$, *then*

$$\left(\int_0^1 |F(x)|^p x^{-1}\{\log(R/x)\}^{-\mu}\,dx\right)^{1/p} \leq \frac{p}{|\mu-1|}\left(\int_0^1 |f(x)|^p x^{p-1}\{\log(R/x)\}^{p-\mu}\,dx\right)^{1/p} .$$

Almost certainly independently, Copson [1] gave integral analogues of some series inequalities he had found many years earlier. Typical of the integral analogues is his Theorem 1, which may be written as follows, again expressed in the notation of this paper.

THEOREM C. *Let* $\phi(x)$ *have a continuous derivative in* $[0,\infty)$ *with* $\phi'(x) \geq 0$, *and let* $\phi(0) = 0$. *Let* $f(x)$ *be continuous in* $[0,\infty)$ *and non-negative, and*

$$F(x) = \int_0^x f(t)\phi'(t)dt .$$

If $p \geq 1$, $0 < b \leq \infty$, $c > 1$ *and the integral on the left converges, then*

$$\left(\int_0^b F(x)^p\phi(x)^{-c}\phi'(x)dx\right)^{1/p} \leq \frac{p}{c-1}\left(\int_0^b f(x)^p\phi(x)^{p-c}\phi'(x)dx\right)^{1/p} .$$

In the present paper Theorem 3 is a generalization of Hardy's Inequality which also includes both Theorems C and K. Theorem 3 is built on Theorem 1, which is an intermediate generalization. Theorem 2 includes the case $\lambda > 1$ of Theorem K(a), thereby showing that this case is a consequence of Theorem 1.

Theorems 4 and 5 are specializations of Theorem 3. In particular, Theorem 4 shows that Theorem C is a consequence of Theorem 3; and Theorem 5 shows that the case $\mu < 1$ of Theorem K(b) is a consequence of Theorem 3. Actually Theorem 5 is a good deal more general than that case.

The "constants" in the right sides of the inequalities obtained may not be the best possible. They do however reduce to the known best possible constant in Hardy's Inequality when the parameters are chosen so as to come down to that inequality.

Unless otherwise indicated, the functions occurring are from subsets of $(0,\infty)$ or $(0,\infty)^2$ to the complex numbers. Powers of $\phi(x)$ are written $\phi(x)^p$; and $\phi(x)$ is increasing means that $\phi(x') \le \phi(x'')$ whenever $x' < x''$.

$H(x,y)$ is homogeneous of degree m on S, a subset of $(0,\infty)^2$, means that $H(tx,ty) = t^m H(x,y)$ whenever (x,y) and (tx,ty) are in S.

2. A GENERALIZED HARDY'S INEQUALITY

The original Hardy's Inequality (integral version; see § 1) is the case $a = 1$, $b = \infty$, $q = 1$, $H(x,y) = 1/x$, $\omega(x) = 1$ of the following theorem, which deals with a moderately general integral transform Hf instead of the integral F which appears in Theorems C and K.

THEOREM 1. _If_ $p \ge 1$, q _is real_, $0 < a < b \le \infty$, $H(x,y)$ _is measurable and homogeneous of degree_ -1 _on_ $(0,b)^2$, _and_

$$C = \int_0^a |H(a,t)|(a/t)^{q/p}dt < \infty ;$$

if also $\omega(x)$ _is decreasing and positive in_ $(0,b)$, $f(x)$ _is measurable on_ $(0,b)$ _and_

$$||f|| = \left(\int_0^b |f(x)|^p x^{q-1}\omega(x)dx\right)^{1/p} < \infty ;$$

then

$$Hf(x) = \int_0^x H(x,y)f(y)dy$$

exists for almost all $x \in (0,b)$ _and_

$$||Hf|| \le C||f|| .$$

Proof. (i) As a function of t, $H(x,t)$ is measurable on $(0,b)$ for almost all x in $(0,b)$. If ξ is such a value of x, homogeneity gives that $H(a,t) = (\xi/a)H(\xi,(\xi/a)t)$; consequently $H(a,t)$ is a measurable function of t on $(0,a)$. This ensures existence (finite or infinite) of the integral for C, so that the hypothesis that this integral be finite is meaningful.

(ii) Suppose that H and f are non-negative. Then $Hf(x)$ exists (finite or infinite) for almost all $x \in (0,b)$. For such x,

$$Hf(x) = \int_0^a H(x,xt/a)f(xt/a)(x/a)dt = \int_0^a H(a,t)f(xt/a)dt .$$

Since Hf is measurable and non-negative on (0,b), we can consider

$$\|Hf\| = \left(\int_0^b \left(\int_0^a H(a,t)f(xt/a)dt\right)^p x^{q-1}\omega(x)dx\right)^{1/p}$$

$$\le \int_0^a \left(\int_0^b H(a,t)^p f(xt/a)^p x^{q-1}\omega(x)dx\right)^{1/p} dt \qquad (1)$$

$$= \int_0^a H(a,t)(a/t)^{q/p}\left(\int_0^b f(xt/a)^p (xt/a)^{q-1}\omega(x)(t/a)dx\right)^{1/p} dt$$

$$= \int_0^a H(a,t)(a/t)^{q/p}\left(\int_0^{bt/a} f(y)^p y^{q-1}\omega(ya/t)dy\right)^{1/p} dt$$

$$\le \int_0^a H(a,t)(a/t)^{q/p}\left(\int_0^b f(y)^p y^{q-1}\omega(y)dy\right)^{1/p} dt \qquad (2)$$

$$= C\|f\| . \qquad (3)$$

In (1) we have used Minkowski's Inequality, and in (2) the hypothesis that ω is decreasing.

By (3) $\|Hf\|$ is finite, and so its integrand is finite almost everywhere in (0,b); so also is Hf(x), since $x^{q-1}\omega(x) > 0$.

(iii) Finally suppose H and f are no longer required to be non-negative. By (ii),

$$H^+f(x) = \int_0^x |H(x,y)||f(y)|dy$$

is finite almost everywhere and $\|H^+f\| \le C\|f\| < \infty$. Thus Hf(x) exists almost everywhere in (0,b); and since $|Hf(x)| \le H^+f(x)$,

$$\|Hf\| \le \|H^+f\| \le C\|f\| .$$

REMARKS. Observe that C is independent of ω as well as of f. It is also independent of a in (0,b), because of the homogeneity of H. At first sight it might appear simpler to take $a = 1$, or $a = b$; but the former excludes $b < 1$ and the latter excludes $b = \infty$, for both of which the theorem holds when formulated as above.

3. AN INEQUALITY OF KADLEC AND KUFNER

These authors prove, in [3: Lemma 3(a)] (see Theorem K(a) in § 1) a case of the following theorem, which we obtain by specializing Theorem 1. Kadlec and Kufner's case of Theorem 2 is that in which $a = 0$, $b = 1$, $\kappa = 2$, $f(x)$ is infinitely differentiable with support in $(0,1)$, and $R \geq e^{2|\mu/(\lambda-1)|}$.

THEOREM 2. *If* $p \geq 1$, $\lambda > 1$, μ *is real*, $\kappa > 1$, $\kappa'^{-1} + \kappa^{-1} = 1$, $b > a \geq 0$, $R \geq b \max\{1, e^{\kappa\mu/(\lambda-1)}\}$, $f(x)$ *is measurable on* (a,b) *and the norm on the right below is finite, then*

$$F(x) = \int_a^x f(t)dt$$

exists for $a < x < b$ *and*

$$\left\{\int_a^b |F(x)|^p x^{-\lambda}\{\log(R/x)\}^{-\mu}\,dx\right\}^{1/p} \leq \frac{p\kappa'}{\lambda-1}\left\{\int_a^b |f(x)|^p x^{p-\lambda}\{\log(R/x)\}^{-\mu}\,dx\right\}^{1/p} . \qquad (4)$$

Proof. (i) Let $k = (\lambda-1)/\kappa$, a positive constant. Calculus shows that the positive function

$$\omega(x) = x^{-k}\{\log(R/x)\}^{-\mu} \quad \text{for } 0 < x < R ,$$

is decreasing wherever $x \leq Re^{-\mu/k}$. Since $R \geq b$, ω is certainly decreasing on $(0,b)$ if $b \leq Re^{-\mu/k}$. If $\mu > 0$ this is ensured by $R \geq be^{\mu/k} = be^{\mu\kappa/(\lambda-1)}$, while if $\mu \leq 0$ it is ensured merely by $R \geq b$. In either event, ω is decreasing and positive in $(0,b)$ under the stated hypothesis on R.

(ii) Suppose that $a = 0$. To apply Theorem 1 with its $a = \frac{1}{2}b$, $q = p - (\lambda-1)/\kappa'$, $H(x,y) = 1/x$ and $\omega(x)$ as in (i),

$$C = \int_0^1 s^{-q/p}ds = \frac{p}{p-q} = \frac{p\kappa'}{\lambda-1},$$

$C\|f\|$ simplifies to the right side of (4) and is therefore finite, and

$$Hf(x) = \frac{1}{x}\int_0^x f(t)dt = \frac{F(x)}{x}.$$

Theorem 1 now gives that $F(x)$ exists for almost all x in $(0,b)$, and therefore for all; and that (4) holds.

(iii) Suppose that $a > 0$. Extending the definition of f by writing $f(x) = 0$ for $0 < x < a$, the theorem follows from (ii).

REMARKS. Kadlec and Kufner also include in [3: Lemma 3(a)] a theorem like our Theorem 2 with $\lambda < 1$. That result is descended from a sister theorem of our Theorem 1 in a similar way. The remaining value $\lambda = 1$ is considered below in Theorem 5. Our notation puts λ, μ, F and f in place of Kadlec and Kufner's $-\beta$, $-\gamma$, f and f' respectively.

4. SOME SIMPLE LEMMAS

LEMMA 1. *If ϕ is increasing and differentiable in [a,b] then ϕ is absolutely continuous on [a,b].*

Proof. The derivative ϕ' is L-integrable on [a,b], by [4: § 11.54]. Also $\phi(x)-\phi(a)$ is the integral of its derivative, for $a \le x \le b$, by [4: § 11.83]; it is therefore absolutely continuous on [a,b].

LEMMA 2. *If ψ is strictly increasing and continuous in $[\alpha,\beta]$, and its inverse ϕ is absolutely continuous on $[a,b] = [\psi(\alpha),\psi(\beta)]$; and if g is (Lebesgue-) measurable on [a,b], then $g(\psi(x))$ is measurable on $[\alpha,\beta]$.*

Proof. We may suppose that g is real-valued. For real γ, $\{x\colon g(\psi(x)) > \gamma \;\&\; \alpha \le x \le \beta\} = \{\phi(u)\colon g(u) > \gamma \;\&\; a \le u \le b\}$; the right side is the ϕ-map of a measurable set S in [a,b]. Since ϕ is absolutely continuous on [a,b], $\phi(S)$ is measurable; as required.

LEMMA 3. *If ϕ, ψ, α, β, a, b are as in Lemma 2 and K(u,v) is measurable on $(a,b)^2$, then $K(\psi(x),\psi(y))$ is measurable on $(\alpha,\beta)^2$.*

Proof. As for Lemma 2.

5. A GENERALIZATION OF THEOREM 1

Theorem 3 (below) is the most general in this paper; the

others are all special cases of it, including Theorem 2. Theorem 2 differs from the others in that it is a special case of Theorem 1 as well. Accordingly it has been proved directly from Theorem 1 rather than from the more complicated Theorem 3. The leading idea in Theorem 3 is to extend Theorem 1 by a monotonic change of variable.

THEOREM 3. *Let* $0 < a < b \le \infty$; *let* $\phi(x)$ *be differentiable in* $(0,b)$ *with* $0 < \phi'(x) < \infty$, $\phi(0) = \phi(0+) = 0$, $\beta = \phi(b-) = \phi(b) \le \infty$; *and let* ψ *be the inverse of* ϕ.

Let $g(x)$ *and* $K(x,y)$ *be measurable on* $(0,b)$ *and* $(0,b)^2$ *respectively, and let* $K(\psi(x),\psi(y))$ *be homogeneous of degree* -1 *on* $(0,\beta)^2$.

If $p \ge 1$, q *is real*, $\omega(x)$ *is decreasing and positive in* $(0,b)$,

$$C = \int_0^a |K(a,t)| \left(\frac{\phi(a)}{\phi(t)}\right)^{q/p} \phi'(t)dt < \infty \ ,$$

and

$$\|g\| = \left(\int_0^b |g(x)|^p \phi(x)^{q-1} \omega(x)\phi'(x)dx\right)^{1/p} < \infty \ ,$$

then

$$Kg(x) = \int_0^x K(x,t)g(t)\phi'(t)dt$$

exists for almost all $x \in (0,b)$ *and*

$$\|Kg\| \le C\|g\| \ .$$

Proof. Let $\rho(x) = \omega(\psi(x))$ and $f(x) = g(\psi(x))$ for $0 < x < \beta$; also let $H(x,y) = K(\psi(x),\psi(y))$ for $0 < x,y < \beta$. Since ψ is increasing in $(0,\beta)$, ρ is decreasing and positive in it. By Lemma 1, ϕ is locally absolutely continuous in $(0,b)$. By Lemma 2, f is measurable on $(0,\beta)$. By Lemma 3, $H(x,y)$ is measurable on $(0,\beta)^2$.

Let $\alpha = \phi(a)$, so that $0 < \alpha < \beta \le \infty$. Since ψ is locally absolutely continuous in $(0,\beta)$, by Lemma 1, the substitutions $t = \psi(s)$ and $x = \psi(s)$ give

$$C = \int_0^\alpha |H(\alpha,s)|(\alpha/s)^{q/p} ds < \infty \tag{5}$$

and

$$\|g\| = \left(\int_0^\beta |f(s)|^p s^{q-1}\rho(s)ds\right)^{1/p} < \infty \ . \tag{6}$$

So Theorem 1 applies with a, b and ω replaced by α, β and ρ, giving existence of

$$\int_0^x H(x,y)f(y)dy = \int_0^x K(\psi(x),\psi(y))g(\psi(y))dy$$
$$= \int_0^{\psi(x)} K(\psi(x),t)g(t)\phi'(t)dt = Kg(\psi(x)) \quad (7)$$

for almost all $x \in (0,\beta)$. Since ψ is locally absolutely continuous in $(0,\beta)$, $Kg(s)$ exists for almost all $s \in (0,\beta)$. Writing $(Hf)(x)$ for the left side of (7),

$$\|Kg\| = \left\{\int_0^b |(Hf)(\phi(t))|^p \phi(t)^{q-1}\omega(t)\phi'(t)dt\right\}^{1/p}$$
$$= \left\{\int_0^\beta |(Hf)(s)|^p s^{q-1}\rho(s)ds\right\}^{1/p} .$$

This, with (5), (6), (7) and Theorem 1, gives the required inequality.

6. AN INEQUALITY OF COPSON

In [1: Theorem 1] Copson gives a case of the following theorem, which we obtain by specializing Theorem 3. Copson's case of Theorem 4 is substantially that in which $\omega(x) = 1$, $a = 0$, f is non-negative and f and ϕ' are continuous in $[a,b]$. He also makes a convergence restriction which seems unnecessary in our context; but on the other hand he does not need our requirement that ϕ' be strictly positive.

THEOREM 4. *If* $-\infty < a < b \le \infty$, $c > 1$, $p \ge 1$, $\phi(x)$ *is differentiable in* (a,b) *with* $0 < \phi'(x) < \infty$, $\phi(a) = \phi(a+) = 0$, $\omega(x)$ *is decreasing and positive in* (a,b), $f(x)$ *is measurable on* (a,b) *and the norm on the right below is finite, then*

$$F(x) = \int_a^x f(t)\phi'(t)dt$$

exists for all $x \in (a,b)$ *and*

$$\left\{\int_a^b |F(x)|^p \phi(x)^{-c}\omega(x)\phi'(x)dx\right\}^{1/p}$$
$$\le \frac{p}{c-1}\left\{\int_a^b |f(x)|^p \phi(x)^{p-c}\omega(x)\phi'(x)dx\right\}^{1/p} .$$

Proof. By translation it is evidently sufficient to prove the case $a = 0$. In Theorem 3 take $q = p-c+1$, $K(x,y) = 1/\phi(x)$ and $g = f$. With ψ and a defined as in that theorem, $K(\psi(x),\psi(y)) = 1/x$ is measurable and homogeneous of degree -1, and

$$C = \int_0^a \frac{1}{\phi(a)^{1-(q/p)}}\, \phi(t)^{-q/p}\phi'(t)dt = \frac{1}{1-(q/p)} = \frac{p}{c-1} \ ;$$

this evaluation uses the absolute continuity of ϕ ensured by Lemma 1 and the convergence of the integral ensured by $q-p = 1-c < 0$. Further,

$$Kg(x) = \int_0^x \frac{1}{\phi(x)}\, f(t)\phi'(t)dt = \frac{F(x)}{\phi(x)} \ .$$

Theorem 3 now gives the existence of $F(x)$ for almost all, and so for all, $x \in (0,b)$; and it gives also the required inequality.

REMARKS. Copson's paper also contains another theorem [1: Theorem 3] of the same kind as his Theorem 1 but with $c < 1$. This is descended in a similar way from a sister theorem of our Theorem 3. Copson's paper contains, further, another similar theorem with $c = 1$, and others with $0 < p < 1$; but there is not space here to consider them. Our notation differs from Copson's in that p, ϕ and ϕ' replace his k, Φ and ϕ respectively.

7. ANOTHER INEQUALITY OF KADLEC AND KUFNER

These authors give in [3: Lemma 3(b)] a case of our remaining theorem, Theorem 5. We obtain this by specializing our Theorem 4. It is thus also a special case of Theorem 3; but unlike Theorem 2 it does not seem to be a special case of Theorem 1.

Theorem 5 is like Theorem 2 with $\lambda = 1$ and $R \geq b$. Kadlec and Kufner's Lemma 3(b) with $\gamma > -1$ (that is, $\mu < 1$) is the case of Theorem 5 in which $a = 0$, $b = 1$, $\omega(x) = 1$ and $f(x)$ is infinitely differentiable with support in (0,1).

THEOREM 5. If $p \geq 1$, $\mu < 1$, $R \geq b > a \geq 0$, $\omega(x)$ is decreasing and positive in (a,b), $f(x)$ is measurable on (a,b) and the norm on the right below is finite, then

$$F(x) = \int_a^x f(t)dt$$

exists for $a < x < b$ and

$$\left(\int_a^b |F(x)|^p x^{-1}\{\log(R/x)\}^{-\mu}\,\omega(x)dx\right)^{1/p}$$
$$\leq \frac{p}{1-\mu}\left(\int_a^b |f(x)|^p x^{p-1}\{\log(R/x)\}^{p-\mu}\omega(x)\,dx\right)^{1/p}.$$

Proof. If $a = 0$, the results follow from Theorem 4 by taking $a = 0$, $c = 2-\mu > 1$, $\phi(x) = 1/\log(R/x)$ in $0 < x < b$, and replacing $f(x)$ by $f(x)/\phi'(x)$.

Suppose $a > 0$. Extending the definition of f by writing $f(x) = 0$ for $0 < x < a$, we may consistently write

$$F(x) = \int_0^x f(t)dt$$

for $0 < x < b$; this also makes $F(x) = 0$ for $0 < x < a$. Now let $a_n \downarrow a$ as $n \uparrow \infty$, and

$$\omega_n(x) = \omega(a_n) \text{ for } 0 < x \leq a_n, \quad \omega_n(x) = \omega(x) \text{ for } a_n < x < b.$$

Each ω_n is finite, positive and decreasing in $(0,b)$. The preceding paragraph now gives the stated conclusions but with ω replaced by ω_n; the finiteness of the norm on the right being ensured by $0 < \omega_n(x) \leq \omega(x)$. Since also $\omega_n(x) \uparrow \omega(x)$ as $n \uparrow \infty$, the required result with ω unreplaced now follows by monotonic convergence.

REMARKS. Kadlec and Kufner also include in [3: Lemma 3(b)] a theorem like our Theorem 5 specialized, but with $\mu > 1$ as well as $\omega(x) = 1$. This is descended from a sister theorem of our Theorem 1 in a similar way.

8. ACKNOWLEDGMENT

I am very grateful to Dr. P. Grisvard, of the UNESCO-supported Centre Internationale pour Mathématiques Pures et Appliquées at Nice, who kindly drew my attention to Reference 3 and obtained a copy of it for me. Without it this work would not have been done.

REFERENCES

1. E.T. Copson, Some integral inequalities. Proc. Roy. Soc. Edinburgh 75A (1975-76), 157-164.

2. G.H. Hardy, J.E. Littlewood and G. Pólya, Inequalities. Cambridge, 1934.

3. J. Kadlec and A. Kufner, Characterization of functions with zero traces by integrals with weight functions II. Časopis Pěst. Mat. 92 (1967), 16-28.

4. E.C. Titchmarsh, The theory of functions. Oxford, 2nd edition, 1939.

E.R. Love, Department of Mathematics, The University of Melbourne, Parkville, Victoria 3052, Australia.

International Series of
Numerical Mathematics, Vol. 71

INEQUALITIES FOR COMPARISON OF MEANS

Zsolt Páles

Abstract. In the present note we investigate the general comparison inequality of means, i.e. the inequality

(0) $$C(M(x_1,\dots,x_n),N(x_1,\dots,x_n)) \le 0 ,$$

where C is a comparative function, M and N are given symmetric means, and n and $x_1,\dots,x_n$ run over the set of positive integers and over a real interval I, respectively. The main result of this paper states that if M and N are upper and lower semiintern repetition invariant means, respectively, and if one of them is infinitesimal, then (0) holds if and only if

$$C(M(\underbrace{x,\dots,x}_{k},\underbrace{y,\dots,y}_{m}),N(\underbrace{x,\dots,x}_{k},\underbrace{y,\dots,y}_{m})) \le 0$$

is satisfied for any nonnegative integers k, m with $k+m>0$ and x, y in I. The most important special cases of this result, the problems of the comparison and complementary comparison, are discussed in detail in this paper.

1. INTRODUCTION

In the theory of means the oldest and the most classical problem is to find necessary and sufficient conditions in order that two given means M and N belonging to a given class be comparable, i.e. that

(1) $$M(x_1,\dots,x_n) \le N(x_1,\dots,x_n)$$

be satisfied for any positive integer n and $x_1,\dots,x_n$ from a given real interval I.

The simplest conditions can be obtained in the class of quasiarithmetic means (see [9]). Namely, Jessen [10] proved that if M and N are quasiarithmetic means then (1) holds if and only if

(2) $$M(x,y) \le N(x,y) \quad (x,y \in I) .$$

The necessity of this condition is obvious. The proof of the sufficiency can be done by the help of Cauchy-type induction.

In 1958 Bajraktarevič [1] generalized the concept of quasiarithmetic means by defining quasiarithmetic means with weight function. The investigation of the comparison inequality (1) for means M and N of this kind was initiated by Bajraktarevič [2] in 1969. However, he obtained sufficient conditions and necessary conditions. The first necessary and sufficient conditions were discovered by Daróczy and Losonczi [7] in the class of homogeneous quasiarithmetic means with continuous weight function. Generalizing the method of [7] and assuming some regularity properties, Losonczi [12] found the criterion of the comparability of nonhomogeneous means of this type. Without any regularity assumptions this problem was completely solved by Daróczy and the author [8] in 1982. We proved that if M and N are quasiarithmetic means with weight function then (1) holds if and only if

$$M(\underbrace{x,\dots,x}_{k},\underbrace{y,\dots,y}_{m}) \le N(\underbrace{x,\dots,x}_{k},\underbrace{y,\dots,y}_{m}) \tag{3}$$

for any natural numbers k, m and x, y in I. This result is the special case of a more general one. Namely, in [8], inequality (1) was considered for deviation means (see [5]), and it was proved that (3) is a necessary and sufficient condition for (1) if M and N are deviation means.

Now compare inequalities (1), (2) and (3). Inequality (1) contains arbitrarily many independent variables, however inequalities (2) and (3) contain only a finite number of independent variables, namely x, y and x, y, k/m, respectively. Inequality (2) does not imply (1) in the classes of quasiarithmetic means with weight function and deviation means. This fact shows that we cannot expect inequality (2) to be a sufficient condition. Now consider inequality (3). It contains three independent variables and implies (1) if M and N are deviation means. Thus, it is natural to raise the following problem: What kind of assumptions are to be assumed in order that (3) implies (1)?

One can think that the proof of the sufficiency of (3) requires some upper and lower estimates concerning M and N, respectively. Such type of estimates are given by the characterization theorem of quasideviation means (see [14]). It is proved in [14] that every quasideviation mean is strongly intern. A simple weakening of this statement is that every deviation mean is intern, i.e., the mean value of any (m+n)-tuple lies between the mean values of the first m and last n components. With the help of this property we shall introduce the lower and upper semiintern means. Applying these concepts one could conjecture the following result: If M and N are upper and lower semiintern means, respectively, then (1) holds if and only if (3) is satisfied. This statement may be true, but I have not been able to prove it until now. I have obtained a somewhat weaker result: If one of the means M and N is an infinitesimal mean, then the above statement is valid (see Theorem 4 below). The infinitesimality can be regarded as a regularity property concerning means. It is satisfied in wide classes of means, e.g., in the class of deviation means (see Theorem 3 below).

In the theory of means the problem of the complementary comparison is also fundamental. The first result in this theory is due to Schweitzer [15], who obtained an upper bound for the ratio of the arithmetic and harmonic means. Schweitzer's result has been generalized in several ways; see, e.g., Specht [16], Cargo and Shisha [4]. In 1969 Beck [3] introduced the concept of comparative functions. Using this notion he obtained general complementary inequalities for quasiarithmetic means. His result generalizes those mentioned above. For deviation means the problem of complementary comparison was considered and discussed in detail in [13]. The results obtained are the generalizations of Beck's result. Using the Main Theorem of this paper we can give a new proof for this result (see Theorem 5 and Corollary 3 below). In the last example of this paper we derive Schweitzer's inequality with the help of Corollary 3.

2. BASIC CONCEPTS

Let I be a fixed real interval. A function

$$M: \bigcup_{n=1}^{\infty} I^n \to \mathbb{R}$$

is called a <u>symmetric mean</u> (or briefly a mean) on I if

(i) M is a symmetric function, i.e.,

$$M(x_1,\dots,x_n) = M(x_{p(1)},\dots,x_{p(n)})$$

for any positive integer n, $x_1,\dots,x_n$ in I and any permutation p of the index set $\{1,\dots,n\}$.

(ii) M satisfies the mean value property, i.e.,

$$\min(x_1,\dots,x_n) \le M(x_1,\dots,x_n) \le \max(x_1,\dots,x_n)$$

for any positive integer n and $x_1,\dots,x_n$ in I.

We remark that the mean value property is called internity in [14]. However, we shall use this word in a different meaning.

A mean M is called <u>repetition invariant</u> if

$$M(\underbrace{x_1,\dots,x_1}_{k},\dots,\underbrace{x_n,\dots,x_n}_{k}) = M(x_1,\dots,x_n)$$

for any positive integers k, n and $x_1,\dots,x_n$ in I.

For the sake of brevity the following simple notation is often used:

If $\underline{x}$ denotes the vector $(x_1,\dots,x_n)$ then $M(\underline{x})$ means $M(x_1,\dots,x_n)$. If $\underline{k} = (k_1,\dots,k_n)$ is an n-tuple of nonnegative integers with $k_1 + \dots + k_n > 0$ then $M(\underbrace{x_1,\dots,x_1}_{k_1},\dots,\underbrace{x_n,\dots,x_n}_{k_n})$ is denoted by $M(x_1,\dots,x_n;k_1,\dots,k_n)$ or by $M(\underline{x};\underline{k})$.

Using the latter notation we can easily prove that a mean M is repetition invariant if and only if

$$M(\underline{x};m\underline{k}) = M(\underline{x};\underline{k})$$

for any n-tuples $\underline{x}$, $\underline{k}$ and for any positive integers n and m.

Now, we introduce the concept of semideviation and deviation. It leads to some new classes of means.

A function

$$E: I^2 \to \mathbb{R}$$

is called a <u>semideviation</u> on I if

$$\operatorname{sgn} E(x,t) = \operatorname{sgn}(x-t)$$

for $x \neq t$ in I.

A semideviation E is called a <u>deviation</u> if the function

$$t \to E(x,t)\,, \qquad t \in I\,,$$

is continuous and strictly monotone decreasing for each fixed x in I.

Let $x_1,\dots,x_n$ be in I, let E be a semideviation on I and consider the sum

$$(4) \qquad E(x_1,t) + \dots + E(x_n,t)$$

for t in I. Since E is a semideviation, it is obvious that (4) is positive if t is less than $x_1,\dots,x_n$. Similarly, (4) is negative if t is greater than $x_1,\dots,x_n$. Hence there exists a maximal $\underline{t}_o$ between $\min\limits_{1\le i\le n} x_i$ and $\max\limits_{1\le i\le n} x_i$ such that (4) is positive if $t < \underline{t}_o$. This value $\underline{t}_o$ is called the <u>lower E-semideviation mean</u> of $x_1,\dots,x_n$ and is denoted by $\underline{M}_E(x_1,\dots,x_n)$. Analogously, the minimal value $\bar{t}_o$ for which (4) is negative if $\bar{t}_o < t$ is called the <u>upper E-semideviation mean</u> of $x_1,\dots,x_n$ and is denoted by $\bar{M}_E(x_1,\dots,x_n)$.

If E is a deviation, then we obtain $\underline{t}_o = \bar{t}_o = t_o$, since E is strictly decreasing in the second variable. Therefore, we write M_E instead of $\underline{M}_E$ and $\bar{M}_E$, and we say that t_o is the <u>E-deviation mean</u> of $x_1,\dots,x_n$. Since E is continuous in the second variable, it follows that (4) is equal to zero for $t = t_o$, i.e., t_o is the unique solution of

$$(5) \qquad E(x_1,t) + \dots + E(x_n,t) = 0\,.$$

It is easy to check that $\underline{M}_E$, $\bar{M}_E$, M_E are repetition invariant means.

Now we define the most important special deviation means. Let $f\colon I \to \mathbb{R}$ be a strictly increasing continuous function and let $p\colon I \to \mathbb{R}$ be a positive function. The mean $M_{f,p}$ defined by

$$M_{f,p}(x_1,\dots,x_n) = f^{-1}\left(\sum_{i=1}^{n} p(x_i)f(x_i) \Big/ \sum_{i=1}^{n} p(x_i)\right)$$

is called a <u>f-quasiarithmetic mean with weight function</u> p.

If we take $p = 1$ then the mean obtained is simply called <u>f-quasiarithmetic mean</u> and is denoted by M_f.

It is easily seen that quasiarithmetic means with weight function are deviation means. Namely, let

$$E(x,t) = p(x)(f(x)-f(t))$$

for x, t in I. Then, for this deviation E, equation (5) has a unique solution which is equal to $M_{f,p}(x_1,\dots,x_n)$. Hence we have $M_E = M_{f,p}$.

EXAMPLES. 1) Let M be defined as follows

$$M(x_1,\dots,x_n) = \begin{cases} x_1 & \text{for } n = 1 \\ \sqrt{\sum_{i<j} x_i x_j / \binom{n}{2}} & \text{for } n \geq 2 . \end{cases}$$

An easy calculation shows that M is not a repetition invariant mean.

2) The well-known arithmetic, geometric and harmonic means are quasiarithmetic means on $(0,\infty)$ (take $f(x) = x$, $\ln x$, $1/x$, respectively).

3) The mean

$$M(x_1,\dots,x_n) = \sum_{i=1}^{n} x_i^{k+1} \Big/ \sum_{i=1}^{n} x_i^k$$

is a quasiarithmetic mean with weight function on $(0,\infty)$ for any real k. It is quasiarithmetic if and only if either $k = 0$ or $k = -1$.

4) Let $E(x,t) = x^2-t^2+x(x-t)$ for $x,t > 0$. Then E is a deviation and

$$M_E(x_1,\dots,x_n) = -\frac{1}{2n}\sum_{i=1}^{n} x_i + \sqrt{\left(\frac{1}{2n}\sum_{i=1}^{n} x_i\right)^2 + \frac{2}{n}\sum_{i=1}^{n} x_i^2} .$$

It is easy to check that M is not a quasiarithmetic mean with weight function.

3. INTERN AND INFINITESIMAL MEANS

A symmetric mean M, defined on the interval I, is called a <u>lower</u> (resp. <u>upper</u>) <u>semiintern mean</u> if

$$\min(M(x_1,\ldots,x_n),M(y_1,\ldots,y_m)) \le M(x_1,\ldots,x_n,y_1,\ldots,y_m)$$

(resp. $M(x_1,\ldots,x_n,y_1,\ldots,y_m) \le \max(M(x_1,\ldots,x_n),M(y_1,\ldots,y_m)))$ for any natural numbers n, m and $x_1,\ldots,x_n$, $y_1,\ldots,y_m$ in I. A lower as well as an upper semiintern mean is called an <u>intern mean</u>.

It is obvious that intern means are automatically repetition invariant. However, in general, semiintern means are not repetition invariant; e.g., the mean defined in example 1) is lower semiintern, but not invariant under repetitions.

The concept of internity is somewhat weaker than strong internity. An intern mean M is strongly intern if both former inequalities are satisfied with the strict inequality sign if $M(x_1,\ldots,x_n) \neq M(y_1,\ldots,y_m)$ (see [14]).

In the following theorems we give the most important semiintern and intern means.

THEOREM 1. <u>Lower and upper semideviation means are lower and upper semiintern means, respectively.</u>

<u>Proof.</u> We prove that if E is a semideviation on I then $\underline{M}_E$ is lower semiintern. (The upper semiinternity of $\bar{M}_E$ can be proved similarly.)

Let $x_1,\ldots,x_n$, $y_1,\ldots,y_m$ be in I and let

$$t_o = \min(\underline{M}_E(x_1,\ldots,x_n),\underline{M}_E(y_1,\ldots,y_m)) .$$

By the definition of $\underline{M}_E$ we have

$$E(x_1,t)+\ldots+E(x_n,t) > 0 , \quad E(y_1,t)+\ldots+E(y_m,t) > 0$$

for $t < t_o$. Adding these inequalities we obtain

$$E(x_1,t)+\ldots+E(x_n,t)+E(y_1,t)+\ldots+E(y_m,t) > 0$$

for $t < t_o$. This implies

$$t_o \le \underline{M}_E(x_1,\ldots,x_n,y_1,\ldots,y_m) .$$

COROLLARY 1. <u>Deviation means are intern means. Especially, quasiarithmetic means and quasiarithmetic means with weight function are intern means.</u>

The latter result has been obtained in [14].

THEOREM 2. *Let* $f_1,\dots,f_k,g\colon I \to \mathbb{R}$ *be strictly increasing continuous functions. Assume that the quasiarithmetic mean* M_g *is less (resp. greater) than the other means* $M_{f_1},\dots,M_{f_k}$, *i.e.*,

(6) $$M_g \le M_{f_1},\dots,M_g \le M_{f_k}$$

(*resp.* $M_g \ge M_{f_1},\dots,M_g \ge M_{f_k}$). *Then the mean* M *defined by*

(7) $$M(\underline{x}) = M_g(M_{f_1}(\underline{x}),\dots,M_{f_k}(\underline{x}))$$

is a lower (resp. upper) semiintern mean.

Proof. Assume that (6) is satisfied and let $\underline{x} = (x_1,\dots,x_n)$ be in I^n and $y = (y_1,\dots,y_m)$ be in I^m. Using (6) and some simple observations on quasiarithmetic means we easily obtain

$$\begin{aligned} M(\underline{x},\underline{y}) &= M_g(M_{f_1}(\underline{x},\underline{y}),\dots,M_{f_k}(\underline{x},\underline{y})) \\ &= M_g(M_{f_1}(M_{f_1}(\underline{x}),M_{f_1}(\underline{y});n,m),\dots,M_{f_k}(M_{f_k}(\underline{x}),M_{f_k}(\underline{y});n,m)) \\ &\ge M_g(M_g(M_{f_1}(\underline{x}),M_{f_1}(\underline{y});n,m),\dots,M_g(M_{f_k}(\underline{x}),M_{f_k}(\underline{y});n,m)) \\ &= M_g(M_g(M_{f_1}(\underline{x}),\dots,M_{f_k}(\underline{x})),M_g(M_{f_1}(\underline{y}),\dots,M_{f_k}(\underline{y}));n,m) \\ &= M_g(M(\underline{x}),M(\underline{y});n,m)\,. \end{aligned}$$

The mean value property of M_g implies

$$M_g(M(\underline{x}),M(\underline{y});n,m) \ge \min\,(M(\underline{x}),M(\underline{y}))\,.$$

This proves the lower semiinternity of M. If M_g is greater than $M_{f_1},\dots,M_{f_k}$, then the proof of the upper semiinternity of M is similar.

EXAMPLE. Denote by A and G the arithmetic mean and the geometric mean on the interval $(0,\infty)$, respectively. Applying Theorem 2 we easily obtain that the means K and L defined by

$$K(\underline{x}) = \tfrac{1}{2}(A(\underline{x}) + G(\underline{x})) \qquad \text{and} \qquad L(\underline{x}) = \sqrt{(A(\underline{x})G(\underline{x}))}$$

are upper and lower semiintern means, respectively.

Now let $x_1 = 1$, $x_2 = 81$, $y_1 = y_2 = 25$. Then $K(x_1,x_2) = 25$ and $K(y_1,y_2) = 25$. However, $K(x_1,x_2,y_1,y_2) = 24 < 25$. Therefore, K is not a lower semiintern mean. Since, by Corollary 1, deviation means are intern means, we have proved that K cannot be generated by any deviation. A similar argument shows that L is not a deviation mean, either.

In what follows we define a regularity property for symmetric means. Its significance will become clear in the proof of the Main Theorem.

A symmetric mean M is said to be <u>infinitesimal</u> if for any $x \le s < t \le y$ there exist natural numbers k and m such that

$$s < M(x,y;k,m) < t .$$

We remark that in [14] a different concept of infinitesimality is used. A mean M is infinitesimal according to [14] if, for x, y in I,

$$\lim_{n\to\infty} \max_{1\le k\le n} |M(x,y;k,n-k) - M(x,y;k-1,n-k+1)| = 0 .$$

One can easily prove that the infinitesimality in this sense implies the infinitesimality in our sense. Moreover, for intern means these two concepts are equivalent to each other.

THEOREM 3. <u>Let</u> $E_1,\dots,E_k$ <u>be deviations on</u> I. <u>Furthermore</u>, <u>let</u> $m\colon I^k \to \mathbb{R}$ <u>be a continuous function such that</u>

$$\min(x_1,\dots,x_k) \le m(x_1,\dots,x_k) \le \max(x_1,\dots,x_k)$$

<u>for any</u> $x_1,\dots,x_k$ <u>in</u> I. <u>Then the mean</u> M <u>defined by</u>

$$M(\underline{x}) = m(M_{E_1}(\underline{x}),\dots,M_{E_k}(\underline{x}))$$

<u>is an infinitesimal mean</u>.

<u>Proof.</u> It is easy to check that the function M is really a symmetric mean on I. To prove the statement we need the following property of deviation means (see [8], [13]):

If E is a deviation, then, for x, y in I, there exists a continuous function $e\colon [0,1] \to I$ such that

$$M_E(x,y;k,m) = e\left(\frac{m}{k+m}\right) .$$

Now let $x \le s < t \le y$ be arbitrary but fixed. Let $e_1,\dots,e_k$ be the corresponding functions to $E_1,\dots,E_k$, respectively. Define f: $[0,1] \to I$ as follows

$$f(u) = m(e_1(u),\dots,e_k(u)), \quad u \in [0,1].$$

Then f is a continuous function and, furthermore, $f(0) = x$, $f(1) = y$. Hence there exists a rational u in (0,1) such that f(u) is in (s,t). If $u = p/q$, then we have

$$M(x,y;q-p,p) = m(M_{E_1}(x,y;q-p,p),\dots,M_{E_k}(x,y;q-p,p))$$
$$= m\left(e_1\left(\tfrac{p}{q}\right),\dots,e_k\left(\tfrac{p}{q}\right)\right) = f\left(\tfrac{p}{q}\right) \in (s,t),$$

i.e., M is infinitesimal.

Applying Theorem 3 for the case $k = 1$, we easily obtain that deviation means, quasiarithmetic means, quasiarithmetic means with weight function and the mean M defined by (7) are infinitesimal means.

4. MAIN RESULTS

The function $C: I^2 \to \mathbb{R}$ is called a comparative function on I if it is continuous on I^2 and if

$$x \mapsto C(x,y) \quad \text{and} \quad y \mapsto C(x,y)$$

are increasing and decreasing functions on I for any fixed y and x in I, respectively.

MAIN THEOREM. *Let C be a comparative function on I and let M and N be upper and lower semiintern repetition invariant symmetric means on I, respectively. Assume that one of the means M and N is infinitesimal. Then the inequality*

$$C(M(x_1,\dots,x_n),N(x_1,\dots,x_n)) \le 0 \tag{8}$$

holds for any positive integer n and $x_1,\dots,x_n$ in I if and only if

$$C(M(x,y;k,m),N(x,y;k,m)) \le 0 \tag{9}$$

is satisfied for any nonnegative integers k, m with $k+m > 0$ and x, y in I.

Proof. The necessity of inequality (9) is obvious. Hence it suffices to show that (9) implies (8).

Assume (9) to be valid and denote by P_n the following statement:

(10) $$C(M(x_1,\dots,x_n;k_1,\dots,k_n),N(x_1,\dots,x_n;k_1,\dots,k_n)) \le 0$$

for any positive integers $k_1,\dots,k_n$ and $x_1,\dots,x_n$ in I. We prove by induction that P_n is valid for any natural number n.

P_1 and P_2 obviously follow from (9). Let $n \ge 3$ be fixed and assume that P_{n-1} and P_{n-2} have been proved. In order to show that P_n is also valid, suppose M to be infinitesimal and choose $\underline{x} = (x_1,\dots,x_n)$ in I^n and $\underline{k} = (k_1,\dots,k_n)$ in $\mathbb{N}^n$. Without loss of generality we may assume that $x_1 \le \dots \le x_n$. If $M(x;k) = x_1$, then, by the properties of C and by (9), we get

$$C(M(\underline{x};\underline{k}),N(\underline{x};\underline{k})) \le C(x_1,x_1) \le 0 .$$

Therefore we can assume that $x_1 < M(\underline{x};\underline{k})$. Choose $\varepsilon > 0$ such that $x_1 < M(\underline{x};\underline{k}) - \varepsilon$. Since M is infinitesimal, there exist p_1, p_n in $\mathbb{N}$ such that

(11) $$M(\underline{x};\underline{k}) - \varepsilon < M(x_1,x_n;p_1,p_n) < M(\underline{x};\underline{k}) .$$

Let $\underline{p} = (p_1,0,\dots,0,p_n)$. Then $M(x_1,x_n;p_1,p_n)$ can be rewritten as $M(\underline{x};\underline{p})$. Further let

$$\frac{m}{q} = \min\left(\frac{k_1}{p_1},\frac{k_n}{p_n}\right) ,$$

where m and q are natural numbers. Since at least one of the values $qk_1 - mp_1$ and $qk_n - mp_n$ is equal to zero, P_{n-1} and P_{n-2} imply

(12) $$C(M(\underline{x};q\underline{k}-m\underline{p}),N(\underline{x};q\underline{k}-m\underline{p})) \le 0 .$$

By P_2 we have

(13) $$C(M(\underline{x};m\underline{p}),N(\underline{x};m\underline{p})) \le 0 .$$

Now we prove that (12) and (13) imply (10). Using the fact that M and N are upper and lower semiintern repetition invariant means, we get

(14) $$M(\underline{x};\underline{k}) = M(\underline{x};q\underline{k}) \le \max(M(\underline{x};q\underline{k}-m\underline{p}),M(\underline{x};m\underline{p}))$$

and

(15) $$N(\underline{x};\underline{k}) = N(\underline{x};q\underline{k}) \ge \min(N(\underline{x};q\underline{k}-m\underline{p}),N(\underline{x};m\underline{p})) .$$

By (11), $M(\underline{x};m\underline{p}) = M(\underline{x};\underline{p}) < M(\underline{x};\underline{k})$. Therefore (14) implies

(16) $$M(\underline{x};\underline{k}) \le M(\underline{x};q\underline{k}-m\underline{p}) .$$

Now, applying the properties of C, (15), (16) and the first inequality in (11), we obtain

$$C(M(\underline{x};\underline{k})-\varepsilon, N(\underline{x};\underline{k})) \le C(M(\underline{x};\underline{k})-\varepsilon, \min(N(\underline{x};q\underline{k}-m\underline{p}), N(\underline{x};m\underline{p})))$$

$$= \max(C(M(\underline{x};\underline{k})-\varepsilon, N(\underline{x};q\underline{k}-m\underline{p})), C(M(\underline{x};\underline{k})-\varepsilon, N(\underline{x};m\underline{p})))$$

$$\le \max(C(M(\underline{x};q\underline{k}-m\underline{p}), N(\underline{x};q\underline{k}-m\underline{p})), C(M(\underline{x};\underline{p}), N(\underline{x};\underline{p}))) \le 0 ,$$

that is

$$C(M(\underline{x};\underline{k})-\varepsilon, N(\underline{x};\underline{k})) \le 0$$

for any sufficiently small $\varepsilon > 0$. Taking the limit $\varepsilon \to 0$ we obtain the desired inequality (10). Thus the proof is complete.

First we apply our Main Theorem to the problem of the comparison of means.

THEOREM 4. *Let M and N be upper and lower semiintern repetition invariant symmetric means, respectively. Assume that one of them is infinitesimal. Then*

$$M(x_1,\dots,x_n) \le N(x_1,\dots,x_n)$$

holds for all natural numbers n and $x_1,\dots,x_n$ in I if and only if

$$M(x,y;k,m) \le N(x,y;k,m)$$

for any natural numbers k, m and x, y in I.

Proof. Let $C(x,y) = x-y$ in the Main Theorem. Then we obtain Theorem 4.

Applying Theorems 1, 2 and 3 one can get several special versions of this result. Now we mention the most interesting one.

COROLLARY 2. *Let E and F be deviations on I. Then*

$$M_E(x_1,\dots,x_n) \le M_F(x_1,\dots,x_n)$$

holds for all n and $x_1,\dots,x_n$ if and only if

(17) $$M_E(x,y;k,m) \le M_F(x,y;k,m)$$

for all k, m *and* x, y.

Inequality (17) is equivalent to

$$E(x,t)F(y,t) \le F(x,t)E(y,t)$$

for $x \le t \le y$. Using this inequality we can easily get necessary and sufficient conditions in order that quasiarithmetic means with weight function be comparable (see [8]).

At last we consider the problem of the complementary comparison of means.

THEOREM 5. *Let* C *be a comparative function on* I, *let* M *and* N *be upper and lower semiintern repetition invariant symmetric means, respectively, and assume that one of them is infinitesimal. If*

$$\sup_{\substack{x,y\in I\\ k,m\ge 0,\, k+m>0}} C(M(x,y;k,m),N(x,y;k,m)) = c^*$$

is finite, then

$$\sup_{\substack{x_1,\dots,x_n\in I\\ n\in\mathbb{N}}} C(M(x_1,\dots,x_n),N(x_1,\dots,x_n)) = c^* .$$

Proof. Applying the Main Theorem to the comparative function $C-c^*$ we easily obtain this result.

COROLLARY 3. *Let* E *and* F *be deviations and let* C *be a comparative function on the compact interval* I. *Then*

$$\sup_{\substack{x_1,\dots,x_n\in I\\ n\in\mathbb{N}}} C(M_E(x_1,\dots,x_n),M_F(x_1,\dots,x_n))$$

$$= \sup_{\substack{x,y\in I\\ k,m\ge 0,\, k+m>0}} C(M_E(x,y;k,m),M_F(x,y;k,m)) .$$

Proof. The right-hand side of the above equation is finite, since C is bounded on I^2. Therefore, Theorem 5 implies this corollary.

Corollary 3 was first proved in [13]. The results obtained by Schweitzer [15], Kantorovich [11], Specht [16], Cargo and Shisha [4] and Beck [3] can easily be derived from it (see [13]).

EXAMPLE. We prove Schweitzer's inequality. Let [a,b] be a subinterval of $(0,\infty)$. Denote by A and H the arithmetic and harmonic means on [a,b], respectively. Then, by Corollary 3, we have for $n \in \mathbb{N}$

$$\sup_{x_1,\dots,x_n\in[a,b]} \frac{A(x_1,\dots,x_n)}{H(x_1,\dots,x_n)} = \sup_{\substack{a\le x,y\le b\\ k,m>0}} \frac{A(x,y;k,m)}{H(x,y;k,m)}$$

$$= \sup_{\substack{a\le x,y\le b\\ k,m>0}} \left\{1 + \frac{km}{(k+m)^2}\left(\frac{x}{y}+\frac{y}{x}-2\right)\right\} = 1 + \frac{1}{4}\left(\frac{a}{b}+\frac{b}{a}-2\right) = \frac{(a+b)^2}{4ab} .$$

REFERENCES

1. M. Bajraktarevič, Sur une équation fonctionelle aux valeurs moyennes. Glasnik Mat. Fiz. Astr. 13 (1958), 243-248.

2. M. Bajraktarevič, Über die Vergleichbarkeit der mit Gewichtsfunktionen gebildeten Mittelwerte. Studia Sci. Math. Hungar. 4 (1969), 3-8.

3. E. Beck, Komplementäre Ungleichungen bei vergleichbaren Mittelwerten. Monatsh. Math. 73 (1969), 289-308.

4. G.T. Cargo and O. Shisha, Bounds on ratios of means. J. Res. Nat. Bur. Standards Sect. B 66 (1962), 169-170.

5. Z. Daróczy, Über eine Klasse von Mittelwerten. Publ. Math. Debrecen 19 (1972), 211-217.

6. Z. Daróczy, A general inequality for means. Aequationes Math. 7 (1972), 16-21.

7. Z. Daróczy and L. Losonczi, Über den Vergleich von Mittelwerten. Publ. Math. Debrecen 17 (1970), 289-297.

8. Z. Daróczy and Zs. Páles, On comparison of mean values. Publ. Math. Debrecen 29 (1982), 107-115.

9. G.H. Hardy, J.E. Littlewood and G. Pólya, Inequalities. Cambridge Univ. Press, London and New York, 2nd Edition, 1952.

10. B. Jessen, Om Uligheder imellem Potensmiddelvaerdier. Mat. Tidsskrift B 1 (1931).

11. L.V. Kantorovic, Functional analysis and applied mathematics (Russian). Uspehi Mat. Nauk. 3 No. 6(28) (1948), 89-185.

12. L. Losonczi, Subadditive Mittelwerte. Arch. Math. 22 (1971), 168-174.

13. Zs. Páles, On complementary inequalities. Publ. Math. Debrecen 30 (1983), 75-88.

14. Zs. Páles, Characterization of quasideviation means. Acta Math. Sci. Hungar. 40 (1982), 243-260.

15. P. Schweitzer, Egy egyenlötlenség az aritmetikai középértékekröl. Mat. Fiz. Lapok 23 (1914), 257-261.

16. W. Specht, Zur Theorie der elementaren Mittel. Math. Z. 74 (1960), 91-98.

Zsolt Páles, Department of Mathematics, Kossuth Lajos University, Debrecen, 4010 Pf. 12, Hungary.

International Series of
Numerical Mathematics, Vol. 71

EXTREME VALUES OF CERTAIN INTEGRALS

Ray Redheffer and Ernst Straus†

In loving memory of Ed Beckenbach and Ernst Straus

Abstract. Let the zeros of a polynomial be prescribed in magnitude but not in phase. It is a familiar fact that certain integrals involving the polynomial are maximized when the zeros all lie on a ray through the origin. Theorems of this type were first proved by Gol'dberg in 1954 and were used by him in the study of the deficiencies of meromorphic functions of genus zero. In 1983 a theorem of the same type was encountered by Kolesnik and Straus in their investigation of the Turán inequalities. Here we give a proof which depends on ideas similar to Gol'dberg's but is simpler in detail. This simplicity enables us to complete the original result of Gol'dberg, showing that a certain condition of convexity is necessary as well as sufficient, and it leads to generalizations that have not been noted hitherto.

1. HISTORICAL PERSPECTIVE

So as not to interrupt the thread of the discussion, we begin by giving the historical background which underlies this study. In an investigation of the Turán inequalities Kolesnik and Straus needed the case $p = -1$ of Theorem 1 below. A simple proof of this was found by the first author, which let to the other results presented here. Later it was observed by Prof. Steinmetz of the University of Karlsruhe that our Theorems 1 and 2 are already contained in the work of Gol'dberg [1] as presented in Hayman [2]. According to the latter reference, the results of Gol'dberg have "far-reaching effects on the problem of deficiencies of meromorphic functions of genus zero". It is of some interest that the same integral inequality arises in such disparate contexts as the deficiencies of meromorphic functions and the Turân inequality; but this relationship also requires that the present approach be examined in the light of Gol'dberg's clear priority.

Comparing our method with that in [2], we note that the underlying idea is similar but the details here are simpler. This simplicity results from two specific features of our exposition. The first is that the global problem of integration is replaced by a local problem which considers an arc $d\theta$ together with its mirror image in the x axis. The second is a geometric lemma (too trivial to be stated formally) to the effect that a set S of points in the complex plane lies on a ray through the origin if, and only if, S lies on one side of every line through the origin. By rotational invariance the line can be taken to be the real axis, and the way to our simplified proof is open.

Our approach leads to a converse to Gol'dberg's theorem which is given in Theorem 3, and to various generalizations, two of which are stated in Theorems 4 and 5. Because of overlap with the methods and results of Gol'dberg, Theorems 1 and 2 as presented in Secs. 2 and 3 should be regarded as expository. As far as we know, however, the other results are new.

In justice to the memory of the second author, it should be stated that this note is only a fragment of a larger work which was never completed. The goal was to subsume all our theorems under a general point of view, in which the principal object of interest would be a group of transformations leaving the contour of integration invariant. Some remarks pertaining to this extension are given in Section 6 but the subject is not developed in detail. The fragment presented here is due in the main to the first author. Had the second author lived to complete his part, the work would be a good deal better than it is.

2. A PRELIMINARY RESULT

Throughout this paper P(z) is a polynomial of degree n over the complex field with leading coefficient $c \neq 0$ and with zeros z_j, thus

$$P(z) = c(z-z_1)(z-z_2)\cdots(z-z_n) .$$

To avoid convergence arguments of an essentially trivial nature we agree that $|z_j| \neq 1$ and we remark that results for unrestricted z_j can be obtained by a continuity argument when the relevant integrals converge. This line of thought applies, for example, to the following theorem when $p > -1$:

THEOREM 1. _Let_ p _be real and let_ $|z_i|$ _be prescribed_, $i = 1,2,\dots,n$. _Then the integral_

$$I = \int_0^{2\pi} |P(e^{i\theta})|^p \, d\theta$$

is maximized when all z_i _lie on a ray through the origin._

The conclusion means that if $|z_i| = a_i$ are given, $i = 1,2,\dots,n$, the choice $z_i = a a_i$ with $|a| = 1$ maximizes I.

For proof, it suffices to show that all z_i lie on one side of each line through the origin, when I is maximum. By rotation invariance we can take the "line through the origin" to be the real axis. Suppose, then, that some z_i lie in the upper half plane and some in the lower half plane. We pick a point z on the unit circle in the upper half plane and consider the contribution to the integral from short arcs containing z and its conjugate $\bar{z}$. Let

$$a = \Pi\,|z-z_i|, \quad A = \Pi\,|\bar{z}-z_i|, \quad B = \Pi\,|z-z_i|, \quad b = \Pi\,|\bar{z}-z_i|,$$

where the products of a, A are over z_i in the upper half plane and those for B, b are over z_i in the lower half plane. One of these half planes, it does not matter which, is regarded as closed. Clearly $A > a$ and $B > b$. The contribution to the integral from these two arcs is

$$|c|(a^p B^p + A^p b^p)\, d\theta \, .$$

If, however, we replace the upper z_i by their conjugates $\bar{z}_i$ this has the effect of interchanging a and A, and the contribution would be

$$|c|(A^p B^p + a^p b^p)\, d\theta \, .$$

Since the latter expression is larger when $p \neq 0$, Theorem 1 follows.

3. THE THEOREMS OF GOL'DBERG

The case $p = 2$ of Theorem 1 is trivial, because the coefficients of $P(z)$ are maximized, in absolute value, when the zeros lie on a ray. The interest of the theorem is that it allows arbitrary p, including negative as well as positive values. Although this fact is at first surprising, it becomes less so in the light of the following Theorem 2. We get Theorem 1 by the choice $\tau'(t) = e^{pt}$ in Theorem 2, and obviously e^{pt} is convex no matter what the sign of p:

THEOREM 2. *Let* $\sigma(r) = \tau(\log r)$ *for* $0 < r < \infty$, *where* $\tau(t)$ *is positive and convex. Then the integral*

$$J = \int_0^{2\pi} \sigma(|P(re^{i\theta})|)\, d\theta$$

is maximum, subject to prescribed $|z_i|$, *when all* z_i *lie on a ray through the origin.*

The proof of Theorem 1 gives Theorem 2, provided

$$\sigma(aB) + \sigma(Ab) \le \sigma(ab) + \sigma(AB)$$

whenever $0 < a < A$, $0 < b < B$. In terms of τ this means

$$\tau(u+V) + \tau(U+v) \le \tau(u+v) + \tau(U+V)$$

where we have written $u = \log a$, $U = \log A$, $v = \log a$, $V = \log B$. The above condition is equivalent to a requirement that

$$\tau(y) + \tau(z) \le \tau(x) + \tau(w)$$

whenever x, y, z, w are real numbers satisfying $x < y < w$, $x < z < w$, $y+z = x+w$. This holds as an equality when τ is linear, hence it holds as an inequality when τ is sublinear as in Theorem 2.

Theorems 1 and 2 apply, and the proof is virtually unchanged, if $|P(z)|$ is replaced by

$$(1) \qquad H(z) = H_1(|z-z_1|)H_2(|z-z_2|)\cdots H_n(|z-z_n|)$$

where the H_j are positive and monotone in the same sense. If this is done, Theorem 1 implies Gol'dberg's special result, [2, Theorem 4.9], and Theorem 2 yields Gol'dberg's general result, [2,

Theorem 4.8]. Conversely, the theorems of [2] contain our Theorems 1 and 2, as was mentioned in the historical introduction, and this applies also when P(z) is replaced by (1). A less superficial generalization than (1) is described in Section 5.

4. THE CONVERSE

We shall establish the following:

THEOREM 3. _Let_ σ _be a positive_ C^2 _function such that the integral_ J _of Theorem_ 2 _has the property there described for all polynomials_ P. _Then_ $\sigma(r)$ _must be a convex function of_ log r.

To see this, let us note that the choice $\sigma(r) = \log r$, corresponding to $\tau(t) = t$, is critical. In this case the integral breaks up into a sum of integrals each of which is rotation invariant, and its value does not depend on the arguments of the z_i. In fact, the opposite conclusion would hold in Theorem 2 if τ were concave instead of convex; the integral would be minimized rather than maximized when the z_i lie on a ray. In general a concave τ requires that τ be negative at some points, but this disadvantage can be overcome by suitably restricting P. For instance the choice $\tau(t) = t^{1/2}$ shows that the integral

$$\int_0^{2\pi} (\log |P(re^{i\theta})|)^{1/2} \, d\theta$$

is minimized, subject to prescribed $|z_i|$, when all z_i lie on a ray through the origin. Here we suppose that the modified polynomial P^* obtained by moving the zeros to a ray satisfies $|P^*(z)| \geq 1$ for $|z| = 1$; then the corresponding property holds for P automatically.

After these preliminaries, we can complete the proof of Theorem 3. Define τ by $\tau(t) = \sigma(e^t)$, so that $\tau \in C^2$ and $\sigma(r) = \tau(\log r)$. If τ is not convex then $\tau''(C) < 0$ at some real value C, and by continuity the same inequality holds throughout a neighborhood of C. Hence, τ is concave in this neighborhood. If $c = e^C$ and $|z_i|$ are sufficiently small, the calculation of J involves only the interval around C on which $\tau(t)$ is concave rather than convex.

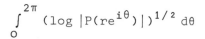

Since the conclusion of the theorem is reversed in that case, as seen above, the theorem cannot hold in the form originally stated. This shows that τ must be convex and completes the proof.

5. GENERALIZATION

We use u, U, v, V to stand for vectors of R^n with componentwise ordering and with scalar product (,). Thus, $u \le U$ means $u_j \le U_j$ for $j = 1,2,\dots,n$ and

$$(U,V) = U_1V_1 + U_2V_2 + \dots + U_nV_n .$$

We shall establish the following:

THEOREM 4. *Let F be a continuous function $R^n \to R$ such that*

$$F(u+V) + F(U+v) \le F(u+v) + F(U+V)$$

holds whenever $0 \le u \le U$, $0 \le v \le V$, $(U,V) = 0$. Then the integral

$$K = \int_0^{2\pi} F(|e^{i\theta}-z_1|,|e^{i\theta}-z_2|,\dots,|e^{i\theta}-z_n|)\, d\theta$$

is maximum, subject to prescribed $|z_j|$, when all z_j lie on a ray through the origin.

Again the proof is simple. With $|z| = 1$ and Im $z > 0$ as in the proof of Theorem 1, let

$$u_j = |z-z_j| , \quad U_j = |\bar{z}-z_j| \qquad (\text{Im } z_j > 0)$$

and $u_j = U_j = 0$ otherwise. Similarly, let

$$V_j = |z-z_j| , \quad v_j = |\bar{z}-z_j| \qquad (\text{Im } z_j \le 0)$$

and $V_j = v_j = 0$ otherwise. Then $0 \le u \le U$, $0 \le v \le V$, $(U,V) = 0$. The contribution to K from short arcs near z and $\bar{z}$ has the integrand $F(u+V) + F(U+v)$. Replacing the upper z_j by their conjugates has the effect of interchanging u and U, and Theorem 4 follows.

Since Theorem 1 has a multiplicative structure and Theorem 4 has an additive structure, it is not obvious at first glance that the latter implies the former. This matter is investigated next. The application of Theorem 4 to I in Theorem 1 leads to the inequality

$$(u_1+V_1)^p(u_2+V_2)^p\cdots(u_n+V_n)^p + (U_1+v_1)^p(U_2+v_2)^p\cdots(U_n+v_n)^p$$
$$\leq (u_1+v_1)^p(u_2+v_2)^p\cdots(u_n+v_n)^p + (U_1+V_1)^p(U_2+V_2)^p\cdots(U_n+V_n)^p$$

for $0 \leq u \leq U$, $0 \leq v \leq V$, $(U,V) = 0$. To check this, let the variables be numbered so that

$$U_1 = U_2 = \dots = U_m = 0, \quad V_{m+1} = V_{m+2} = \dots = V_n = 0$$

and note that the hypothesis $0 \leq u \leq U$, $0 \leq v \leq V$ gives corresponding relations for u and v. If we set

$$A = U_{m+1}U_{m+2}\cdots U_n, \quad B = V_1V_2\cdots V_m$$

and similarly for a and b, it is seen that the desired inequality reduces to the same inequality as was used in the proof of Theorem 1. This shows that Theorem 4 does imply Theorem 1, but that a direct proof is simpler. In a like manner Theorem 4 implies Theorem 2 and the generalization noted in connection with (1), but here too a direct proof is preferable. The interest of Theorem 4 lies in the insight which it gives, not in its application to Theorem 1 or 2.

6. CONCLUDING REMARKS

All these results have analogs in which the integration is over the real axis rather than over the unit circle. With suitable conventions regarding convergence (which are not spelled out here) the analog of Theorem 4 is:

THEOREM 5. *Let F be as in Theorem 4. Then the integral*

$$L = \int_{-\infty}^{\infty} F(|x-z_1|, |x-z_2|, \dots, |x-z_n|)\, dx$$

is maximum, subject to prescribed Im z_j, *when all* z_j *lie on a line perpendicular to the real axis.*

The proof is similar to the proof of Theorem 4.

The conformal mapping $w = (1+iz)/(i+z)$ maps the upper half plane onto the unit circle. Applying this mapping to Theorem 1, 2, 4 gives results for the half plane which are entirely different from Theorem 5. In a like manner, applying the mapping to Theorem

5 gives theorems for the circle which are different from Theorems 1, 2, 4. The essence of the matter is that the extremum of the integral is characterized when z_j are subject to a group of transformations that takes the path of integration into itself. The orbits of the z_j under this group are circles centered at the origin in Theorems 1, 2, 4 and they are horizontal lines in Theorem 5.

Finally we mention that the results extend to higher dimensions; we have taken z and z_j to be complex only for simplicity of notation. The extension to higher dimensions involves no new ideas and is not developed here.

REFERENCES

1. A.A. Gol'dberg, On an inequality for log convex functions. Do. Akad. Nauk. Ukrain. R.S.R. (1957), 227-230.

2. W.K. Hayman, Meromorphic Functions. Oxford Math. Monographs, Oxford 1964, 106-109.

3. G. Kolesnik and E.G. Straus, On the sum of powers of complex numbers. Turán Memorial Volume, Hungarian Academy of Sciences, 1983.

Ray Redheffer, Department of Mathematics, University of California at Los Angeles, Los Angeles, CA 90024, U.S.A.

International Series of
Numerical Mathematics, Vol. 71

REMARK ON AN INEQUALITY OF N. OZEKI

Dennis C. Russell

Abstract. The inequality under discussion gives an upper bound for the minimum difference between n real numbers in terms of the sum of the p-th powers of their absolute values. A simple proof is given for $p \geq 1$ and the inequality is extended to the case $0 < p < 1$.

Given different real numbers $a_1, a_2, \dots, a_n$ $(n \geq 2)$, write

$$d := \min_{1 \leq i < j \leq n} |a_i - a_j| .$$

D.S. Mitrinović and G. Kalajdžić [1] have discussed the problem of finding an upper bound for d in terms of positive powers of a_k or $|a_k|$. They give several solutions (by different authors), one of which is

$$(1) \qquad d^2 \leq \frac{12}{n(n^2-1)} \left\{ \sum_{k=1}^{n} a_k^2 - \frac{1}{n} \left(\sum_{k=1}^{n} a_k \right)^2 \right\} .$$

We may assume that $a_1 < a_2 < \dots < a_n$. Then $|a_i - a_j| \geq d|i-j|$ for any i, j, and hence, for $p > 0$,

$$(2) \qquad \sum_{1 \leq i<j \leq n} |a_j - a_i|^p \geq d^p \sum_{1 \leq i<j \leq n} |j-i|^p = d^p \sum_{k=1}^{n-1} (n-k)k^p .$$

In the case $p = 2$, (2) reduces to (1).

A further inequality is quoted in [1, § 5], namely

$$(3) \qquad \sum_{k=1}^{n} |a_k|^p \geq C_{np} d^p , \quad C_{np} := \begin{cases} 2(1^p + 2^p + \dots + m^p), & n = 2m+1; \\ 2^{1-p}(1^p + 3^p + \dots + (2m-1)^p), & n = 2m; \end{cases}$$

for example, $C_{n2} = n(n^2-1)/12$, so that (1) implies (3) in the case $p = 2$. The inequality (3) is given without proof by N. Ozeki [2, p.203]; although he writes $p > 0$, the context suggests that he tacitly assumes p to be a positive integer (in which case (3) is true). In [1, p.6], Mitrinović and Kalajdžić specifically assume p to be a positive real number and proceed to supply a proof of (3), but their proof covers only the case $p \geq 1$. Indeed, for _even_ n and $0 < p < 1$, there are real n-vectors $\underline{a}$ for which (3) (with C_{np} as stated) is false: e.g. take $n = 2$, $p = \frac{1}{2}$, $\underline{a} = (-1,2)$; or $n = 4$, $p = \frac{1}{2}$, $\underline{a} = (-4,-1,2,5)$. We can modify the constant C_{np} in (3) to give an inequality valid for all $p > 0$, and we state this in the Theorem which follows. In the case $0 < p < 1$ and n odd, the proof follows closely that in [1, p.7], suitably corrected. In the other cases (which includes $p \geq 1$) a simple inductive proof suffices.

THEOREM. _Let_ $p > 0$, $n \in \{2,3,\ldots\}$, $e_p := \min\{2^{1-p},1\}$, _and_

(5) $$c_{np} := \begin{cases} 2(1^p + 2^p + 3^p + \ldots + [\frac{1}{2}(n-1)]^p), & n \text{ odd}, \\ e_p(1^p + 3^p + 5^p + \ldots + (n-1)^p), & n \text{ even}. \end{cases}$$

_If_ $a_1,\ldots,a_n$ _are different real numbers, and_ $d := \min_{i \neq j} |a_i - a_j|$, _then_

(6) $$\min_{x \in \mathbb{R}} \sum_{k=1}^{n} |a_k - x|^p \geq c_{np} d^p .$$

Proof. Since $a_i - a_j = (a_i - x) - (a_j - x)$, it suffices, in place of (6), to prove

(7) $$\sum_{k=1}^{n} |a_k|^p \geq c_{np} d^p .$$

We may also suppose that $a_1 < a_2 < \ldots < a_n$. Then $a_n - a_1 \geq (n-1)d$ and hence

(8) $$|a_1|^p + |a_n|^p \geq e_p(|a_1| + |a_n|)^p \geq e_p(a_n - a_1)^p \geq e_p(n-1)^p d^p .$$

Suppose that (7) holds for a set of (n-2) numbers. Then, applying (7) to $a_2 < \ldots < a_{n-1}$, we get

(9) $$\sum_{k=2}^{n-1} |a_k|^p \geq c_{n-2,p} \min_{2 \leq i < j \leq n-1} |a_i - a_j|^p \geq c_{n-2,p} d^p .$$

Now (by definition (5)) $c_{n-2,p} + e_p(n-1)^p = c_{np}$ in the cases

(10) $$p > 0 \quad (n \text{ even}) \qquad \text{or} \qquad p \geq 1 \quad (n \text{ odd}) .$$

Thus by adding (8) and (9) together we see that, in these cases, (7) holds for a set of n numbers. For $n = 2$ and $p > 0$, (7) is the same as (8); while for $n = 3$ and $p \geq 1$ ($e_p = 2^{1-p}$), (7) is implied by (8). It follows by induction that (7) holds for any combination of $n \geq 2$, $p > 0$ satisfying (10).

The remaining case is $0 < p < 1$, n odd, and, following [1, (10)], we consider

$$f(\underline{a}) := \sum_{k=1}^{n} |a_k|^p , \quad \underline{a} = (a_1, \dots, a_n);$$

by replacing a_k by a_k/d, we can suppose without loss of generality that

$$d(\underline{a}) := \min_{1 \leq i < j \leq n} |a_i - a_j| = 1 .$$

As before, we can also suppose that $a_1 < a_2 < \dots < a_n$ and, if $n = 2m+1$, then by symmetry we may if necessary reflect all the points a_k about 0, in order to ensure that at least m+1 of the values a_k are non-negative. Now we define s as the subscript for which

$$a_{s-1} < 0 \leq a_s \leq a_{m+1} , \quad \text{so that } s-1 \leq m \leq 2m+1-s .$$

If $a_s > 1$ then $\underline{b} := (a_1, \dots, a_{s-1}, 1, a_{s+1}, \dots, a_n)$ satisfies $d(\underline{b}) = 1$ and $f(\underline{a}) \geq f(\underline{b})$. Hence it suffices to take $0 \leq a_s < 1$, so that we now have

$$t := a_s \in [0,1] \quad \text{and} \quad s-1 \leq m \leq 2m+1-s . \tag{11}$$

Now since $d(\underline{a}) = 1$ we see that

$$a_{s-r_1} \leq t - r_1 \leq 0 \qquad (1 \leq r_1 \leq s-1) ,$$

$$a_{s+r_2} \geq t + r_2 \geq 0 \qquad (0 \leq r_2 \leq 2m+1-s)$$

and then

$$\begin{aligned} f(\underline{a}) &\geq \sum_{r_1=1}^{s-1} (r_1 - t)^p + \sum_{r_2=0}^{2m+1-s} (r_2 + t)^p \\ &\geq \sum_{r_1=1}^{m} (r_1 - t)^p + \sum_{r_2=0}^{m} (r_2 + t)^p =: h(t) \end{aligned} \tag{12}$$

because of (11) and the fact that

$$r_2 + t \geq r_1 - t \quad \text{for} \quad s-1 \leq r_1 \leq m \leq r_2 \leq 2m+1-s .$$

Now if $0 < p < 1$ and $0 < t < 1$ we have

$$h''(t) = p(p-1)\left\{\sum_{r_1=1}^{m}(r_1-t)^{p-2} + \sum_{r_2=0}^{m}(r_2+t)^{p-2}\right\} < 0$$

and h is continuous on [0,1]. Hence h is <u>concave</u> on [0,1]. But, by (12), $h(1)-h(0) = (m+1)^p - m^p > 0$ and the concavity then gives, for $0 \le t \le 1$,

$$f(\underline{a}) \ge h(t) \ge \min\{h(1),h(0)\} = h(0) = 2\sum_{r=1}^{m} r^p = c_{np},$$

which proves (7) for $0 < p < 1$, n odd, and thus completes the proof. □

Note that when $p = 2$, the minimum on the left of (6) occurs when $x = \bar{a} := \frac{1}{n}(a_1 + \dots + a_n)$, and (6) then reduces exactly to (1).

REFERENCES

1. D.S. Mitrinović and G. Kalajdzić, On an inequality. Univ. Beograd Publ. Electrotehn. Fak. Ser. Mat. Fiz. No. 678 (1980), 3-9 [Math. Rev. 82m: 26019].

2. N. Ozeki, On the estimation of inequalities by maximum and minimum values. J. College Arts Sci. Chiba Univ. 5 (1968), 199-203 (in Japanese) [Math. Rev. 40: 7408].

Dennis C. Russell, Department of Mathematics, York University, Toronto-Downsview, Ontario M3J 1P3, Canada.

International Series of
Numerical Mathematics, Vol. 71
© 1984 Birkhäuser Verlag Basel

SUR UNE INEGALITE DE JENSEN-STEFFENSEN

Petar M. Vasić et Josip E. Pečarić

Résumé. Il est notoire que beaucoup d'inégalités peuvent être présentées dans la forme des inégalités du type de Rado et Popoviciu, c'est-à-dire dans la forme (1). Dans cet article on a démontré que le cas est le même pour les inégalités de Jensen-Steffensen pour les fonctions convexes.

La suite des inégalités

$$n(A_n-G_n) \geq (n-1)A_{n-1}-G_{n-1}) \geq \dots \geq 2(A_2-G_2) \geq A_1-G_1 = 0$$

se trouve dans la monographie [1] (A_k et G_k désignent la moyenne arithmétique et la moyenne géometrique respectivement des nombres $a_1,\dots,a_k$).

Après cela plusieurs travaux ont été publiés dans lesquels on forme la suite des inégalités

(1) $$F_n(x_1,\dots,x_n) \geq F_{n-1}(x_1,\dots,x_{n-1}) \geq \dots \geq F_2(x_1,x_2) \geq F_1(x_1) = 0$$

en partant de l'inégalité

(2) $$F_n(x_1,\dots,x_n) \geq 0 ,$$

où F_n est une fonction donnée.

Un tel résultat est le suivant:

Soit

$$F_k(x_1,\dots,x_k) = G_k(x_1,\dots,x_k,p_1,\dots,p_k)$$

(3) $$= \frac{1}{P_k}\sum_{i=1}^{k} p_i f(x_i) - f\left(\frac{1}{P_k}\sum_{i=1}^{k} p_i x_i\right)$$

$$\left(P_k = \sum_{i=1}^{k} p_i ; \quad k = 1,\dots,n\right) .$$

Si $f: [a,b] \to R$ est une fonction convexe, $x_i \in [a,b]$ $(i=1,\dots,n)$,

$p_i > 0$ $(i = 1,\dots,n)$, on a (1) (voir P.M. Vasić et Ž. Mijalković [2]). Evidemment, l'inégalité $F_2(x_1,x_2) \geq 0$ est la définition des fonctions convexes. D'autre part, l'inégalité $F_n(x_1,\dots,x_n) \geq 0$ est l'inégalité classique de Jensen.

Une généralisation de l'inégalité de Jensen est l'inégalité de Jensen-Steffensen:

THEOREME 1. *Soient* x_i, p_i $(i = 1,\dots,n)$ *des nombres réels tels que* $a \leq x_1 \leq \dots \leq x_n \leq b$ *et*

$$(4) \qquad 0 \leq P_k \leq P_n \quad (k = 1,\dots,n-1), \qquad P_n > 0 .$$

Alors, pour chaque fonction convexe $f\colon [a,b] \to R$ *l'inégalité*

$$F_n(x_1,\dots,x_n) \geq 0 ,$$

où F *est donnée par* (3), *est vraie.*

Dans cette Note nous avons obtenu une suite des inégalités de la forme (1) se rapportant au résultat de Jensen-Steffensen. Notre résultat est:

THEOREME 2. *Soient* x_i, p_i $(i = 1,\dots,n)$ *des nombres réels tels que* $a \leq x_1 \leq \dots \leq x_n \leq b$ *et*

$$(4) \qquad 0 \leq P_k \leq P_n \quad (k = 1,\dots,n-1), \qquad P_n > 0 .$$

Pour chaque fonction $f\colon [a,b] \to R$ *convexe sur* $[a,b]$ *on a*

$$(5) \qquad H_n(x_1,\dots,x_n) \geq H_{n-1}(x_1,\dots,x_{n-1}) \geq \dots \geq H_2(x_1,x_2) \geq H_1(x_1) = 0$$

où la fonction H_k *est définie par*

$$H_k(x_1,\dots,x_k) = G_k(x_1,\dots,x_k,p_1,\dots,p_{k-1},\overline{P_k}) \quad (\overline{P_k} = P_n - P_{k-1})$$

et G_k *est définie par* (3).

DEMONSTRATION 1. L'inégalité

$$(6) \qquad H_k(x_1,\dots,x_k) \geq H_{k-1}(x_1,\dots,x_{k-1})$$

est équivalente à la suivante

$$(7) \qquad (f(x_k) - f(x_{k-1}))\frac{\overline{P_k}}{P_n} \geq f(x_k) - f(x_{k-1}) .$$

Dans la dernière inégalité nous avons utilisé la notation suivante

$$\overline{x_k} = \frac{1}{P_n}\left(\sum_{i=1}^{k-1} p_i x_i + x_k \overline{P_k}\right) .$$

Pour les fonctions convexes le résultat suivant est connu (voir, par exemple, [6]):

Si $x_1 \geq y_1$, $x_2 \geq y_2$, $x_1 \neq x_2$, $y_1 \neq y_2$, alors, on a

$$\frac{f(x_1)-f(x_2)}{x_1-x_2} \geq \frac{f(y_1)-f(y_2)}{y_1-y_2} \tag{8}$$

De même, on a (voir [6]):

$$x_1 \leq \frac{1}{P_k} \sum_{i=1}^{k} p_i x_i \leq x_k .$$

Partant du résultat cité, nous avons

$$\overline{x_k} \leq x_k , \qquad \overline{x_{k-1}} \leq x_{k-1} .$$

Sans nuire à la généralité, nous pouvons supposer

$$x_1 < \dots < x_n , \quad 0 < P_k < P_n \quad (k = 1,\dots,n-1) .$$

Alors, si l'on pose

$$x_1 \to x_k , \quad x_2 \to x_{k-1} , \quad y_1 \to \overline{x_k} , \quad y_2 \to \overline{x_{k-1}} ,$$

de (8) il vient (7), c'est-à-dire (6).

REMARQUE 1. Dans le cas où $P_k = 0$ pour quelque k nous pouvons poser $p_1 \to p_1+\varepsilon$, $\varepsilon > 0$. Les conditions utilisées dans notre démonstration sont remplies et en mettant $\varepsilon \to 0$ nous obtenons de nouveau (6).

DEMONSTRATION 2. Soit las fonction a_k définie par

$$a_k(t) = H_k(x_1,\dots,x_{k-1},t) \quad (t \in [x_{k-1},x_k];\ k = 2,\dots,n) .$$

Supposons que f est une fonction convexe et continue sur [a,b]. Puisque on peut approcher uniforme f sur [a,b] par les convexes polynomes, sans nuire à la généralité, on peut supposer que f est une fonction différentiable et que f' est une fonction non-décroissante. Alors,

$$a_k'(t) = f'(t)\,\frac{\overline{P_k}}{P_n} - f'\left(\frac{1}{P_n}\left(\sum_{i=1}^{k-1} p_i x_i + t\overline{P_k}\right)\right)\frac{\overline{P_k}}{P_n} \geq 0 ,$$

car

$$x_1 \leq \dots \leq x_{k-1} \leq t , \qquad \text{et } \frac{1}{P_n}\left(\sum_{i=1}^{k-1} p_i x_i + t\overline{P_k}\right) \leq t .$$

De là, il suit que a_k est une fonction non-décroissante sur $[x_{k-1},x_k]$, et on a $a_k(x_k) \geq a_k(x_{k-1})$, ou bien, l'inégalité (6).

La fonction arbitraire sur $[a,b]$ peut être discontinue seulement dans les points a et b. Etant donné que $f(a+0) \leq f(a)$ et $f(b-0) \leq f(b)$, on peut démontrer, sans diffculté, que (5) reste vraie pour les fonctions convexes arbitraires (sans la supposition que f soit continue).

REMARQUE 2. Les Démonstrations 1 et 2 sont, en effet, deux démonstrations nouvelles, très simples, de l'inégalité de Jensen-Steffensen.

DEMONSTRATION 3. L'inégalité de Jensen-Steffensen est pour la première fois démontrée par Steffensen [7]. Il est parti du résultat suivant:

Supposons que f et g sont des fonctions intégrables sur $[a,b]$. Alors,

$$\int_{b-\lambda}^{b} f(t)dt \leq \int_a^b f(t)g(t)dt \leq \int_a^{a+\lambda} f(t)dt , \tag{9}$$

si f est décroissante et que $0 \leq g(t) \leq 1$ sur $[a,b]$, et

$$\lambda = \int_a^b g(t)dt .$$

Posons

$$G(x) = \int_a^x g(t)dt , \quad \text{et} \quad S(x) = \int_a^{a+G(x)} f(t)dt - \int_a^x f(t)g(t)dt .$$

Evidemment,

$$\lambda' \equiv (a+G(x)) - (a+G(y)) = \int_y^x g(t)dt \leq x-y .$$

Parce-que $a+G(y) \leq y$, et f est une fonction décroissante, nous avons

$$(10) \qquad \int_{a+G(y)}^{a+G(x)} f(t)dt \geq \int_{y}^{y+\lambda'} f(t)dt .$$

En utilisant l'inégalité (9) (avec [y,x] au lieu de [a,b]), nous obtenons

$$(11) \qquad \int_{y}^{y+\lambda'} f(t)dt \geq \int_{y}^{x} f(t)g(t)dt .$$

En confrontant (10) et (11), on a

$$\int_{a+G(y)}^{a+G(x)} f(t)dt \geq \int_{y}^{x} f(t)g(t)dt$$

ou bien

$$(12) \qquad S(x) \geq S(y) \quad (x \geq y) .$$

A la même manière, comme dans la Démonstration 2, nous pouvons supposer que f est une fonction différentiable. En posant $a = x_1$, $g(t) = \frac{\overline{P_k}}{P_n}$ pour $x_{k-1} < t \leq x_k$ $(k = 2,\dots,n)$ et $f(x) \to -f'(x)$, de (12) nous concluons que a_k est une fonction décroissante.

REMARQUE 3. Le résultat du théorème 1 reste valable si les conditions $x_1 \leq \dots \leq x_n$ sont remplies par $x_1 \geq \dots \geq x_n$.

BIBLIOGRAPHIE

1. G.H. Hardy, J.E. Littlewood and G. Pólya, Inequalities, Cambridge 1934.

2. P.M. Vasić and Ž. Mijalković, On an index set function connected with Jensen inequality. Univ. Beograd. Publ. Elektroteh. Fak. Ser. Mat. Fiz. N°544 - N°576 (1976), 110-112.

3. D.S. Mitrinović, P.S. Bullen and P.M. Vasić, Sredine i sa njima povezane nejednakosti, Ibid. N°600, 1-232.

4. P.M. Vasić and J.E. Pečarić, On the Jensen inequality. Ibid. N°634 - N°677 (1979), 50-54.

5. P.S. Bullen, The Steffensen inequality. Ibid. N°320 - N°328 (1970), 59-63.

6. D.S. Mitrinović (in cooperation with P.M. Vasić), Analytic inequalities. Berlin - Heidelberg - New York, 1970.

7. J.F. Steffensen, On certain inequalities and methods of approximation. J. Instr. Actuaries 51 (1919), 274-297.

8. J.E. Pečarić, On the Jensen-Steffensen inequality. Univ. Beograd. Publ. Elektrotehn. Fak. Ser. Mat. Fiz. N°634 - N°677 (1979), 101-107.

Petar M. Vasić, Department of Mathematics, Faculty of Electrical Engineering, University of Beograd, 11001 Beograd, Yugoslavia.

Josip E. Pečarić, Faculty of Civil Engineering, Bulevar revolucije 73, Beograd, Yugoslavia.

Inequalities in
Analysis and Approximation

Conifer Seedlings

Winter Willows

International Series of
Numerical Mathematics, Vol. 71

DIFFERENCE CALCULUS WITH APPLICATIONS TO DIFFERENCE EQUATIONS

Ravi P. Agarwal

Abstract. We present discrete analogues of Taylor's formula, l'Hospital's rule, Kneser's theorem etc., and use these to study qualitative properties of solutions of higher order difference equations. The proofs are based on some simple inequalities.

1. INTRODUCTION

In the last few years there has been considerable interest in studying discrete analogues of qualitative properties known for nth order ordinary differential equations; see, e.g., [1-3,5 and references therein]. The purpose of this paper is to study oscillatory behaviour of nth order nonlinear difference equations. The results obtained are discrete analogues of some known results given in [4, 9-11, 14, 15, 17]; they generalize or improve the results discussed in [1, 2, 12, 16 and references therein].

2. DIFFERENCE CALCULUS

In what follows, I is the discrete set $\{0,1,\dots\}$ and Δ is the forward difference operator $\Delta x(t) = x(t+1)-x(t)$. The set $\{t_1, t_1+1, \dots\}$ where $t_1 \in I$ is denoted by I_{t_1}. For all $t \in R$, $(t)^{(m)} = t(t-1)\dots(t-m+1)$ is the usual factorial notation with $(t)^{(0)} = 1$. For all $t_1 > t_2 \in I$ and any $f(s)$ defined on I, $\sum_{s=t_1}^{t_2} f(s) = 0$.

LEMMA 1. (*Taylor's formula*). *Let* $u(t)$ *be some function defined on* I_{t_1}. *Then, for* $0 \le k \le n-1$, $t \ge t_1$

$$\Delta^k u(t) = \sum_{i=k}^{n-1} \frac{(t-t_1)^{(i-k)}}{(i-k)!} \Delta^i u(t_1) + R(t,n,k) \tag{2.1}$$

where

$$R(t,n,k) = \frac{1}{(n-k-1)!} \sum_{s=t_1}^{t-n+k} (t-s-1)^{(n-k-1)} \Delta^n u(s).$$

Proof. The proof is by backward induction in k. Since

$$\Delta^m u(t) = \Delta^m u(t_1) + \sum_{s=t_1}^{t-1} \Delta^{m+1} u(s), \quad 0 \le m \le n-1 \tag{2.2}$$

the identity (2.1) is trivial for $k = n-1$. Next, assume (2.1) is true for $k = m+1$, then from (2.2)

$$\Delta^m u(t) = \Delta^m u(t_1) + \sum_{s=t_1}^{t-1} \Bigg[\sum_{i=m+1}^{n-1} \frac{(s-t_1)^{(i-m-1)}}{(i-m-1)!} \Delta^i u(t_1)$$

$$+ \frac{1}{(n-m-2)!} \sum_{s_1=t_1}^{s-n+m+1} (s-s_1-1)^{(n-m-2)} \Delta^n u(s_1) \Bigg]$$

$$= \Delta^m u(t_1) + \sum_{i=m+1}^{n-1} \sum_{s=t_1}^{t-1} \frac{\Delta(s-t_1)^{(i-m)}}{(i-m)!} \Delta^i u(t_1)$$

$$+ \frac{1}{(n-m-2)!} \sum_{s=t_1+n-m-1}^{t-1} \sum_{s_1=t_1}^{s-n+m+1} (s-s_1-1)^{(n-m-2)} \Delta^n u(s_1)$$

$$= \Delta^m u(t_1) + \sum_{i=m+1}^{n-1} \frac{(s-t_1)^{(i-m)}}{(i-m)!} \Bigg|_{s=t_1}^{t} \Delta^i u(t_1)$$

$$+ \frac{1}{(n-m-2)!} \sum_{j=0}^{t-n+m-t_1} \sum_{s=t_1+j}^{t-n+m} (t-1-s-1)^{(n-m-2)} \Delta^n u(t_1+j)$$

$$= \Delta^m u(t_1) + \sum_{i=m+1}^{n-1} \frac{(t-t_1)^{(i-m)}}{(i-m)!} \Delta^i u(t_1)$$

$$- \frac{1}{(n-m-1)!} \sum_{j=0}^{t-n+m-t_1} \sum_{s=t_1+j}^{t-n+m} \Delta(t-s-1)^{(n-m-1)} \Delta^n u(t_1+j)$$

$$= \sum_{i=m}^{n-1} \frac{(t-t_1)^{(i-m)}}{(i-m)!} \Delta^i u(t_1) - \frac{1}{(n-m-1)!} \sum_{j=0}^{t-n+m-t_1} [(n-m-2)^{(n-m-1)} - (t-t_1-j-1)^{(n-m-1)}] \Delta^n u(t_1+j)$$

$$= \sum_{i=m}^{n-1} \frac{(t-t_1)^{(i-m)}}{(i-m)!} \Delta^i u(t_1) + \frac{1}{(n-m-1)!} \sum_{s=t_1}^{t-n+m} (t-s-1)^{(n-m-1)} \Delta^n u(s)$$

and hence (2.1) is true for $k = m$.

LEMMA 2. (*Taylor's formula*). *Let* $u(t)$ *be some function defined on* I_{t_1}. *Then, for* $p \geq t$ *and* $n \geq 1$

$$u(t) = \sum_{i=0}^{n-1} \frac{(p+i-1-t)^{(i)}}{(i)!} (-1)^i \Delta^i u(p) - \frac{(-1)^{n-1}}{(n-1)!} \sum_{s=t}^{p-1} (s+n-1-t)^{(n-1)} \Delta^n u(s). \quad (2.3)$$

Proof. Identity (2.3) is obvious for $n = 1$. Assuming it is true for $n = k$, then from

$$\sum_{s=t}^{p-1} (s+k-1-t)^{(k-1)} \Delta^k u(s) = \frac{1}{k} \sum_{s=t}^{p-1} \Delta(s+k-1-t)^{(k)} \Delta^k u(s)$$

$$= \frac{1}{k} \{ (s+k-1-t)^{(k)} \Delta^k u(s) \Big|_{s=t}^{p} - \sum_{s=t}^{p-1} (s+k-t)^{(k)} \Delta^{k+1} u(s) \}$$

$$= \frac{1}{k} (p+k-1-t)^{(k)} \Delta^k u(p) - \frac{1}{k} \sum_{s=t}^{p-1} (s+k-t)^{(k)} \Delta^{k+1} u(s)$$

we find (2.3) for $n = k+1$.

REMARK. It cannot be expected that the Lagrange analogue of the remainder terms in (2.1) or (2.3) holds true.

LEMMA 3. (*ℓ'Hospital's rule*). *Let* $u(t)$ *and* $v(t)$ *be defined on* I *and* $v(t) > 0$, $\Delta v(t) < 0$ *for all sufficiently large* t *in* I. *Then, if* $u(t) \to 0$ *and* $v(t) \to 0$

$$\liminf \frac{\Delta u(t)}{\Delta v(t)} \leq \liminf \frac{u(t)}{v(t)} \leq \limsup \frac{u(t)}{v(t)} \leq \limsup \frac{\Delta u(t)}{\Delta v(v)} . \quad (2.4)$$

Proof. Let t_1 be sufficiently large so that for all $t \geq t_1$, $v(t) > 0$ and $\Delta v(t) < 0$. We assume that

$$\frac{\Delta u(t)}{\Delta v(t)} \geq \alpha \qquad \text{for} \quad t \geq t_1 .$$

Then, $\Delta u(t) \leq \alpha \Delta v(t)$ and by summation

$$u(t+p) - u(t) \leq \alpha(v(t+p)-v(t)) \qquad \text{for} \quad t \geq t_1,\ p > 0.$$

Letting $p \to \infty$, we find $-u(t) \leq -\alpha v(t)$, or

$$\frac{u(t)}{v(t)} \geq \alpha \qquad \text{for } t \geq t_1.$$

Since the same holds with inequalities reversed, (2.4) follows.

LEMMA 4. (_ℓ'Hospital's rule_). _Let_ $u(t)$ _and_ $v(t)$ _be defined on_ I _and_ $v(t) > 0$, $\Delta v(t) > 0$ _for all sufficiently large_ t. _Then_, _if_ $v(t) \to \infty$

$$\lim_{t\to\infty} \frac{\Delta u(t)}{\Delta v(t)} = r \ \underline{\text{implies}} \ \lim_{t\to\infty} \frac{u(t)}{v(t)} = r. \tag{2.5}$$

Proof. $\lim_{t\to\infty} \frac{\Delta u(t)}{\Delta v(t)} = r$ (finite) implies that for all sufificiently large $t > t_1$ in I

$$(r-\varepsilon)\Delta v(t) \leq \Delta u(t) \leq (r+\varepsilon)\Delta v(t).$$

Summing the above inequality, we find

$$(r-\varepsilon)\left[v(t+p)-v(t)\right] \leq u(t+p)-u(t) \leq (r+\varepsilon)\left[v(t+p)-v(t)\right] \quad \text{for } t \geq t_1,\ p > 0$$

or

$$(r-\varepsilon)\left[1 - \frac{v(t)}{v(t+p)}\right] \leq \frac{u(t+p)}{v(t+p)} - \frac{u(t)}{v(t+p)} \leq (r+\varepsilon)\left[1 - \frac{v(t)}{v(t+p)}\right] \text{for } t \geq t_1,\ p>0.$$

Letting $p \to \infty$, we get (2.5)

If r is infinite say $+\infty$ (the case $-\infty$ can be treated similarly) then, for arbitrary $k > 0$, there exists t_1 sufficiently large so that

$$\frac{\Delta u(t)}{\Delta v(t)} \geq k \quad \text{for} \quad t \geq t_1$$

or

$$\Delta u(t) \geq k \Delta v(t) \quad \text{for} \quad t \geq t_1.$$

Summing the above inequality

$$u(t+p) - u(t) \geq k\left[v(t+p)-v(t)\right] \quad \text{for } t \geq t_1,\ p > 0$$

or

$$\frac{u(t+p)}{v(t+p)} - \frac{u(t)}{v(t+p)} \geq k\left[1 - \frac{v(t)}{v(t+p)}\right] \text{for } t \geq t_1,\ p > 0.$$

Taking $p \to \infty$, we get $\lim_{t\to\infty} \frac{u(t)}{v(t)} \geq k$.

LEMMA 5. Let $u(t)$ be some function defined on I_{t_1}. Then,

$$\sum_{s=t_1}^{t-1} (s)^{(n-1)}\Delta^n u(s) = \sum_{k=1}^{n} (-1)^{k+1}\Delta^{k-1}(s)^{(n-1)}\Delta^{n-k}u(s+k-1)\Big|_{s=t_1}^{t} . \quad (2.6)$$

Proof. The identity (2.6) is a repeated application of

$$\sum_{s=t_1}^{t-1} u(s)\Delta v(s) = u(s)v(s)\Big|_{s=t_1}^{t} - \sum_{s=t_1}^{t-1} \Delta u(s)v(s+1).$$

LEMMA 6. Let $1 \leq q \leq n-1$ and $u(t)$ be some function defined on I. Then

(a) $\liminf_{t\to\infty} \Delta^q u(t) > 0$ implies $\lim_{t\to\infty} \Delta^i u(t) = \infty$, $0 \leq i \leq q-1$

(b) $\limsup_{t\to\infty} \Delta^q u(t) < 0$ implies $\lim_{t\to\infty} \Delta^i u(t) = -\infty$, $0 \leq i \leq q-1$.

Proof. $\liminf_{t\to\infty} \Delta^q u(t) > 0$ implies that for all sufficiently large t $t \geq t_1$, $\Delta^q u(t) \geq c > 0$. Now, from (2.2) for $m = q-1$, we find $\Delta^{q-1}u(t) \geq \Delta^{q-1}u(t_1) + c_1(t-t_1)$ and hence $\lim_{t\to\infty} \Delta^{q-1}u(t) = \infty$. Rest of the proof follows by induction. The case (b) is treated similarly.

LEMMA 7. (Kneser [8], Kiguardze [6,7], Onose [9,10], Philos [13]). Let $u(t)$ be some function defined on I and let $u(t) > 0$ with $\Delta^n u(t)$ of constant sign on I_{t_1} and not identically zero. Then, there exists an integer ℓ, $0 \leq \ell \leq n$ with $n + \ell$ odd for $\Delta^n u(t) \leq 0$ or $n+\ell$ even for $\Delta^n u(t) \geq 0$ and such that

$\ell \leq n-1$ implies $(-1)^{\ell+j}\Delta^j u(t) > 0$ for $t \in I_{t_1}$, $\ell \leq j \leq n-1$

$\ell \geq 1$ implies $\Delta^i u(t) > 0$ for all large $t \in I$, $1 \leq i \leq \ell-1$.

Proof. Case (i) $\Delta^n u(t) \leq 0$ on $t \in I_{t_1}$. First, we shall prove that $\Delta^{n-1}u(t) > 0$ on $t \in I_{t_1}$. If not, then there exists some $t_2 \geq t_1$ such that $\Delta^{n-1}u(t_2) \leq 0$. Since $\Delta^{n-1}u(t)$ is decreasing and not identically zero on I_{t_2}, there exists $t_3 \geq t_2$ such that $\Delta^{n-1}u(t) \leq \Delta^{n-1}u(t_3) < \Delta^{n-1}u(t_2) \leq 0$ for all $t \geq t_3$. But, from lemma 6, $\lim_{t\to\infty} u(t) = -\infty$ which is a contradiction to $u(t) > 0$. Thus, $\Delta^{n-1}u(t) > 0$ on I_{t_1} and there exists smallest integer

ℓ, $0 \le \ell \le n-1$ with $n+\ell$ odd and

$$(-1)^{\ell+j}\Delta^j u(t) > 0 \quad \text{on} \quad t \in I_{t_1}, \ \ell \le j \le n-1. \tag{2.7}$$

Next, let $\ell > 1$ and

$$\Delta^{\ell-1} u(t) < 0 \quad \text{on} \quad I_{t_1} \tag{2.8}$$

then, once again from lemma 6, it follows that

$$\Delta^{\ell-2} u(t) > 0 \quad \text{on} \quad I_{t_1}. \tag{2.9}$$

Inequalities (2.7)-(2.9) can be unified to

$$(-1)^{(\ell-2)+j}\Delta^j u(t) > 0 \quad \text{on} \quad I_{t_1}, \quad \ell-2 \le j \le n-1$$

which is a contradiction to the definition of ℓ. So, (2.8) fails and $\Delta^{\ell-1}u(t) \ge 0$ on I_{t_1}. From (2.7), $\Delta^{\ell-1}u(t)$ is nondecreasing and hence $\lim_{t\to\infty} \Delta^{\ell-1}u(t) > 0$. If $\ell > 2$, we find from lemma 6 that $\lim_{t\to\infty} \Delta^i u(t) = \infty$, $1 \le i \le \ell-2$. Thus, $\Delta^i u(t) > 0$ for all large t, $1 \le i \le \ell-1$.

Case (ii). $\Delta^n u(t) \ge 0$ on $t \in I_{t_1}$. Let t be such that $\Delta^{n-1}u(t) \ge 0$, then since $\Delta^{n-1}u(t)$ is nondecreasing and not identically zero, there exists some $t_5 \ge t_4$ such that $\Delta^{n-1}u(t) > 0$, $t \ge t_5$. Thus, $\lim_{t\to\infty} \Delta^{n-1}u(t) > 0$ and from lemma 6, $\lim_{t\to\infty} \Delta^i u(t) = \infty$, $1 \le i \le n-2$ and so $\Delta^i u(t) > 0$ for all large t, $1 \le i \le n-1$. This proves the lemma for $\ell = n$. In case $\Delta^{n-1}u(t) < 0$ for all $t \in I_{t_1}$, we find from lemma 6 that $\Delta^{n-2}u(t) > 0$ for all $t \in I_{t_1}$. Rest of the proof is same as in Case (i).

COROLLARY 1. *Let u(t) be some function defined on I and let u(t) > 0 with $\Delta^n u(t) \le 0$ on I_{t_1} and not identically zero. Then, there exists sufficiently large $t_2 \ge t_1$ in I_{t_1} such that*

$$u(t) \ge \frac{1}{(n-1)!}\Delta^{n-1}u(2^{n-\ell-1}t)(t-t_2)^{(n-1)}, \quad t \ge t_2. \tag{2.10}$$

Proof. From lemma 7, we find $(-1)^{n+j-1}\Delta^j u(t) > 0$ for $t \in I_{t_1}$, $\ell \le j \le n-1$

and $\Delta^j u(t) > 0$ for all large $t \in I_{t_1}$ (say $t \geq t_2$ i.e. $t \in I_{t_2}$), $1 \leq j \leq \ell-1$. Using these inequalities, we obtain

$$-\Delta^{n-2}u(t) = -\Delta^{n-2}u(\infty) + \sum_{s=t}^{\infty} \Delta^{n-1}u(s)$$

$$\geq \sum_{s=t}^{2t} \Delta^{n-1}u(s) \geq \Delta^{n-1}u(2t)(t)^{(1)}$$

$$\Delta^{n-3}u(t) = \Delta^{n-3}u(\infty) - \sum_{s=t}^{\infty} \Delta^{n-2}u(s)$$

$$\geq \sum_{s=t}^{2t} (s)^{(1)}\Delta^{n-1}u(2s) \geq \sum_{s=t}^{2t} (s-t)^{(1)}\Delta^{n-1}u(2s)$$

$$\geq \Delta^{n-1}u(2^2 s) \cdot \frac{1}{2!}(t)^{(2)}$$

$$\cdots \qquad \cdots$$

$$\Delta^{\ell}u(t) \geq \Delta^{n-1}u(2^{n-\ell-1}t) \cdot \frac{1}{(n-\ell-1)!}(t)^{(n-\ell-1)}.$$

Next, we get

$$\Delta^{\ell-1}u(t) = \Delta^{\ell-1}u(t_0) + \sum_{s=t_2}^{t-1} \Delta^{\ell}u(s)$$

$$\geq \sum_{s=t_2}^{t-1} \frac{1}{(n-\ell-1)!}(s-t_2)^{(n-\ell-1)}\Delta^{n-1}u(2^{n-\ell-1}s)$$

$$\geq \frac{1}{(n-\ell)!}\Delta^{n-1}u(2^{n-\ell-1}t)(t-t_2)^{(n-\ell)}.$$

Hence, after $(\ell-1)$ summations, we obtain (2.10).

COROLLARY 2. Let $u(t)$ be as in Corollary 1 and bounded. Then

(a) $\lim_{n\to\infty} \Delta^k u(t) = 0$, $1 \leq k \leq n-1$

(b) $(-1)^{k+1}\Delta^{n-k}u(t) \geq 0$, $1 \leq k \leq n-1$.

Proof. Part (a) follows from lemma 6. Also, for (b) we note that in the conclusion of lemma 7, ℓ cannot be greater than 1 and $n+\ell$ is odd.

COROLLARY 3. Let $u(t)$ be as in Corollary 1 . Then, exactly one of the following is true

(a) $\lim_{t\to\infty} \Delta^k u(t) = 0$, $1 \leq k \leq n-1$

(b) there is an odd integer k, $1 \le k \le n-1$ such that $\lim_{t\to\infty} \Delta^{n-j}u(t) = 0$ for $1 \le j \le k-1$, $\lim_{t\to\infty} \Delta^{n-k}u(t) \ge 0$ (finite), $\lim_{t\to\infty} \Delta^{n-k-1}u(t) > 0$ and $\lim_{t\to\infty} \Delta^{j}u(t) = \infty$, $0 \le j \le n-k-2$.

Proof. The proof is contained in lemma 7 and Corollary 2.

LEMMA 8. Let u(t) be as in Corollary 1. Then,

$$\lim_{t\to\infty} \frac{\Delta^i u(t)}{u(t)} = 0, \quad 1 \le i \le n-1$$

unless $\lim_{t\to\infty} \Delta^i u(t) = 0$, $0 \le i \le n-1$. The exceptional case may arise only when n is odd.

Proof. We first assume that the case (a) of Corollary 3 holds. Then, from lemma 6 in the conclusion of lemma 7, ℓ cannot be greater than 1. Thus, u(t) is monotone nondecreasing or nonincreasing on I_{t_1} according as n is even or odd. Thus, (2.11) follows unless $\lim_{t\to\infty} u(t) = 0$, which is possible only when n is odd.

Next, we assume the case (b) of Corollary 3. Then, (2.11) is obvious for $n-k \le i \le n-1$. If $i \le n-k-1$, then for $\lim_{t\to\infty} \Delta^{n-k}u(t) = 0$ we have from lemma 4 that

$$0 = \lim_{t\to\infty} \frac{\Delta^{n-k}u(t)}{\Delta^{n-k-1}u(t)} = \lim_{t\to\infty} \frac{\Delta^{n-k-1}u(t)}{\Delta^{n-k-2}u(t)} = \ldots = \lim_{t\to\infty} \frac{\Delta u(t)}{\Delta u(t)}$$

also, if $\lim_{t\to\infty} \Delta^{n-k}u(t) = c > 0$, then $\lim_{t\to\infty} \Delta^{n-k-1}u(t) = \infty$ and from lemma 4

$$0 = \lim_{t\to\infty} \frac{\Delta^{n-k+1}u(t)}{\Delta^{n-k}u(t)} = \lim_{t\to\infty} \frac{\Delta^{n-k}u(t)}{\Delta^{n-k-1}u(t)} = \ldots = \lim_{t\to\infty} \frac{\Delta u(t)}{\Delta u(t)} .$$

Thus, as long as $\lim_{t\to\infty} \Delta^{n-k}u(t) \ge 0$, we have

$$\lim_{t\to\infty} \frac{\Delta^i u(t)}{\Delta^{i-1}u(t)} = 0, \; 1 \le i \le n-k.$$

Next, from (2.12) we find

$$\lim_{t\to\infty} \frac{\Delta^i u(t)}{u(t)} = \lim_{t\to\infty} \frac{\Delta^i u(t)}{\Delta^{i-1}u(t)} \times \lim_{t\to\infty} \frac{\Delta^{i-1}u(t)}{\Delta^{i-2}u(t)} \times \ldots \times \lim_{t\to\infty} \frac{\Delta u(t)}{u(t)} = 0, \; 1 \le i \le n-k-1.$$

Some of the above lemmas are proved in our earlier work [1,2] also.

3. APPLICATIONS TO DIFFERENCE EQUATIONS.

Here, we shall consider nth order nonlinear difference equation

$$\Delta^n x(t) + \sum_{i=1}^{m} f_i(t)F_i(x(t),\Delta x(t),\ldots,\Delta^{n-1}x(t)) = 0, \quad t \in I \tag{3.1}$$

and assume that f_i, F_i; $1 \leq i \leq m$ are defined in their domain of definition and for all initial values of the type

$$\Delta^i x(t_1) = A_i, \quad 0 \leq i \leq n-1, \quad t_1 \in I \tag{3.2}$$

solutions of (3.1), (3.2) exist for all $t \geq t_1$, $t \in I$.

By a solution of (3.1) we shall always refer to a nontrivial solution. A solution $x(t)$ of (3.1) is called nonoscillatory if it is eventually of fixed sign; and it is called oscillatory if there is no end of t_1 and $t_2 (t_1 < t_2)$ in I such that if $x(t_1) > 0 (x(t_1) < 0)$ then $x(t_2) < 0 (x(t_2) > 0)$. For example, the equation $\Delta^2 x(t) + \frac{8}{3}\Delta x(t) + \frac{4}{3}x(t) = 0$, $t \in I$ has an oscillatory solution $x(t) = (-1)^t$ and a nonoscillatory solution $x(t) = (\frac{1}{3})^t$.

THEOREM 1. <u>In equation</u> (3.1), <u>we assume</u>

(i) $f_i(t) \geq 0$ <u>for all</u> $t \in I_{t_1}$, $1 \leq i \leq m$

(ii) $\operatorname{sgn} F_i(x_1,x_2,\ldots,x_n) = \operatorname{sgn} x_1$ <u>and</u>

$F_i(-x_1,-x_2,\ldots,-x_n) = -F_i(x_1,x_2,\ldots,x_n)$ <u>for all</u> $(x_1,x_2,\ldots,x_n) \in R^n$, $1 \leq i \leq m$

(iii) <u>there is an index</u> j <u>such that</u> $F_j(x_1,x_2,\ldots,x_n)$ <u>is continuous and</u>

(a) $F_j(\lambda x_1,\lambda x_2,\ldots,\lambda x_n) = \lambda^{2\alpha+1}F_j(x_1,x_2,\ldots,x_n)$ for all $(x_1,x_2,\ldots,x_n) \in R^n$ <u>and</u> $\lambda \in R$, α <u>is some nonnegative integer</u>

(b) $\sum_{s=t_1}^{\infty} f_j(s) = \infty$.

<u>Then</u>, (1) <u>if</u> n <u>even</u>, <u>every solution of</u> (3.1) <u>is oscillatory</u>

(2) <u>if</u> n <u>odd</u>, <u>each solution of</u> (3.1) <u>is oscillatory or tends monotonically to zero together with</u> $\Delta^i x(t)$, $1 \leq i \leq n-1$.

<u>Proof</u>. Let $x(t)$ be nonoscillatory solution of (3.1), which must then

eventually be of one sign. Since, condition (ii) implies $-x(t)$ is again a solution of (3.1), without loss of generality we can assume that $x(t) > 0$ for $t \geq t_1 > 0$. For this solution $x(t)$ our hypotheses implies $\Delta^n x(t) \leq 0$ for $t \geq t_1$.

If n is even, we find from lemma 8, $\lim_{t\to\infty} \frac{\Delta^i x(t)}{x(t)} = 0$, $1 \leq i \leq n-1$.

Since F_j is continuous, for any $\varepsilon > 0$ there exists $t_2 \geq t_1$ such that

$$|F_j(1, \frac{\Delta x(t)}{x(t)}, \ldots, \frac{\Delta^{n-1} x(t)}{x(t)} - F_j(1,0,\ldots,0)| < \varepsilon, \quad t \geq t_2.$$

From (ii), $F_j(1,0,\ldots,0) > 0$ and we may assume $0 < \varepsilon < F_j(1,0,\ldots,0)$. From lemma 7, we have $\Delta x(t) > 0$, $\Delta^{n-1} x(t) > 0$ also as a consequence $x(t) > c_n > 0$ for all $t \in I_{t_2}$. Define $y(t) = \frac{\Delta^{n-1} x(t)}{x(t)}$, then we find

$$\Delta y(t) = \frac{\Delta^{n-1} x(t+1)}{x(t+1)} - \frac{\Delta^{n-1} x(t)}{x(t)}$$

$$\leq \frac{\Delta^{n-1} x(t+1)}{x(t)} - \frac{\Delta^{n-1} x(t)}{x(t)} = \frac{\Delta^n x(t)}{x(t)} .$$

Thus, from (3.1) and the hypotheses

$$\Delta y(t) \leq -f_j(s) \frac{F_j(x(s), \Delta x(s), \ldots, \Delta^{n-1} x(s))}{x(s)} .$$

Summing the above inequality, we find from (iii)

$$y(t) - y(t_3) \leq - \sum_{s=t_2}^{t-1} f_j(s) x^{2\alpha}(s) F_j(1, \frac{\Delta x(s)}{x(s)}, \ldots, \frac{\Delta^{n-1} x(s)}{x(s)})$$

$$\leq - \left[F_j(1,0,\ldots,0) - \varepsilon\right] c_n^{2\alpha} \sum_{s=t_2}^{t-1} f_j(s).$$

In the above inequality right side tends to $-\infty$ as $t \to \infty$ whereas left side remains bounded. This contradiction proves (1).

If n is odd, then the case (b) of Corollary 3 is impossible because we get a contradiction as in the case n even. Next, we assume

$$\lim_{t\to\infty} x(t) = c > 0, \quad \lim_{t\to\infty} \Delta^i x(t) = 0, \quad 1 \leq i \leq n-1$$

then from the continuity of F_j and (ii), we find $\lim_{t\to\infty} F_j(x(t), \Delta x(t), \ldots,$

$\Delta^{n-1}x(t)) = F_j(c,0,\ldots,0) > 0$. Thus, $F_j > 0$ for all $t \geq t_3$. From (3.1) and (ii), we have

$$\Delta^n x(t) \leq -f_j(t)F_j(x(t),\Delta x(t),\ldots,\Delta^{n-1}x(t)).$$

Summing the above inequality, we obtain

$$-\Delta^{n-1}x(t) + \Delta^{n-1}(t_3) \geq \sum_{s=t_3}^{t-1} f_j(s)F_j(x(s),\Delta x(s),\ldots,\Delta^{n-1}x(s)).$$

If we let t tend to infinity in the above inequality, we have a contradiction $\Delta^{n-1}(t_3) \geq \infty$. This completes the proof of (2).

THEOREM 2. <u>In theorem 1, condition (iii)(a) can be replaced by</u>

(a)' <u>for any</u> k, $2 \leq k \leq n$ <u>and any</u> $c \geq 0$, $\liminf F_j(x_1,x_2,\ldots,x_n) > 0$ <u>or</u> ∞ <u>as</u> $x_1 \to \infty,\ldots, x_{k-1} \to \infty$, $x_k \to c$, $x_{k+1} \to 0,\ldots,x_n \to 0$.

<u>Proof</u>. Assuming $x(t) > 0$ on I_{t_1} as in the proof of theorem 1. Then, $\Delta^n x(t) \leq 0$ for $t \geq t_1$ and from lemma 7, $\Delta^{n-1}x(t) > 0$ on I_{t_1}. From (3.1), we find

$$\Delta^n x(t) \leq -f_j(t)F_j(x(t),\Delta x(t),\ldots,\Delta^{n-1}x(t))$$

and hence

$$\Delta^{n-1}x(t_1) > \sum_{s=t_1}^{t-1} f_j(s)F_j(x(s),\Delta x(s),\ldots,\Delta^{n-1}x(s)). \tag{3.3}$$

We distinguish two cases :

<u>Case 1</u>. There exists k, $0 \leq k \leq n-1$ such that $\lim_{t\to\infty} \Delta^i x(t) = \infty$ for $0 \leq i \leq k-1$, $\lim_{t\to\infty} \Delta^k x(t) = c > 0$ and $\lim_{t\to\infty} \Delta^i x(t) = 0$ for $k+1 \leq i \leq n-1$. Then, from (a)' $\liminf_{t\to\infty} F_j(x(t),\Delta x(t),\ldots,\Delta^{n-1}x(t)) \geq \varepsilon > 0$. So, there exists a $t_2 \geq t_1$ such that $F_j(x(t), \Delta x(t),\ldots,\Delta^{n-1}x(t)) \geq \varepsilon$ for all $t \geq t_2$. Replacing t_1 by t_2 in (3.3), we find

$$\Delta^{n-1}x(t_2) > \varepsilon \sum_{s=t_2}^{t-1} f_j(s)$$

and this leads to a contradiction.

Case 2. $\lim_{t\to\infty} x(t) = c > 0$ and $\lim_{t\to\infty} \Delta^i x(t) = 0$ for $1 \leq i \leq n-1$. If $c < \infty$, then since F_j is continuous for every $0 < \varepsilon < F_j(c,0,\ldots,0)$ there exists $t_2 \geq t_1$ such that $F_j(c,0,\ldots,0) - \varepsilon < F_j(x(t),\Delta x(t),\ldots,\Delta^{n-1}x(t))$ for all $t \geq t_2$ and we get from (3.3).

$$\Delta^{n-1}x(t_2) > \left[F_j(c,0,\ldots,0) - \varepsilon\right] \sum_{s=t_2}^{t-1} f_j(s)$$

which is again a contradiction. If $c = \infty$, then also from (a)' we have a contradiction. This compltes the proof.

THEOREM 3. <u>In addition to</u> (i) <u>and</u> (ii) <u>of theorem 1, we assume</u> (iii)' <u>there is an index j such that</u> $F_j(x_1,x_2,\ldots,x_n)$ <u>is continuous and</u>

$$\sum^{\infty}(s)^{(n-1)}f_j(s) = \infty. \tag{3.4}$$

<u>Then</u>, (1) <u>if</u> n <u>even, every bounded solution of (3.1) is oscillatory</u>

(2) <u>if</u> n <u>odd, each bounded solution of (3.1) is oscillatory or tend to zero monotonically</u>.

<u>Proof</u>. Assuming $x(t) > 0$ is bounded on I_{t_1} as in the proof of theorem 1. Then, $\Delta^n x(t) \leq 0$ for $t \geq t_1$ and from Corollary 2 $\lim_{t\to\infty} \Delta^i x(t) = 0$, $1 \leq k \leq n-1$. also $\Delta x(t) \geq 0$ if n is even whereas for n odd $\Delta x(t) \leq 0$. Since $x(t)$ is bounded we find, for n even $x(\infty) = c > 0$ and for n odd either $x(\infty) = c > 0$ or $x(\infty) = 0$. Thus, to complete the proof we need to consider the case $x(\infty) = c > 0$ whether n is even or odd.

Since F_j is continuous we find, $\lim_{t\to\infty} F_j(x(t),\Delta x(t),\ldots,\Delta^{n-1}x(t)) = F_j(c,0,\ldots,0) > \varepsilon > 0$. Hence, there exists $t_2 \geq t_1$ such that $F_j(c,0,\ldots,0) - \varepsilon < F_j(x(t),\Delta x(t),\ldots,\Delta^{n-1}x(t))$ for all $t \geq t_2$.

From the equation (3.1), we have

$$\Delta^n x(t) + f_j(t)F_j(x(t),\Delta x(t),\ldots,\Delta^{n-1}x(t)) \leq 0, \quad t \geq t_1$$

and hence

$$\Delta^n x(t) + \left[F_j(c,0,\ldots,0) - \varepsilon\right] f_j(t) < 0, \quad t \geq t_2. \tag{3.5}$$

Multiplying (3.5) by $(t)^{(n-1)}$ and using lemma 5, we obtain

$$\sum_{k=1}^{n} (-1)^{k+1}\Delta^{k-1}(s)^{(n-1)}\Delta^{n-k}x(s+k-1)\Big|_{s=t_2}^{t} + [F_j(c,0,\ldots,0)-\varepsilon]\sum_{s=t_2}^{t-1}(s)^{(n-1)}f_j(s)<0. \tag{3.6}$$

Using Corollary 2 in (3.6), we get

$$[F_j(c,0,\ldots,0)-\varepsilon]\sum_{s=t_2}^{t-1}(s)^{(n-1)}f_j(s) < k + (-1)^{n+2}!x(t+n-1) \tag{3.7}$$

where k is some finite constant. Using (3.4) in (3.7), we have a contradiction to our assumption that x(t) is bounded. Thus, the proof is completed.

Our next result is for the even order equation and in (3.1) we shall take n as 2N where $N \geq 1$.

THEOREM 4. In addition to (i) and (ii) of theorem 1, we assume

(iii)" $E \neq \phi$ where E denotes the set of all indices i for which the function $F_i(x_1,x_2,\ldots,x_{2N})$ is continuous and nondecreasing with respect to each variable $x_2,x_4,\ldots,x_{2N}$ and nonincreasing with respect to $x_3,x_5,\ldots,x_{2N-1}$ as well as the function $\frac{F_i(x,0,\ldots,0)}{x}$ in nonincreasing on $(0,\infty)$

(iv) there exists a positive function $\phi(t)$, $t \in I$ such that

$$\sum^{\infty}\left[\phi(s)\sum_{i\in E}\frac{f_i(2^{2N-2}s)F_i(k(s)^{(2N-1)},0,\ldots,0)}{k(s)^{(2N-1)}} - \frac{1}{4}\,\frac{(2N-2)!(\Delta\phi(s))^2}{(s/2)^{(2N-2)}\phi(s)}\right] = \infty \tag{3.8}$$

for every $k \geq 1$.

Then all bounded solutions of (3.1) are oscillatory.

Proof. Assuming $x(t) > 0$ is bounded on I_{t_1} as in the proof of theorem 1. Then, $\Delta^{2N}x(t) \leq 0$ for $t \geq t_1$ and from Corollary 2, $\lim_{t\to\infty}\Delta^i x(t) = 0$, $1 \leq i \leq 2N-1$ and $(-1)^{i+1}\Delta^{2N-i}x(t) \geq 0$, $1 \leq i \leq 2N-1$. We define the transformation

$$y(t) = -\frac{\Delta^{2N-1}x(2^{2N-2}t)}{x(t)}\,\phi(t) \leq 0,\ t \geq t_1$$

and obtain on using nonincreasing nature of $\Delta^{2N-1}x(t)$ that

$$\Delta y(t) \geq -\frac{\Delta^{2N}x(2^{2N-2}t)}{x(t)}\phi(t) + \frac{\Delta\phi(t)y(t+1)}{\phi(t+1)} - \frac{y(t+1)\phi(t)\Delta x(t)}{x(t)\phi(t+1)} . \tag{3.9}$$

Using the equation (3.1), the hypotheses (iii)" and the above inequalities in (3.9), to obtain

$$\Delta y(t) \geq \phi(t)\sum_{i\in E}\frac{f_i(2^{2N-2}t)F_i(x(t),0,\ldots,0)}{x(t)} + \frac{\Delta\phi(t)y(t+1)}{\phi(t+1)} - \frac{y(t+1)\phi(t)\Delta x(t+1)}{x(t)\ \phi(t+1)},$$
$$t \geq t_1. \tag{3.10}$$

Since $\Delta x(t+1) > 0$ and $\Delta^{2N}x(t) \leq 0$, Corollary 1 is applicable and we find

$$\Delta x(t+1) \geq \frac{1}{(2N-2)!}\Delta^{2N-1}x(2^{2N-2}\overline{t+1})(t+1-t_2)^{(2N-2)}, \quad t \geq t_2 \geq t_1$$

or

$$\Delta x(t+1) \geq -\frac{y(t+1)x(t+1)}{\phi(t+1)}\frac{(t+1-t_2)^{(2N-2)}}{(2N-2)!}, \quad t \geq t_2. \tag{3.11}$$

Substituting (3.11) in (3.10), we get

$$\Delta y(t) \geq \phi(t)\sum_{i\in E}\frac{f_i(2^{2N-2}t)F_i(x(t),0,\ldots,0)}{x(t)}$$
$$+\left[\frac{\Delta\phi(t)y(t+1)}{\phi(t+1)} + \frac{y^2(t+1)\phi(t)x(t+1)}{\phi^2(t+1)x(t)}\frac{(t+1-t_2)^{(2N-2)}}{(2N-2)!}\right], \quad t \geq t_2. \tag{3.12}$$

Since $\frac{x(t+1)}{x(t)} \geq 1$, the term inside the brackets in (3.12) (calling it Z) is

$$Z \geq \phi(t)\frac{(t+1-t_2)^{(2N-2)}}{(2N-2)!}\left[\frac{y(t+1)}{\phi(t+1)} + \frac{1}{2}\frac{(2N-2)!}{(t+1-t_2)^{(2N-2)}}\frac{\Delta\phi(t)}{\phi(t)}\right]^2$$
$$-\frac{1}{4}\frac{(2N-2)!}{(t+1-t_2)^{(2N-2)}}\frac{(\Delta\phi(t))^2}{\phi(t)}, \quad t \geq t_2.$$

Using this in (3.12), we get

$$\Delta y(t) \geq \phi(t)\sum_{i\in E}\frac{f_i(2^{2N-2}t)F_i(x(t),0,\ldots,0)}{x(t)}$$
$$-\frac{1}{4}\frac{(2N-2)!}{(t+1-t_2)^{(2N-2)}}\frac{(\Delta\phi(t))^2}{\phi(t)}, \quad t \geq t_2. \tag{3.13}$$

Next, from lemma 1, we have

$$x(t) \leq \sum_{i=0}^{2N-1} \frac{(t-t_2)^{(i)}}{(i)!} \Delta^i u(t_2), \quad t \geq t_2.$$

and hence there exists some $k \geq 1$ such that

$$x(t) \leq k(t)^{(2N-1)}, \quad t \geq t_2. \tag{3.14}$$

Using (3.14) in (3.13), we obtain

$$\Delta y(t) \geq \phi(t) \sum_{i \in E} \frac{f_i(2^{2N-2}t)F_i(k(t)^{(2N-1)},0,\ldots,0)}{k(t)^{(2N-1)}}$$

$$- \frac{1}{4} \frac{(2N-2)!}{(t/2)^{(2N-2)}} \frac{(\Delta\phi(t))^2}{\phi(t)}, \; t \geq 2t_2 - 2. \tag{3.15}$$

Summing up the above inequality, we find from (3.8) that y(t) is eventually positive, which is a contradiction. Hence the result follows.

REFERENCES

1. R. P. Agarwal, Properties of solutions of higher order nonlinear difference equations, An. Sti. Univ. "Al. I. Cuza" Iasi, to appear.

2. R. P. Agarwal, Properties of solutions of higher order nonlinears difference equations II, An. Sti. Univ. "Al. I. Cuza" Iasi, to appear.

3. R. P. Agarwal and E. Thandapani, On some new discrete inequalities, Appl. Math. Comp. 7(1980), 205-224.

4. R. P. Agarwal, Oscillation and asymptotic behaviour of solutions of differential equations with nested arguments, Boll. U. M. I.

5. P. Hartman, Difference equations : Disconjugacy, Principal solutions, Green's functions, complete monotonicity, Trans. Amer. Math. Soc. 246 (1978), 1-30.

6. I. T. Kiguardze, On the oscillation of the equation $\frac{d^m u}{dt^m} + a(t)u^n \text{sgn } u^{\,n}=0$. Mat. Sb. 65(1964), 172-187.

7. I. T. Kiguardze, The problem of socillation of solution of nonlinear differential equations. Differencial 'nye Uravnenija 1(1965), 995-1006.

8. A Kneser, Untersuchungen uber die reellen Nullstellen der Integrale linearer Differentialgleichungen, Math. Ann. 42(1893), 409-435.

9. H. Onose, Oscillatory property of certain non-linear ordinary differential equations, Proc. Japan Acad. 44(1968), 232-237.

10. H. Onose, Oscillatory property of certain non-linear ordinary differential equations II, Proc. Japan Acad. 44(1968), 876-878.

11. H. Onose, Some oscillatory criteria for nth order nonlinear delay-differential equations Hiroshima Math. J. 1(1971), 171-176.

12. W. T. Patula, Growth and oscillation properties of second order linear difference equations, SIAM J. Math. Anal. 10(1979), 55-61.

13. Ch. G. Philos, Oscillatory and asymptotic behaviour of all solutions of differential equations with deviating arguments, Proc. Royal Soc. Edinburgh, 81A(1978), 195-210.

14 V. A. Staikos, Oscillatory property of certain delay-differential equations, Bull. Soc. Math. Grece 11(1970), 1-5.

15. V. A. Staikos and A. G. Petsoulas, Some oscillation criteria for second order nonlinear delay-differential equations, J. Math. Anal. Appl. 30(1970), 695-701.

16. B. Szmanda, Oscillation theorems for nonlinear second-order difference equations, J. Math. Anal. Appl. 79(1981), 90-95.

17. E. Thandapani and R. P. Agarwal, Asymptotic behaviour and oscillation of solutions of differential equations with deviating arguments,Boll. U. M. I. 17(1980), 82-93.

Ravi P. Agarwal, Department of Mathematics, National University of Singapore, Kent Ridge, Singapore 0511.

International Series of
Numerical Mathematics, Vol. 71

POSITIVITY IN SUMMABILITY

Wolfgang Beekmann and Karl Zeller

Abstract. Given a triangular summability matrix A we define a sequence s to be positive if its A-transform has positive entries. We are interested in matrices with the property that the ordinary sections of a sequence or certain weighted means of them determine positive operators in the convergence domain. Especially important is the autopositive case where the weighted means are given by the matrix itself.

1. NOTATION

We consider infinite matrices which are triangles:

$$a_{nk} = 0 \quad (k > n), \quad a_{nn} \neq 0 \qquad (n = 0,1,\ldots) .$$

Such a matrix A defines a transformation

$$t = As: \quad t_n = \sum_{k=0}^{n} a_{nk} s_k \qquad (n = 0,1,\ldots)$$

and a summability method A:

$$\text{A-lim } s_k := \lim t_n$$

(if the latter exists). The corresponding sequences $s = (s_k)$ constitute the summability field c_A . This is a BK-space with norm $\|s\| := \sup |t_n|$, cf. [18], § 22 I. We say (for $s \in c_A$) that

$$s \text{ is positive, if } t_n \geq 0 \qquad (n = 0,1,\ldots).$$

For simplicity we consider regular methods only, i.e. A-lim s_k = lim s_k for every convergent sequence. A method B is said to be stronger than A (with consistency), if $c_A \subseteq c_B$ and B-lim s_k = A-lim s_k for every $(s_k) \in c_A$. This is equivalent to the regularity of BA^{-1}.

2. POSITIVE SECTIONS

Let A be a regular triangle. We say that it has positive sections, if the matrix operators

$$I_n : c_A \to c_A , \quad s = (s_k) \to I_n s := (s_0, \ldots, s_n, 0, 0, \ldots)$$

are positive, i.e. $I_n s$ is positive for positive s. This is equivalent to

$$A I_n A^{-1} \geqq 0 ,$$

and in case $A \geqq 0$ it is equivalent to $\bar{a}_{nk} \leqq 0$ for $0 \leqq k < n$, where $(\bar{a}_{nk}) = A^{-1}$, see [17]. Examples are the Cesàro methods $A = C_p$ with $0 \leqq p \leqq 1$.

A positive linear functional L in c_A is given by

$$Ls = g \lim t_n + \sum_{n=0}^{\infty} g_n t_n \quad (g \geqq 0,\ g_n \geqq 0,\ \sum g_n < \infty).$$

With e^k denoting the k-th unit vector we put

$$f_k := Le^k = \sum_n g_n a_{nk} \qquad (k = 0, 1, \ldots) ,$$

then, if A has positive sections and L is positive, we have

$$\sum_{k=\nu}^{\infty} f_k \bar{a}_{k\nu} \geqq 0 \quad (\nu = 0, 1, \ldots) , \text{ shortly: } fA^{-1} \geqq 0 ,$$

whenever the sums exist. We notice that also the sums

$$fI_n s := \sum_{k=0}^{n} f_k s_k$$

define positive linear functionals in c_A (consider $L \circ I_n$), and if

$$0 \leqq f_k \searrow , \text{ i.e. } fI_n C_1^{-1} \geqq 0,$$

then the sums

$$\sum_{k=0}^{n} f_k s_k e^k$$

define positive linear operators in c_A. The latter results are useful for comparison theorems, cf. [4].

3. POSITIVE D-SECTIONS

Let A be a regular triangle and D a triangle with inverse $D^{-1} = (\bar{d}_{nk})$. We say that A has positive D-sections if

$$D_n s := (d_{n0}s_0,\ldots,d_{nn}s_n,0,0,\ldots) = \sum_{k=0}^{n} d_{nk}s_k e^k$$

determine positive linear operators in c_A. Examples are $A = C_p$ with $D = C_q$ $(1 \leq p \leq q)$, see [1] and [17]. The property is important for comparison theorems in connection with factors. Suitable factors give new positive operators (and lead to stronger summability methods B where $b_{nk} = f_{nk}a_{nk}$).

PROPOSITION 1. _Suppose_ A _is a regular triangle having positive_ D-_sections_. _Then_, _for_ $n = 0,1,\ldots,$

$$\sum_{k=0}^{n} f_k d_{nk} s_k$$

defines a positive linear functional in c_A _if_ $f_k = Le^k$ _for some positive linear functional_ L _in_ c_A. _If_

$$\sum_{k=0}^{n} f_k \bar{d}_{kj} \geq 0 \qquad (j = 0,\ldots,n),$$

then each sum

$$\sum_{k=0}^{n} f_k s_k e^k$$

defines a positive linear operator in c_A.

Proof. For the first part we consider $L \circ D_n$. For the second part we point out that the operator in question is the sum of the positive operators given by

$$\left(\sum_{k=0}^{n} f_k \bar{d}_{kj}\right) D_j s = \sum_{k=0}^{n} \sum_{r=0}^{j} f_k \bar{d}_{kj} d_{jr} s_r e^r \quad (j=0,\ldots,n).$$

The condition in the first assertion leads to $fA^{-1} \geq 0$, as shown in § 2. The second condition can be described by

$$fI_n D^{-1} \geq 0.$$

4. THE AUTOPOSITIVE CASE

Given a regular triangle A we say that A is autopositive, if it has positive A-sections. Examples are the Cesàro matrices C_p for $p \geqq 1$, and also some more general Nörlund matrices, see [11]. The coincidence $D = A$ simplifies the considerations and leads to stronger results.

PROPOSITION 2. *Suppose that A is a triangle which is regular and autopositive. If the sequence (f_k) is such that*

$$f_k = Le^k$$

for a positive linear functional L in c_A, then the mapping

$$\operatorname{diag}(f_k) : (s_k) \to (f_k s_k)$$

is a positive linear operator in c_A.

If each of the two sequences (f'_k) and (f''_k) are generated as above by positive linear functionals in c_A, then the same is true for the product sequence $(f'_k f''_k)$.

Proof. From proposition 1 we have

$$w_n(s) := \sum_{k=0}^{n} a_{nk} f_k s_k \geqq 0 \quad (n = 0,1,\ldots) \text{ for positive } s.$$

Hence the operator $\operatorname{diag}(f_k)$ is positive. To show that $(f_k s_k) \in c_A$ for $s \in c_A$, we prove $w_n(s) \to 0$ for $s \in c_A$: It can be seen from the definition in § 3 that A-positivity of A is equivalent to $AA_nA^{-1} \geqq 0$. In particular, we have

$$a_{mn} a_{nn} \bar{a}_{nn} \geqq 0 \text{ , hence } A \geqq 0.$$

Consequently $u := Ae > 0$, $u_n \to 1$ and $0 \leqq w_n(e) \to 0$, since A is regular and $\sum |f_k| < \infty$. By positivity of $\operatorname{diag}(f_k)$ we now get, for $s = A^{-1}t \in c_A$ and $\tilde{t} = \operatorname{diag}(u_k^{-1})t$,

$$|w_n(s)| \leqq w_n(A^{-1}\operatorname{diag}(u_k)e) \sup|\tilde{t}_k| = w_n(e) \sup|\tilde{t}_k| \to 0.$$

The second assertion follows by putting $Ls := L''(f'_k s_k)$, where $L''e^k = f''_k$.

5. CESÀRO METHODS

As mentioned above the Cesàro triangle C_p is regular for $p \geq 0$ and autopositive for $p \geq 1$. The elements of C_p and C_p^{-1} are (for $0 \leq k \leq n$)

$$\binom{n-k+p-1}{n-k} \Big/ \binom{n+p}{n} \quad \text{and} \quad \binom{n-k-p-1}{n-k} \cdot \binom{k+p}{k},$$

respectively. We define the p-th differences of a given (bounded) sequence $f = (f_k)$ by

$$h_k := \sum_{j=k}^{\infty} \binom{j-k-p-1}{j-k} f_j \qquad (k = 0,1,\ldots).$$

Of course, $h = (h_k) \geq 0$ if and only if $fC_p^{-1} \geq 0$

THEOREM 3. Let $A := C_p$ for some $p \geq 1$ and let B be a regular triangle with

$$b_{nk} = f_{nk} a_{nk}$$

such that, for each $n = 0,1,\ldots,\lim_k f_{nk}$ exists and the p-th differences h_{nk} of $(f_{nk})_{k=0,1,\ldots}$ satisfy

$$h_{nk} \geq 0 \quad \text{and} \quad \sum_j h_{nj} \binom{j+p}{j} < \infty \qquad (n,k = 0,1,\ldots)$$

Then $BA^{-1} \geq 0$, and B is stronger than A with consistency.

Proof. There are linear functionals L_n in c_A with $L_n e^k = f_{nk}$, see [12], § 4; these are positive by hypothesis. Hence (proposition 2) the rows of B define positive linear functionals B_n in c_A, whence (see § 2) $BA^{-1} \geq 0$. The norms of the B_n's being bounded by $B_n A^{-1} e = B_n e$ are uniformly bounded. Since A is perfect and B is regular, B is stronger than A with consistency.

6. REMARKS

Our investigations are rooted in the well-known mean value theorem of M. Riesz which led to a set of inequalities in connection with Cesàro and general matrix transformations, see e.g. Bosanquet [6] and, also for a discussion of different starting points and results, [7]. Later this has been incorporated into functional analysis (sectional boundedness and sectional convergence, see [15] and [18], Chapt. 24). Subsequent investigations led to sectional summability in summability fields (see [2],[11], [12],[16]) or in more general settings (see e.g. [8],[9],[13]) and to positivity considerations (see e.g. [3],[4],[5],[17]) which partially exhibit methods having been used implicitely in earlier investigations.

Using decomposition we can treat non-positive operators. Also there is no compelling reason to restrict the positivity considerations to the summability field c_A , we could employ larger or smaller sets than c_A , thus obtaining modifications of our results. The basic concepts are also adaptable to other matrices than triangles, for instance for matrices defining the Abel means or the Riesz means. As to the latter, there is a result (presented at the previous conference, see [14]) concerning autopositivity, which is important for the Second Theorem of Consistency (related to our theorem 3).

REFERENCES

1. R. Askey, G. Gasper and M. E.-H.Ismail, A positive sum from summability theory. J. Approx. Th. 13 (1975), 413-42o.

2. W. Balser, Über Abschnittslimitierbarkeit. J. reine angew. Math. 281 (1976), 211-218.

3. W. Beekmann, Total-Vergleich normaler Matrizen mit diagonalpositiver Inversen. Math. Z. 125 (1972), 361-371.

4. W. Beekmann und K. Zeller, Normvergleich bei abschnittsbeschränkten Limitierungsverfahren. Math. Z. 126 (1972), 116-122.

5. W. Beekmann und K. Zeller, Positive Operatoren in der Limitierungstheorie. ISNM 25 (1974), 559-564.

6. L. S. Bosanquet, Note on convergence and summability factors (III). Proc. London Math. Soc. 16 (1941), 146-148

7. L. S. Bosanquet, An inequality for sequence transformations. Mathematika 13 (1966), 26-41.

8. M. Buntinas, Convergent and bounded Cesàro sections in FK-spaces. Math. Z. 121 (1971), 191-200.

9. M. Buntinas, On Toeplitz sections in sequence spaces. Math. Proc. Camb. Phil. Soc. 78 (1975), 451-460.

10. G. H. Hardy and M. Riesz, The general theory of Dirichlet's series, Cambridge University Press, Cambridge 1915, reprinted 1952.

11. J. C. Kurtz, Uniform summability and Toeplitz bases. Proc. Camb. Phil. Soc. 73 (1973), 73-81.

12. G. G. Lorentz and K. Zeller, Abschnittslimitierbarkeit und der Satz von Hardy-Bohr. Archiv Math. 15 (1964), 208-213.

13. G. Meyers, On Toeplitz sections in FK-spaces. Studia Math. 51 (1975), 23-33.

14. H. Türke and K. Zeller, Riesz mean value theorem extended. General Inequalities 3, Birkhäuser Verlag, Basel, 1983.

15. A. Wilansky and K. Zeller, Abschnittsbeschränkte Matrixtransformationen; starke Limitierbarkeit. Math. Z. 64 (1956), 258-269.

16. K. Zeller, Approximation in Wirkfeldern von Summierungsverfahren. Archiv Math. 4 (1953), 425-431.

17. K. Zeller, Abschnittsabschätzungen bei Matrixtransformationen. Math. Z. 80 (1963), 355-357.

18. K. Zeller and W. Beekmann, Theorie der Limitierungsverfahren. Springer-Verlag, Berlin-Heidelberg-New York, 2nd edition, 1970.

Wolfgang Beekmann, Fachbereich Mathematik und Informatik, Fernuniversität-Gesamthochschule-, D-5800 Hagen, West Germany.

Karl Zeller, Mathematische Fakultät, Universität Tübingen, D-7400 Tübingen, West Germany.

International Series of
Numerical Mathematics, Vol. 71

INEQUALITIES FOR RATIOS OF INTEGRALS

Dieter K. Ross

Abstract. A unified method is presented for finding bound estimates on ratios of consecutive moments of a non-negative function having a continuous first or second derivative of one sign. One extension of this work leads to a converse of the Cauchy-Schwarz-Buniakowski inequality and to a best possible constant.

Other results are found that strengthen and generalize some integral inequality of Ting and of Sendov and Skordev concerning moments of concave non-negative functions.

1. INTRODUCTION

Integral inequalities can often be deduced by using simple properties of real functions such as the positive semi-definiteness of quadratic forms or by making use of the monotonicity or convexity of particular functions. A typical example of the former type is the arithmetic-geometric mean, for this result leads to the Cauchy-Schwarz-Buniakowski inequality, see Mitrinović [4]; whereas the integral inequalities of Ross [6] and Eliezer and Daykin [3] illustrate the latter.

The main idea in this paper is to make use of m integrations by parts in order to replace the m-th derivative of a function f by f itself. This technique leads to an identity which can be converted to an inequality provided that $f^{(m)}(t)$ remains one signed for all $t \in I = \{t: 0 \leqslant t \leqslant x\}$. An obvious advantage of this method is that any number of inequalities can be deduced, all of which become equalities when $f(t)$ is a polynomial in t of degree $m-1$, $m \geqslant 1$.

The author wishes to thank the Sir Ian Potter Foundation and Cadbury-Schweppes of Melbourne. Without their support attendance at this Conference would not have been possible.

The motivation for this work is a paper of Ting [8] concerning the determination of tight bounds on the radii of gyration of convex bodies. Ting proved, by using an elaborate geometric argument, that for a continuous non-negative function f,

$$\frac{(n-1)x}{(n+1)} \leqslant \frac{\int_0^x t^{n-1}f(t)dt}{\int_0^x t^{n-2}f(t)dt} \leqslant \frac{nx}{(n+1)}, \quad \text{for } n = 2,3,4,\ldots,$$

provided that $f''(t) \leqslant 0$ on I. Here the lower and upper bounds are attained for different functions f; the lower bound when $f(t) = x-t$ and the upper bound when $f(t) = t$. A benefit of the m-fold integration technique described earlier is that it leads to a stronger set of inequalities because *both* the bounds are attained whenever $f''(t) \equiv 0$ on I.

Throughout this paper, I denotes the interval $0 \leqslant t \leqslant x$, where $x > 0$.

2. PRELIMINARY RESULTS

Lemma 1. *Let Δ^n be the forward difference operators defined by*

$$\Delta^0 U_\alpha = U_\alpha, \quad \Delta^1 U_\alpha = U_{\alpha+1} - U_\alpha, \quad \Delta^n U_\alpha = \Delta(\Delta^{n-1}U_\alpha), \quad \text{for } n=1,2,3,\ldots .$$

Then

$$(x-t)^n t^\alpha = (-1)^n x^{\alpha+n} \Delta^n(t^\alpha/x^\alpha), \quad t > 0. \tag{1}$$

This lemma may now be used to prove the following important theorem.

THEOREM 2 *Let f be a continuous function of t defined on the interval I and let $\alpha > -1$ be a fixed constant. Then*

$$\int_0^x dx_1 \int_0^{x_1} dx_2 \ldots \int_0^{x_n} t^\alpha f(t)dt = (-)^n x^{\alpha+n} \int_0^x \frac{\Delta^n}{n!}(t^\alpha/x^\alpha)f(t)dt\ , \tag{2}$$

for each integer $n = 0,1,2,\ldots$ with the understanding that the left hand side of (2) is to be interpreted as the single integral $\int_0^x t^\alpha f(t)dt$ when $n = 0$.

Proof. Let $h(t)$ be a continuous function of t defined on I. Then by Cauchy's formula for repeated integration, as referred to by Oldham and Spanier [5],

$$\int_0^x dx_1 \int_0^{x_1} dx_2 \dots \int_0^{x_n} h(t)dt = \int_0^x \frac{(x-t)^n}{n!} h(t)dt,$$

for $n = 0,1,2,\dots$.

The required result follows by setting $h(t) = t^\alpha f(t)$ and using Lemma 1.

Lemma 3 *Let* $(\alpha)_0 = 1$ *and* $(\alpha)_m = \alpha(\alpha-1)\dots(\alpha-m+1)$, *for* $m=1,2,3,\dots$. *Then*

$$\Delta^n(\alpha)_m = \Gamma(\alpha+1)\Gamma(m+1)/\Gamma(\alpha-m+n+1)\Gamma(m-n+1), \text{ for } n=0,1,2,\dots,$$

where Γ *is Euler's gamma function*.

Proof. This follows in a straight-forward manner by induction on n.

THEOREM 4. *Let* f *be a function with a continuous m-th derivative defined on* I, *let* n,m *be integers satisfying* $n \geqslant 0$, $n+1 \geqslant m \geqslant 0$, *and let* α *be a real constant with* $\alpha > m$. *Then*

$$\int_0^x dx_1 \int_0^{x_1} dx_2 \dots \int_0^{x_n} t^\alpha f^{(m)}(t)dt = \delta_{m,n+1}\, x^\alpha f(x) + (-)^{n+m} x^{\alpha+n} \frac{\Delta^n}{n!} \int_0^x D^m[t^\alpha/x^\alpha]\, f(t)dt, \qquad (4)$$

where D^m *denotes the m-th derivative with respect to* t *and* $\delta_{i,j}$ *denotes the Kronecker delta*.

Proof. (a) For $m = 0$ the above identity is an immediate consequence of Theorem 2 mentioned above, whereas for $n = 0$ with $m = 1$ equation (4) reduces to the obvious identity

$$\int_0^x t^\alpha f'(t)dt = x^\alpha f(x) - x^\alpha \int_0^x D\,[t^\alpha/x^\alpha]\, f(t)dt, \quad \alpha > 0.$$

(b) The next step is to show that (4) holds for all values of $n \geqslant m \geqslant 1$. The proof which follows makes use of an induction on m. To begin suppose that equation (4) is valid when $m = k$, where k is one of the integers $0,1,2,\dots,n$ then it remains valid when $m = k+1$. This is easily verified by replacing the function f in (4), with $m = k$, by f' and integrating the last integral by parts. Thus

$$\int_0^x dx_1 \int_0^{x_1} dx_2 \ldots \int_0^{x_n} t^\alpha f^{(k+1)}(t)dt = (-)^{n+k} x^{\alpha+n-k} \delta_{k,n} f(x) +$$

$$(-)^{n+k+1} x^{\alpha+n} \frac{\Delta^n}{n!} \int_0^x D^{k+1} [t^\alpha/x^\alpha] f(t)dt,$$

where the factor

$$x^{\alpha+n} \frac{\Delta^n}{n!} [D^k(t^\alpha/x^\alpha)f(t)]_0^x$$

which occured in the integration by parts was replaced by

$$x^{\alpha+n-k} \frac{\Delta^n}{n!} (\alpha)_k = x^{\alpha+n-k} \delta_{k,n}$$

because $\alpha > k$.

(c) In order to complete the proof of (4) it is necessary to consider the remaining case where $n+1 = m$. Here it is sufficient to show that the derivative with respect to x of the right hand side of (4) leads to the same expression but with n replaced by $n-1$ provided that $n \geqslant m$. In essence this depends on the simple identity

$$\frac{\partial}{\partial x} [x^{\alpha+n} \frac{\Delta^n}{n!} (t^\alpha/x^\alpha)] = x^{\alpha+n-1} \frac{\Delta^{n-1}}{(n-1)!} (t^\alpha/x^\alpha), \quad t > 0 ,$$

which follows directly from Lemma 1. Thus

$$\int_0^x dx_1 \int_0^{x_1} dx_2 \ldots \int_0^{x_{n-1}} t^\alpha f^{(m)}(t)\, dt = (-)^{n+m} x^{\alpha+n-m} \frac{\Delta^n}{n!} (\alpha)_m f(x) +$$

$$(-)^{n-1+m} x^{\alpha+n-1} \frac{\Delta^{n-1}}{(n-1)!} \int_0^x D^m [t^\alpha/x^\alpha] f(t)dt ,$$

and the required result follows once again by a direct application of Lemma 3 with $n = m$.

<u>Corollary</u> 5. <u>Let</u> p <u>be a function with a continuous m-th derivative defined on</u> I. <u>Then under the condition stated in Theorem</u> 4

$$\int_0^x dx_1 \int_0^{x_1} dx_2 \ldots \int_0^{x_n} t^\alpha p(t) f^{(m)}(t)dt = \delta_{m,n+1}\, x^\alpha p(x) f(x) +$$

$$(-)^{n+m} x^{\alpha+m} \frac{\Delta^n}{n!} \int_0^x D^m[t^\alpha p(t)/x^\alpha] f(t)dt. \tag{5}$$

For the sake of completeness it should be mentioned that extensions of (5) exist in the case where $n+1 < m$, which involve the first $(m-n-1)$ derivatives of f as well as f itself. The most straight-forward way of finding this identity is to differentiate (4) with respect to x; however, this new result is not found of interest here. It should also be noted that Corollary 5 with m=0 is a trivial integration by parts formula. The first set of non-trivial inequalities therefore stem from the case m=1.

3. INEQUALITIES BASED ON FIRST DERIVATIVES, m = 1.

THEOREM 6. *Let f be a non-negative function of t with a continuous first derivative defined on I. Then*

$$x^{\alpha}f(x) \lessgtr \alpha \int_0^x t^{\alpha-1}f(t)dt \qquad (6)$$

and

$$\frac{\int_0^x t^{\alpha}f(t)dt}{\int_0^x t^{\alpha-1}f(t)dt} \lessgtr \frac{\alpha x}{(\alpha+1)}, \quad \text{provided that } \alpha > 0 \qquad (7)$$

according as $f'(t) \lessgtr 0$, *for all* $t \in I$, *with equality in both cases* (6) *and* (7) *if and only if* $f(t) =$ *constant* .

Proof. The first inequality (6) is a direct consequence of equation (4), with m=1 and n=0 and is equivalent to a simple integration by parts of the obvious inequality

$$\int_0^x t^{\alpha}f'(t)dt \gtrless 0$$

according as $f'(t) \gtrless 0$, for all $t \in I$.

The second inequality (7) is again a direct consequence of equation (4), but with m = 1 and n = 1, and is reminiscent of an inequality for the ratios of the moments of a non-negative concave function as established by Ting [8].

THEOREM 7. *Let f be a non-negative function of t with a continuous first derivative defined on I. Then*

$$\alpha(\alpha+2) \int_0^x t^{\alpha-1}f(t)dt \int_0^x t^{\alpha+1}f(t)dt \leqslant (\alpha+1)^2\left[\int_0^x t^{\alpha}f(t)dt\right]^2 , \qquad (8)$$

for $\alpha > 1$, *provided that* $f'(t) \geqslant 0$, *for all* $t \in I$. *This inequality is a best possible under the stated conditions, since equality is attained when* $f(t) = 1$.

Clearly this is a converse of the classical Cauchy-Schwarz-Buniakowski inequality of the type referred to by Bellman [2],

Mitrinović [4] and Wang [9].

Proof. This is based on the identity (4) with m=1 and n=2 coupled with a well known identity involving persymmetric determinants, viz that

$$\begin{vmatrix} a_\alpha & a_{\alpha+1} \\ a_{\alpha+1} & a_{\alpha+2} \end{vmatrix} = \begin{vmatrix} a_\alpha & \Delta a_\alpha \\ \Delta a_\alpha & \Delta^2 a_\alpha \end{vmatrix}, \tag{9}$$

where Δ is the forward difference operator referred to in Lemma 1. For a generalisation of this identity (9), see Aitken [1].

Now it is interesting to note that the inequality in (8) is not reversed when $f'(t) \leqslant 0$, for all $t \in I$. This observation stems from the fact that the sign of the right hand side of (9) is determined when $a_\alpha \, \Delta^2 \, a_\alpha \leqslant 0$ but is not determined when $a_\alpha \, \Delta^2 \, a_\alpha > 0$ without specifying further conditions on the a_α.

Remark 8. Other inequalities which depend on the sign of $f'(t)$, for all $t \in I$, can be obtained by setting $m=1$ with $n=3,4,5,\ldots$. Thus

$$(-1)^{n+1}\Delta^n \int_0^x [\alpha t^{\alpha-1} f(t)/x^\alpha]dt \gtrless 0, \tag{10}$$

according as $f'(t) \gtrless 0$, for all $t \in I$.

A special case of (10), but with n=1, leads to the conclusion that $\int_0^x [\alpha t^{\alpha-1} f(t)/x^\alpha]dt$ is monotonic in α $(\alpha > 1)$, provided that $f(t) \geqslant 0$ and that $f'(t)$ does not change sign for any $t \in I$. A simple calculation shows that, for $\alpha > 1$,

$$\frac{\alpha x^n}{(\alpha+n)} \int_0^x t^{\alpha-1} f(t)dt \lesseqgtr \int_0^x t^{\alpha+n-1} f(t)dt \leqslant x^n \int_0^x t^{\alpha-1} f(t)dt, \; n = 0,1,2,\ldots, \tag{11}$$

where the left-hand inequality applies according as $f'(t) \gtrless 0$, for all $t \in I$ with equality if and only if $f(t) =$ constant, and where the right hand inequality applies for all $f(t) \geqslant 0$ with equality if and only if $f(t) = 0$, for all $t \in I$. The less than inequalities in (11) are analogous to those given by Sendov and Skordev [7] who assumed the weaker conditions $f(t) \geqslant 0$ with f mid-point convex for all $t \in I$.

4. INEQUALITIES BASED ON SECOND DERIVATIVES, m=2.

In classical mechanics there arises a problem concerning the radii of gyration of a convex body. In such a study Ting [8] was able to prove that the ratio of the moments of a non-negative convex function f(t) satisfies the inequalities

$$\frac{(\alpha-1)x}{(\alpha+1)} \leqslant \frac{\int_0^x t^{\alpha-1}f(t)dt}{\int_0^x t^{\alpha-2}f(t)dt} \leqslant \frac{\alpha x}{(\alpha+1)}, \text{ for } \alpha > 2 \qquad (12)$$

with equality in the former case when $f(t) = x-t$ and in the latter case when $f(t) = t$. Ting verifies (12) in case the graph of the function f bounds a right-angled triangle or a trapezoidal figure and then he uses those results to prove that the same inequality (12) hold for any non-negative convex function f. The next four theorems show that the inequalities (12) can be strengthened and that elementary proofs based on the identity (4) exist.

THEOREM 9. _Let f be a non-negative function of t with a continuous second derivative defined on I._ Then

$$\frac{x(\alpha-1)}{(\alpha+1)}\left[1 + \frac{x^{\alpha-1}f(x)}{\alpha(\alpha-1)\int_0^x t^{\alpha-2}f(t)dt}\right] \gtrless \frac{\int_0^x t^{\alpha-1}f(t)dt}{\int_0^x t^{\alpha-2}f(t)dt}, \text{ for } \alpha > 2 \qquad (13)$$

according as $f''(t) \gtrless 0$, _for all_ $t \in I$. _Equality is attained if and only if_ $f(t) = \lambda t + \mu$, λ,μ constant.

Proof. This is based on the identity (4) with m = 2 and n = 1 in which case $x^\alpha f(x) \gtrless x^{\alpha+1} \Delta \int_0^x [\alpha(\alpha-1)t^{\alpha-2}f(t)/x^\alpha]dt$, according as $f''(t) \gtrless 0$, for all $t \in I$. The inequality (13) is simply a rearrangement of the required result (13). Clearly equality occurs when f(t) is linear in t.

Corollary 10. _Let f be a non-negative function of t with a continuous second derivative defined on I, and let f(t) have lower and upper bounds m and M respectively so that_

$$0 < m \leqslant f(t) \leqslant M, \quad \text{for all } t \in I.$$

Then

$$\frac{x(\alpha-1)}{(\alpha+1)}\left[1 + \frac{m}{\alpha M}\right] \leqslant \frac{\int_0^x t^{\alpha-1}f(t)dt}{\int_0^x t^{\alpha-2}f(t)dt}, \ \alpha > 2 \qquad (14)$$

provided that $f''(t) \leqslant 0$, for all $t \in I$, and

$$\frac{x(\alpha-1)}{(\alpha+1)}\left[1+\frac{M}{\alpha m}\right] \geqslant \frac{\int_0^x t^{\alpha-1}f(t)dt}{\int_0^x t^{\alpha-2}f(t)dt}, \quad \alpha > 1 \tag{15}$$

provided that $f''(t) \geqslant 0$, for all $t \in I$.

THEOREM 11. *Let f be a non-negative function of t with a continuous second derivative defined on I. Then*

$$\frac{\int_0^x t^{\alpha-1}f(t)dt}{\int_0^x t^{\alpha-2}f(t)dt} \gtrless \frac{\alpha x}{(\alpha+1)}\left[1-\frac{f(0)x^{\alpha-1}}{\alpha^2(\alpha-1)\int_0^x t^{\alpha-2}f(t)dt}\right], \text{ for } \alpha > 1 \tag{16}$$

according as $f''(t) \gtrless 0$, for all $t \in I$. Equality is attained if and only if $f(t) = \lambda t + \mu$, λ, μ constant.

Proof. Let $m=1$, $n=1$ and replace α by $(\alpha-1)$ in (4) so that

$$\int_0^x dx_1 \int_0^{x_1} t^{\alpha-1}f'(t)dt = \alpha\int_0^x t^{\alpha-1}f(t)dt - x(\alpha-1)\int_0^x t^{\alpha-2}f(t)dt; \tag{17}$$

now, let $m=0$, $n=1$ and replace α by $(\alpha-2)$ in (4) so that

$$\int_0^x dx_1 \int_0^{x_1} t^{\alpha-2}f(t)dt = -\int_0^x t^{\alpha-1}f(t)dt + x\int_0^x t^{\alpha-2}f(t)dt, \tag{18}$$

then, on making use of the identity

$$-f(0) + \int_0^t sf''(s)\,ds = t\,f'(t) - f(t)$$

and subtracting (17) from (18), it appears that

$$\frac{f(0)x^{\alpha}}{(\alpha-1)\alpha} + (\alpha+1)\int_t^x t^{\alpha-1}f(t)dt \gtrless \alpha x\int_0^x t^{\alpha-2}f(t)dt$$

according as $f''(t) \gtrless 0$, for all $t \in I$. The result follows.

Corollary 12. *Let f be a non-negative function of t with a continuous second derivative defined on I, and let $f(t)$ have lower and upper bounds m and M respectively so that*

$$0 < m \leqslant f(t) \leqslant M, \text{ for all } t \in I.$$

Then

$$\frac{f(0)}{mx\alpha(\alpha-1)} + \frac{\int_0^x t^{\alpha-1}f(t)dt}{\int_0^x t^{\alpha-2}f(t)dt} \geqslant \frac{\alpha x}{\alpha+1}, \quad \alpha > 1 \tag{19}$$

provided that $f''(t) \geqslant 0$, for all $t \in I$, and

$$\frac{f(0)}{Mx\alpha(\alpha-1)} + \frac{\int_0^x t^{\alpha-1}f(t)dt}{\int_0^x t^{\alpha-2}f(t)dt} \leqslant \frac{\alpha x}{\alpha+1}, \quad \alpha > 1 \tag{20}$$

provided that $f''(t) \leqslant 0$, *for all* $t \in I$.

Corollary 13. *Let* f *be a non-negative function of* t *with a continuous second derivative defined on* I. *Then*

$$x^{\alpha}[\alpha f(x) + f(0)] \gtrless \alpha(\alpha+1) \int_0^x t^{\alpha-1}f(t)dt, \quad \alpha > 0 \tag{21}$$

according as $f''(t) \gtrless 0$, *for all* $t \in I$. *Equality is attained if and only if* $f(t) = \lambda t + \mu$, λ,μ constant.

This is the extension of the inequality given in Theorem 6 to functions which are concave (or convex) on I.

Proof. This can be deduced by eliminating $\int_0^x t^{\alpha-2}f(t)dt$ from (13) and (16).

A simple converse of the Cauchy-Schwarz-Buniakowski inequality can now be deduced from Theorems 9 and 11.

THEOREM 14. *Let* f *be a non-negative convex function of* t *with a continuous second derivative defined on* I. *Then*

$$\alpha(\alpha+3) \int_0^x t^{\alpha-1}f(t)dt \int_0^x t^{\alpha+1}f(t)dt \leqslant (\alpha+2)^2 \left[\int_0^x t^{\alpha}f(t)dt\right]^2, \ \alpha > 0,$$

with equality if and only if $f(t) = 0$, *for all* $t \in I$.

A better and stronger result than the above can be obtained by appealing to the identity in (4) with m=2, n=2 and by making use of the properties of the persymmetric determinant referred to in Theorem 7.

THEOREM 15. *Let* f *be a non-negative convex function of* t *with a continuous second derivative defined on* I. *Then*

$$\alpha(\alpha+3)\int_0^x t^{\alpha-1}f(t)dt \int_0^x t^{\alpha+1}f(t)dt \leqslant (\alpha+1)(\alpha+2)\left[\int_0^x t^{\alpha}f(t)dt\right]^2, \tag{22}$$

$\alpha > 0$, *with equality when* $f(t) = (x-t)$ *or* $f(t) = 0$ *in* I.

THEOREM 16. *Let* f *be a non-negative function of* t *with a continuous second derivative defined on* I. *Then*

$$\frac{nx^{\alpha+n}f(x)}{(\alpha+n)(\alpha+n+1)} + \frac{\alpha(\alpha+1)x^n}{(\alpha+n)(\alpha+n+1)} \int_0^x t^{\alpha-1}f(t)dt \gtrless \int_0^x t^{\alpha+n-1}f(t)dt$$

$$\gtrless \frac{(\alpha+1)x^n}{(\alpha+n+1)} \int_0^x t^{\alpha-1}f(t)dt - \frac{nx^{\alpha+n}f(0)}{\alpha(\alpha+n)(\alpha+n+1)}, \tag{24}$$

$\alpha > 1$, $n=0,1,2,\ldots$, *according as* $f''(t) \gtrless 0$, *for all* $t \in I$. *Equality is attained if and only if* $f(t) = \lambda t + \mu$, λ,μ constant.

This is a strengthened version of a two-sided inequality derived by Sendov and Skordev [7] who limited their discussion to the case where $f''(t) \leq 0$ with $\alpha = 1$.

Proof. The first inequality is based on the identity (4), with m=2, and n=1. On the other hand, the right hand inequality follows directly from an application of Theorem 11 from which it can be deduced that

$$(\alpha+1)x^{1-\alpha} \int_0^x t^{\alpha-1}f(t)dt - xf(0)/\alpha$$

is monotonic in α depending on the sign of $f''(t)$, for all $t \in I$.

5. SOME FURTHER RESULTS

All of the inequalities found so far depend on the identity (4) where it is assumed that $n+1 \geq m \geq 0$. In case $n+1 \leq m$ a similar identity exist which takes the form:

THEOREM 17. *Let* f *be a function with a continuous* m-*th derivative defined on* I, *let* n,m *be positive integers satisfying* $n+1 \leq m$ *and let* α *be a real constant with* $\alpha > m$. *Then*

$$\int_0^x dx_1 \int_0^{x_1} dx_2 \ldots \int_0^{x_n} t^\alpha f^{(m)}(t)dt = \sum_{\gamma=0}^{m-n-1} (-)^\gamma x^{\alpha-\gamma} f^{(m-n-\gamma-1)}(x) \cdot \frac{\Delta^n}{n!}(\alpha)_{n+\gamma} + (-)^{m+n} x^{\alpha+n} \frac{\Delta^n}{n!} \int_0^x D^m[t^\alpha/x^\alpha]f(t)dt \quad .(25)$$

It should be noted that if $\sum_{\gamma=0}^{m-n-1}$ is interpreted as zero when $n+1 < m$ then (25) is equivalent to (4).

Proof. This follows in a straight-forward manner by induction on m. The starting value for m is n+1 in which case (25) reduces to (4).

It is now possible to find extensions of some of the earlier mentioned inequalities although now best possible constants are not obtained.

THEOREM 18. _Let_ f _be an m-times completely monotone function with a continuous m-th derivative defined on_ I _and let_ α _be a real constant with_ $\alpha > m$. _Then_

$$\Delta^n \int_0^x (\alpha)_m [t^{\alpha-m}/x^{\alpha}] f(t)\, dt \geqslant 0, \text{ for } n=0,2,4,\ldots,m-1 \text{ and } m=1,3,5\ldots, \tag{26}$$

$$(\alpha-m)(\alpha+1)\int_0^x t^{\alpha-m-1} f(t)dt \int_0^x t^{\alpha-m+1} f(t)dt \leqslant 4\alpha(\alpha-m+1)[\int_0^x t^{\alpha-m} f(t)dt]^2. \tag{27}$$

Once again the latter result is a converse of the Cauchy-Schwarz-Buniakowski inequality like that referred to previously. If n is set equal to 2 in (26) a further and rather obvious generalisation of the Sendov and Skordev type of inequality can be found.

Other interesting results can be obtained by setting $p(t) = e^{-\beta t}$ in equation (5). Thus it appears that

$$\frac{\alpha x}{\alpha+1+\beta x} \lesseqgtr \frac{\int_0^x t^{\alpha} e^{-\beta t} f(t)dt}{\int_0^x t^{\alpha-1} e^{-\beta t} f(t)dt},$$

provided that $\alpha > 1$, $\beta > 0$ and according as $f'(t) \lesseqgtr 0$, for all $t \in I$. In addition it can be shown that

$$\alpha\beta x \int_0^x t^{\alpha-1} e^{-\beta t} f(t)dt \int_0^x t^{\alpha+1} e^{-\beta t} f(t)dt \leqslant (\alpha+1+\beta x)^2 [\int_0^x t^{\alpha} e^{-\beta t} f(t)dt]^2 \tag{29}$$

provided that $\alpha > 1$, $\beta > 0$ and $f'(t) \leqslant 0$, for all $t \in I$. The case where $f'(t) \geqslant 0$, for all $t \in I$ is of no interest here for it leads to a weaker result than the inverse Cauchy-Schwarz-Buniakowski inequality given in Mitrinović [4; p. 43].

REFERENCES

1. A.C.Aitken, Determinants and Matrices. Oliver and Boyd, Edinburgh and London, 1959.

2. R. Bellman, Converses of Schwarz's Inequality. Duke Math. J. 23 (1956), 429-434.

3. C.J. Eliezer and D.E. Daykin, Generalizations and Applications of Cauchy-Schwarz Inequalities. Quart.J.Math. Oxford (2) 18 (1967), 357-360.

4. D. S. Mitrinović, Analytic Inequalities. Springer-Verlag, Berlin, Heidelberg and New York, 1970.

5. K.B. Oldham and J. Spanier, The Fractional Calculus. Acad, Press, New York and London 1974.

6. D.K. Ross, Iequalities for Special Functions. SIAM Rev. 15 (1973), 665-670.

7. B. Sendov and D. Skordev; see Elementary Inequalities, by D.S. Mitrinović. P. Noordhoff Ltd. Groningen, The Netherlands, 1964.

8. T.W. Ting, Upper and Lower Bounds of the Radii of Gyration of Convex Bodies. Trans. Amer. Math. Soc. 128 (1967), 336-357.

9. Chung-Lie Wang, On Developments of Inverses of the Cauchy and Hölder Inequalities. SIAM Rev. 21 (1979), 550-557.

Dieter K. Ross, Applied Mathematics Department, La Trobe University, Victoria, 3083, Australia.

International Series of
Numerical Mathematics, Vol. 71

ON THE NONEXTENDABILITY OF KOLMOGOROFF'S REARRANGEMENT THEOREM

Brent Smith and Frederic Wong

Abstract. Let $\hat{f} = (\hat{f}(n))_{n=0}^{\infty} \in \ell^2(\mathbb{N})$ satisfy $\hat{f}(n) = 0$ unless n is a power of 2. Kolmogoroff's rearrangement theorem says that if $f = \sum_{n=0} \hat{f}(n)e^{in\theta}$ then $\sum \hat{f}(n_k)e^{in_k\theta} \to f(\theta)$ a.e. for every rearrangement n_k. We show there is no generalization of this theorem applicable to all of $\ell^2(\mathbb{N})$.

Let a_n be disjoint finite subsets of $\mathbb{N}$ with $a_1 \cup a_2 \cup \ldots = \mathbb{N}$. Let $A_n(\theta) = \sum_{m \in a_n} e^{im\theta}$.

THEOREM. There exist a rearrangement n_k of $\mathbb{N}$, and an $\hat{f} \in \ell^2(\mathbb{N})$ such that $\sum_{k=1}^{N} A_{n_k} * f$ diverges a.e. as $N \to \infty$, (as usual $g * f(t) = \frac{1}{2\pi}\int g(t-u)f(u)du$.)

We remark that for the special case $a_n = \{n\}$ this is a theorem of Z. Zahorski [3]. See also [2].

To prove the theorem we first work a special case. Let $b_n = \{2^n+1,\ldots,2^{n+1}\}$, $B_n(\theta) = \sum_{m \in b_n} e^{im\theta}$

LEMMA. There exist a rearrangement n_k of $\mathbb{N}$, and an $\hat{f} \in \ell^2(\mathbb{N})$ such that $\sum_{k=1}^{N} B_{n_k} * f$ diverges a.e. as $N \to \infty$.

The lemma (p. 166, [4]) is equivalent to being able to find $F_1 \supset F_2 \supset \ldots \supset F_N$ subsets of $[0,2\pi]$ and a rearrangement n_k of $\{1,2,\ldots,N\}$ such that

$$(1) \qquad \frac{1}{2\pi}\int_0^{2\pi} (|B_{n_1} * \chi F_1|^2 + |B_{n_2} * \chi F_2|^2 + \ldots + |B_{n_N} * \chi F_N|^2)dx$$

is arbitrarily large. (1) is, of course, the same as

(2) $\frac{1}{4\pi}\int_0^{2\pi}|(D_{2^{n_1+1}} - D_{2^{n_1}}) * \chi_{F_1}|^2 + \cdots + |(D_{2^{n_N+1}} - D_{2^{n_N}}) * \chi_{F_N}|^2 dx$

where $D_k = D_k(x) = \sum_{n=-k}^{k} e^{inx}$ is the standard Dirichlet kernel.

Instead of making (1) large, we will make a martingale version of (1) large. To go from the martingale version back to the Fourier version is only tedium.

Uniformly partition [0,1] into 2^k, $0 \leq k < \infty$, intervals $I_{k1}, \ldots, I_{k2^k}$, then the standard dyadic grid on [0,1] is $\{I_{k1}, \ldots, I_{k2^k} | 0 \leq k < \infty\}$. Put the standard dyadic grid on [0,1] and name its elements I. Use $\tilde{I}$ to denote the next larger member of the grid above I. Let $E_I(F) = \frac{|F \cap I|}{|I|}$ for $F \subset [0,1]$.

PROPOSITION. <u>For any</u> N, <u>there exist</u> $F_1 \supset F_2 \supset \cdots \supset F_N$ <u>subsets of</u> [0,1] <u>and a rearrangement</u> n_k <u>of</u> $\{1, 2, \ldots, N\}$ <u>such that</u>

(3) $$\int_0^1 \left\{ \sum_{|I|=2^{-n_1}} (E_I - E_{\tilde{I}})(F_1)\chi_I(x) \right\}^2 + \cdots + \left\{ \sum_{|I|=2^{-n_N}} (E_I - E_{\tilde{I}})(F_N)\chi_I(x) \right\}^2 dx$$
$$\geq \text{Const} \log N \text{ for some constant}$$

<u>Proof.</u> The F_k's are defined by stages. Basically there will be a stage s F over and under each stage (s-1) F, and the 2-norm^2 contribution of the over and under stage s F's = the contribution of the stage (s-1) F.

We use n(k), F(k) for n_k, F_k. Let $L(\ell) = \{I : |I| = 2^{-\ell}\}$. We name the elements of $L(\ell)$ as we go from left to right $I_\ell(1), \ldots, I_\ell(2^\ell)$.

Stage 0 consists only of $F(N/2) = I_1(1)$. So $n(N/2) = 1$.

$|\sum_{I \in L(1)} (E_I - E_{\tilde{I}})(F(N/2)) \cdot \chi_I| = \frac{1}{2}$. The 2-norm^2 contribution of stage 0 is $\frac{1}{4}$.

Stage 1 consists of an over F(N/4) and an under F(3N/4) for the stage 0 F(N/2). n(N/4) = 2, n(3N/4) = 3. $F(N/4) = I_2(1) \cup I_2(2) \cup I_2(3)$. $F(3N/4) = I_3(1) \cup I_3(3)$. Now

$$\left| \sum_{I \in L(2)} (E_I - E_{\tilde I})(F(N/4))\, \chi_I \right| = \tfrac12 \text{ on } I_2(3) \cup I_2(4).$$

and

$$\left| \sum_{I \in L(3)} (E_I - E_{\tilde I})(F(3N/4)) \cdot \chi_I \right| = \tfrac12 \text{ on } I_3(1) \cup \ldots \cup I_3(4).$$

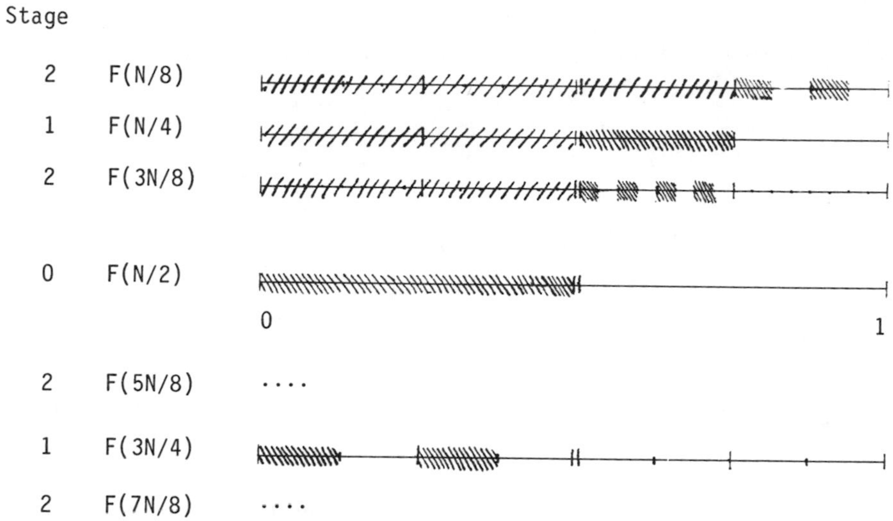

Diagram. Where F in L_k makes its contribution to $\sum_{I \in L_k} (E_I - E_{\tilde I})(F)\, \chi_I$ is indicated by \\\\ .

Hence, the total contribution to 2-norm^2 from F(N/4) is $\frac12 \cdot \frac14$ and the total from F(3N/4) is $\frac12 \cdot \frac14$. And so stage 1 delivers the same $\frac14$ contribution as stage 0. See diagram.

Now we define the stage M F's, which are

$F(\frac{N}{2^{M+1}})$, $F(\frac{3N}{2^{M+1}})$, ..., $F\left(\frac{(2^{M+1}-1)N}{2^{M+1}}\right)$. Put F(N) = [0,1], F(0) = $\emptyset$.

Let $n(\frac{N}{2^{M+1}}) = 2^M$, ..., $n\left(\frac{(2^{M+1}-1)N}{2^{M+1}}\right) = 2^{M+1}-1$. And so we are using $L(2^M), \ldots, L(2^{M+1}-1)$. Let

$$F\left(\frac{2k+1}{2^{M+1}} \cdot N\right) = \begin{cases} F\left(\frac{2k+2}{2^{M+1}} \cdot N\right) \cup \\ \text{the odd indexed members of } L(2^M+k) \\ \text{which are } \subset \overline{C\left(F\left(\frac{2k+2}{2^{M+1}} \cdot N\right)\right)} \cap \\ F\left(\frac{2k}{2^{M+1}} \cdot N\right). \end{cases}$$

The 2-norm^2 contribution from each of

$$\sum_{I \in L(2^M+k)} (E_I - E_I)\left(F\left(\frac{2k+1}{2^{M+1}} \cdot N\right)\right) \cdot \chi_I \text{ is}$$

$$\tfrac{1}{4}\left|C\left(F\left(\frac{2k+2}{2^{M+1}} \cdot N\right)\right) \cap F\left(\frac{2k}{2^M} \cdot N\right)\right| = \tfrac{1}{4} \cdot \frac{1}{2^M}.$$

And so the total contribution from level M is again ¼.

Theorem Proof. In order to make (1) arbitrarily large we require only an infinite subset of $\{b_n\}$. Now we must make (1) large with A_n's in the place of the B_n's. However this follows if we use an infinite subset $\{b_{n_j}\}$ that respects $\{a_n\}$; this means: $a_n \cap b_{n_j} \neq \emptyset$, $a_n \cap b_{n_\ell} \neq \emptyset$ implies $j=\ell$. More explicitly, let S_1 = subset of $\{a_n\}$ that covers $b_1 = b_{n_1}$. Choose b_{n_2} so $b_{n_2} \cap S_1 = \emptyset$. Let S_2 = subset of $\{a_n\}$ that covers b_{n_2}. Choose b_{n_3} so $b_{n_3} \cap (S_1 \cup S_2) = \emptyset$. Let S_3 = subset of $\{a_n\}$ that covers b_{n_3}, and so on. We use the infinite subset $\{b_{n_j}\}$ of $\{b_n\}$ to define the F's for the partial sum operators formed using the blocks $\{S_j\}$ from $\{a_n\}$.

Our puspose in this paper is to indicate that the following theorem [1] is in some sense an optimal rearrangement result.

$\{n_k\}$ is a dyadic rearrangement of $\mathbb{N}$ if it is generated as follows. We are allowed to change the order of the elements of set A with respect to those of set B if under binary notation A has the form

$A = \{b_N \dots b_3\, b_2\, b_1\, 0\, c_1\, c_2 \dots c_M$: $b_N, \dots b_2, b_1$ are fixed; $c_1, \dots, c_M$ are free$\}$ and B has the form $B = \{b_N \dots b_1\, 1\, c_1 c_2 \dots c_M$: $b_N, \dots, b_1$ are the same as in A and $c_1, \dots, c_M$ are free$\}$.

THEOREM. Let $\hat{f} \in \ell^2(\mathbb{N})$ and let n_k be a dyadic rearrangement of $\mathbb{N}$.

Then

$$\sum_{k=1}^{N} \hat{f}(n_k) e^{in_k\theta} \to f(\theta) \text{ a.e. as } N \to \infty .$$

REFERENCES

1. B. Smith and F. Wong, Fourier pointwise convergence under dyadic rearrangements. Preprint.

2. P. L. Ul'yanov, Divergent fourier series, Russian Mathematical Surveys, vol. 16, no. 3, (May-June 1961) 1-35.

3. Z. Zahorski, Une serie de fourier permutée d'une fonction de classe L^2 divergente presque partout, C.R. Acad. Sci., Paris 251(1960) 501-503.

4. A. Zygmund, Trigonometric Series, II, Cambridge University Press (1959).

5. Y. Katznelson, An Introduction to Harmonic Analysis, John Wiley, 1968.

Brent Smith, Department of Mathematics, California Institute of Technology, Pasadena, CA 91125, U.S.A.

Frederic Wong, Department of Mathematics, California Institute of Technology, Pasadena, CA 91125, U.S.A.

International Series of
Numerical Mathematics, Vol. 71

APPROXIMATION OF FUNCTIONS BY WHITTAKER'S CARDINAL SERIES

R.L. Stens

Abstract. According to the classical cardinal series theorem any integrable, entire function of exponential type $\leq \sigma$ can be represented by its cardinal series with nodes $k\pi/\sigma$. It is shown that for continuous functions having an absolute integrable Fourier transform or satisfying certain smoothness conditions this representation holds in the limit for $\sigma \to \infty$. Similarly, the derivatives of a function as well as its Hilbert transform can be approximated by the derivatives and the Hilbert transform of the cardinal series, respectively. Estimates for the approximation error are given; these are shown to be best possible.

1. INTRODUCTION

Let B^p_σ, $1 \leq p < \infty$, be the set of all functions $f \in L^p(\mathbb{R})$ having an extension into the whole complex plane as an entire function of exponential type $\leq \sigma$ (cf. [1, Sec. 70]). The Whittaker cardinal series theorem then states that every $f \in B^p_\sigma$ has the representation

$$(1.1) \quad f(t) = (S_\sigma f)(t) := \sum_{k=-\infty}^{\infty} f\left(\frac{k\pi}{\sigma}\right) \operatorname{sinc}(\sigma t - k\pi) \qquad (t \in \mathbb{R}),$$

where $\operatorname{sinc}(t) := (\sin t)/t$ for $t \neq 0$ and $\operatorname{sinc}(0) := 1$. Moreover, the derivatives $f^{(r)}(t)$, $(r \in \mathbb{N} := \{1,2,3,\ldots\})$ can be obtained by differentiating the series in (1.1) termwise, i.e.,

$$(1.2) \quad f^{(r)}(t) = (S^{(r)}_\sigma f)(t) := \sum_{k=-\infty}^{\infty} f\left(\frac{k\pi}{\sigma}\right) \sigma^r \operatorname{sinc}^{(r)}(\sigma t - k\pi) \ (t \in \mathbb{R}).$$

Replacing the sinc-function by its Hilbert transform sinc~,

given by

$$\operatorname{sinc}^{\sim}(u) = \frac{1-\cos u}{u} \equiv \frac{\sin^2 u/2}{u/2},$$

one has the following representations for the Hilbert transform $f^{\sim}$ of f and its derivatives ($r \in \mathbb{N}_0 := \{0,1,2,\dots\}$)

$$(1.3)\quad (f^{\sim})^{(r)}(t) = (\tilde{S}_\sigma^{(r)} f)(t) := \sum_{k=-\infty}^{\infty} f\left(\frac{k\pi}{\sigma}\right) \sigma^r \operatorname{sinc}^{\sim (r)}(\sigma t - k\pi) \quad (t \in \mathbb{R}).$$

If $f \in L^p(\mathbb{R})$ does not have an extension as an entire function of exponential type, then (1.1)-(1.3) may hold at least in the limit for $\sigma \to \infty$, i.e.,

$$(1.4)\qquad f^{(r)}(t) = \lim_{\sigma\to\infty} (S_\sigma^{(r)} f)(t),$$

$$(1.5)\qquad (f^{\sim})^{(r)}(t) = \lim_{\sigma\to\infty} (\tilde{S}_\sigma^{(r)} f)(t).$$

These representations hold e.g. if $f \in L^2(\mathbb{R}) \cap C(\mathbb{R})$ and if the r th absolute moment of the Fourier transform is finite (cf. e.g. [2;3;4;7;8;15;19]).

The aim of this paper is to give several sufficient conditions upon f for (1.4),(1.5) to hold, as well as to estimate the errors $\| f^{(r)} - S_\sigma^{(r)} f \|$ and $\| (f^{\sim})^{(r)} - \tilde{S}_\sigma^{(r)} f \|$. Furthermore, it will be shown that in many cases the error bounds are best possible.

2. NOTATIONS AND AUXILIARY RESULTS

Let $L^p(\mathbb{R})$, $1 \le p < \infty$, be the space of complex-valued, measurable functions f defined on the real line $\mathbb{R}$ for which the norm

$$\| f \|_p := \left\{ \frac{1}{\sqrt{2\pi}} \int_{-\infty}^{\infty} |f(u)|^p \, du \right\}^{1/p}$$

is finite, and let $C(\mathbb{R})$ denote the space of all bounded, uniformly continuous functions on $\mathbb{R}$ with supremum norm $\|\cdot\|$, whereas $C^{(r)}(\mathbb{R})$ is the subset of all $f \in C(\mathbb{R})$ having an r th derivative $f^{(r)}$ belonging to $C(\mathbb{R})$.

As a measure of smoothness of the functions in question we use the modulus of continuity with respect to the second order difference, namely

$$\omega_2(\delta;f) := \sup\{\| f(\cdot+h)+f(\cdot-h)-2f(\cdot)\| : |h|\leq\delta\} \qquad (\delta>0),$$

and we introduce the Lipschitz class of order $\alpha>0$ by

$$\mathrm{Lip}_2(\alpha;C(\mathbb{R})) := \{f\in C(\mathbb{R}) : \omega_2(\delta;f) = \mathcal{O}(\delta^\alpha),\ \delta\to 0+\}.$$

The classes $\mathrm{Lip}_2(\alpha;L^p(\mathbb{R}))$ are defined with obvious modifications. For $1<\alpha\leq 2$ one has $f\in\mathrm{Lip}_2(\alpha;C(\mathbb{R}))$ if and only if $f'\in\mathrm{Lip}_2(\alpha-1;C(\mathbb{R}))$, and for $\alpha>2$ the class $\mathrm{Lip}_2(\alpha;C(\mathbb{R}))$ consists of constant functions only. The same applies to the classes $\mathrm{Lip}_2(\alpha;L^1(\mathbb{R}))$ with the exception that $\mathrm{Lip}_2(\alpha;L^1(\mathbb{R}))$ only contains the zero-function if $\alpha>2$. So it is convenient to restrict the matter to $0<\alpha\leq 1$. Let us finally mention that for $0<\alpha<1$ the Lipschitz classes can equivalently be defined via a modulus of continuity based upon the first order difference. For these facts see e.g. [13].

The Fourier transform $f^\wedge$ of $f\in L^1(\mathbb{R})$ is given by

$$f^\wedge(v) := \frac{1}{\sqrt{2\pi}}\int_{-\infty}^{\infty} f(u)e^{-ivu}\,du \qquad (v\in\mathbb{R}),$$

and the same symbol will be used for the Fourier-Plancherel transform of $f\in L^2(\mathbb{R})$, defined by

$$\lim_{R\to\infty}\left\| f^\wedge(v) - \frac{1}{\sqrt{2\pi}}\int_{-R}^{R} f(u)e^{-ivu}\,du\right\|_2 = 0.$$

We will need a generalization of [15, La. 4.1]; the proof follows along the same lines.

LEMMA 1. a) _For each_ $r\in\mathbb{N}_0$ _there exists a constant_ $c>0$[1])

1) c always denotes a positive constant which may be different at each occurrence.

such that for all $y \in \mathbb{R}$, $N \in \mathbb{N}$ there holds

$$\sum_{k=-N}^{N} |\operatorname{sinc}^{(r)}(y-k\pi)| \leq c(1+\log N),$$

$$\sum_{k=-N}^{N} |\operatorname{sinc}^{\sim(r)}(y-k\pi)| \leq c(1+\log N).$$

b) Let $f \in C(\mathbb{R})$ satisfy

$$(2.1) \qquad |f(t)| \leq c/(1+|c|^{\gamma}) \qquad (t \in \mathbb{R})$$

for some $0<\gamma\leq 1$, and let $r \in \mathbb{N}_0$. Then there exists $c>0$ such that for each $y \in \mathbb{R}$, $N \in \mathbb{N}$ and $\sigma>0$

$$\sum_{|k|>N} |f(\tfrac{k\pi}{\sigma}) \operatorname{sinc}^{(r)}(y-k\pi)| \leq c\,\sigma^{\gamma} N^{-\gamma/2},$$

$$\sum_{|k|>N} |f(\tfrac{k\pi}{\sigma}) \operatorname{sinc}^{\sim(r)}(y-k\pi)| \leq c\,\sigma^{\gamma} N^{-\gamma/2}.$$

Let us now consider the de la Vallée Poussin means defined for $f \in C(\mathbb{R})$ by

$$(J_\sigma f)(t) := \frac{\sigma/2}{\sqrt{2\pi}} \int_{-\infty}^{\infty} f(t-u)\theta(\sigma u/2)du \qquad (\sigma>0),$$

$$\theta(u) := \frac{3}{\sqrt{2\pi}} \operatorname{sinc}(u/2)\operatorname{sinc}(3u/2).$$

LEMMA 2. a) For $f \in L^p(\mathbb{R})$, $1 \leq p < \infty$, one has $J_\sigma f \in B_\sigma^p$ for all $\sigma>0$.

b) If $f \in C(\mathbb{R})$ satisfies (2.1) for some $0<\gamma\leq 1$, then

$$|(J_\sigma f(t)| \leq c/(1+|t|^{\gamma}) \qquad (t \in \mathbb{R}).$$

c) If $f \in C^{(r)}(\mathbb{R})$ for some $r \in \mathbb{N}_0$, then $(J_\sigma f)^{(r)} = J_\sigma f^{(r)}$ and

$$\|J_\sigma f - f\| \leq c\,\sigma^{-r}\,\omega_2(\sigma^{-1};f^{(r)}) \qquad (\sigma>0).$$

For the proof see [1, Sec. 90; 15].

3. ERROR ESTIMATES FOR THE APPROXIMATION OF A FUNCTION AND ITS DERIVATIVES BY CARDINAL SERIES

For a measurable function f defined on $\mathbb{R}$ the r th ($r \in \mathbb{N}_0$) absolute truncated moment is definded by

$$m_\sigma(f;r) := \frac{1}{\sqrt{2\pi}} \int_{|u|>\sigma} |u|^r |f(u)|\, du.$$

Instead of $m_0(f;r)$ we simply write $m(f;r)$.

THEOREM 1. *If* $f \in L^2(\mathbb{R}) \cap C(\mathbb{R})$ *is such that* $m(f^\wedge;r) < \infty$ *for some* $r \in \mathbb{N}_0$, *then* $f \in C^{(r)}(\mathbb{R})$ *and*

$$\|S_\sigma^{(r)} f - f^{(r)}\| \leqslant m_\sigma(f^\wedge;r) + \sigma^r\, m_\sigma(f^\wedge;0); \tag{3.1}$$

in particular there holds (1.4) *uniformly for* $t \in \mathbb{R}$.

The proof follows from [19, Thm. 1] with $h(v) = e^{ivt}$ and $x = 0$. Since $\sigma^r m_\sigma(f^\wedge;0) \leqslant m_\sigma(f^\wedge;r)$, the right hand side of (3.1) is bounded by $2m_\sigma(f^\wedge;r)$; this is the bound obtained in [19] (cf. also [2;3;7;8]).

This result gives an estimate for $\|S_\sigma^{(r)} f - f^{(r)}\|$ in terms of the Fourier transform $f^\wedge$. If one is interested in estimates involving only properties of the function f itself, then one may restrict the matter to the class $L^1(\mathbb{R}) \cap C(\mathbb{R})$ and make use of the fact that smoothness of the function f gives information on the behaviour of $f^\wedge(v)$ for $v \to \pm\infty$. More precisely, if $f \in C^{(r)}(\mathbb{R})$ has an r th derivative $f^{(r)} \in \mathrm{Lip}_2(\alpha;L^1(\mathbb{R}))$, then $|f^\wedge(v)| = O(|v|^{-r-\alpha})$, $v \to \pm\infty$. (This follows for $0 < \alpha < 1$ from [6, (5.1.3), Prop. 5.1.14]. For $\alpha \geqslant 1$ one has to replace the first order modulus of continuity in [6, (5.1.3)] by the second order one; the proof of this generalization in obvious.) Now, the behaviour of $f^\wedge$ at $\pm\infty$ can be used to estimate $m_\sigma(f^\wedge;0)$ and $m_\sigma(f^\wedge;r)$ for $\sigma \to \infty$. This leads to (cf. [7;8;19])

COROLLARY 1. *If* $f \in L^1(\mathbb{R}) \cap C^{(r+s-1)}(\mathbb{R})$ *is such that* $f^{(r+s)} \in \mathrm{Lip}_2(\alpha; L^1(\mathbb{R}))$ *for some* $r \in \mathbb{N}_0$, $s \in \mathbb{N}$ *and* $0 < \alpha \leq 1$, *then*

$$(3.2) \qquad \| S_\sigma^{(r)} f - f^{(r)} \| = O(\sigma^{-s-\alpha+1}) \qquad (\sigma \to \infty).$$

The disadvantage of Cor. 1 is that f must have a derivative of order at least r+1, i.e. $s = 1$, in order to obtain meaningful estimates. In particular, if $r = 0$, then f must have a first order derivative belonging to some Lipschitz class; otherwise the right hand side of (3.2) tends to infinity. This can be avoided by the following entirely different approach.

THEOREM 2. *Let the function* f *belong to* $C^{(r+s)}(\mathbb{R})$, $r, s \in \mathbb{N}_0$, *and satisfy the growth condition* (2.1) *for some* $\gamma > 0$. *Then*

$$(3.3) \; \| S_\sigma^{(r)} f - f^{(r)} \| \leq c\{(1 + \log\sigma)\sigma^{-s}\omega_2(\sigma^{-1}; f^{(r+s)}) + \sigma^{-s-1}\} \; (\sigma > 1).$$

Proof. Without loss of generality we may assume that $0<\gamma\leq 1$. For arbitrary $N \in \mathbb{N}$ one has in view of La. 2 and (1.2) that

$$\| S_\sigma^{(r)} f - f^{(r)} \| \leq \| S_\sigma^{(r)} f - S_\sigma^{(r)}(J_\sigma f) \| + \| J_\sigma f^{(r)} - f^{(r)} \|$$

$$\leq \sigma^r \, \| \sum_{k=-N}^{N} | f(\tfrac{k\pi}{\sigma}) - (J_\sigma f)(\tfrac{k\pi}{\sigma}) | \; | \mathrm{sinc}^{(r)}(\sigma \cdot - k\pi) | \, \|$$

$$+ \sigma^r \| \sum_{|k| > N} \{ | f(\tfrac{k\pi}{\sigma}) | + | (J_\sigma f)(\tfrac{k\pi}{\sigma}) | \} \; | \mathrm{sinc}^{(r)}(\sigma \cdot - k\pi) | \, \|$$

$$+ \| J_\sigma f^{(r)} - f^{(r)} \| := A_1 + A_2 + A_3,$$

say. Now one has by La. 2c) and La. 1a) that

$$A_1 \leq c \; \sigma^r \; \sigma^{-r-s} \; \omega_2(\sigma^{-1}; f^{(r+s)})(1 + \log N),$$

by La. 2b) and La. 1b) that $A_2 \leq c \; \sigma^r \; \sigma^\gamma \; N^{-\gamma/2}$, and by La. 2c) applied to $f^{(r)}$ that

$$A_3 \leq c\ \sigma^{-s}\ \omega_2(\sigma^{-1};f^{(r+s)}).$$

These inequalities together yield (3.3) by choosing $N \in \mathbb{N}$ such that $\sigma^{\beta} \leq N < \sigma^{\beta}+1$ with $\beta = 2(r+s+\gamma+1)/\gamma$.

Note that the assumptions imply that $m(f^{\wedge};r) < \infty$. A particular case of Thm.2 is contained in

COROLLARY 2. a) *Let* $f \in C^{(r)}(\mathbb{R})$, $r \in \mathbb{N}_0$, *be such that* (2.1) *holds for some* $\gamma > 0$. *If* $f^{(r)}$ *satisfies a Dini-Lipschitz condition, i.e.,*

$$\lim_{\delta\to 0+} \omega_2(\delta;f^{(r)}) \log 1/\delta = 0, \tag{3.4}$$

then (1.4) *holds uniformly for* $t \in \mathbb{R}$.
b) *If* $f \in C^{(r+s)}(\mathbb{R})$, $r,s \in \mathbb{N}_0$, *satisfies* (2.1) *for some* $\gamma > 0$, *and if* $f^{(r+s)} \in \mathrm{Lip}_2(\alpha;C(\mathbb{R}))$, $0 < \alpha \leq 1$, *then*

$$\| S_{\sigma}^{(r)}f - f^{(r)}\| = \mathcal{O}(\sigma^{-s-\alpha}\log\sigma) \qquad (\sigma \to \infty). \tag{3.5}$$

Condition (3.4) can equivalently be expressed by the modulus of continuity with respect to the first order difference. Contrary to Cor. 1, s may now be zero, i.e., f need have a derivative of order r only, and not of order r+1 as in Cor. 1. The particular case r = 0 was already proved in [15] and, under the stronger assumption of f having compact support, also in [5;18].

It should be mentioned that although the order in (3.5) is better than that in (3.2), it may still happen that for a particular function f Cor. 1 delivers a better estimate than Cor. 2 does. This depends on the fact that f may have a derivative of a certain order which belongs to $L^1(\mathbb{R})$ but not to $C(\mathbb{R})$ (for an example cf. Sec. 5).

There exist also error bounds for $\| S_{\sigma}^{(r)}f - f^{(r)}\|$ for functions f having an analytic extension into a strip of the complex plane which contains the real axis. In this respect see [9;17;19].

4. ERROR ESTIMATES FOR THE APPROXIMATION OF THE HILBERT TRANSFORM AND ITS DERIVATIVES

The Hilbert transform or conjugate function $f^\sim$ of $f \in L^p(\mathbb{R})$, $1 \leq p < \infty$, is defined by the Cauchy principal value

$$f^\sim(t) := \lim_{\delta\to 0+} \frac{1}{\pi} \int_{|u|>\delta} \frac{f(t-u)}{u}\, du \equiv \mathrm{PV}\, \frac{1}{\pi} \int_{-\infty}^{\infty} \frac{f(t-u)}{u}\, du$$

existing for almost all $t \in \mathbb{R}$. It defines a bounded linear operator from $L^p(\mathbb{R})$ into itself, provided $1 < p < \infty$. (cf. [6, Chap. 8]). We will need the following Privalov-type result.

LEMMA 3. *If* $f \in C^{(r)}(\mathbb{R})$, $r \in \mathbb{N}_0$, *satisfies* (2.1) *for some* $\gamma > 0$, *and if* $f^{(r)} \in L^p(\mathbb{R}) \cap \mathrm{Lip}_2(\alpha; C(\mathbb{R}))$ *for some* $1 \leq p < \infty$, $0 < \alpha \leq 1$, *then*

$$(f^\sim)^{(r)} \in \mathrm{Lip}_2(\alpha; C(\mathbb{R})). \tag{4.1}$$

Proof. For $r = 0$ and $0 < \alpha < 1$ this result is due to Titchmarsh [20, Sec. 5.15], and for $r = 0, \alpha = 1$ it was proved by Ogieveckiĭ and Boičun [14]. If $r \geq 1$, then the assumptions imply that $f, f^{(r)} \in L^q(\mathbb{R})$ for some $1 \leq q < \infty$ and that (cf. e.g. [10, La. 2.4]) $f^\sim \in C^{(r-1)}(\mathbb{R})$, $(f^\sim)^{(r)} \in L^q(\mathbb{R})$, as well as

$$(f^\sim)^{(r)}(t) = (f^{(r)})^\sim(t) \qquad \text{a.e.} \tag{4.2}$$

The general case now follows from the case $r = 0$.

As counterparts of Thm. 1 and Cor. 1 one has (cf. [8;19]):

THEOREM 3. *If* $f \in L^1(\mathbb{R}) \cap C(\mathbb{R})$ *is such that* $m(f^\wedge; r) < \infty$ *for some* $r \in \mathbb{N}_0$, *then* $f^\sim \in C^{(r)}(\mathbb{R})$ *and*

$$\| \tilde{S}_\sigma^{(r)} f - (f^\sim)^{(r)} \| \leq m_\sigma(f^\wedge; r) + \sigma^r\, m_\sigma(f^\wedge; 0); \tag{4.3}$$

in particular, there holds (1.5) *uniformly for* $t \in \mathbb{R}$.

COROLLARY 3. *If* $f \in L^1(\mathbb{R}) \cap C^{(r+s-1)}(\mathbb{R})$ *is such that* $f^{(r+s)} \in \mathrm{Lip}_2(\alpha; L^1(\mathbb{R}))$ *for some* $r \in \mathbb{N}_0$, $s \in \mathbb{N}$ *and* $0 < \alpha \leq 1$, *then*

$$(4.4) \qquad \|\tilde{S}_\sigma^{(r)} f - (f^\sim)^{(r)}\| = O(\sigma^{-s-\alpha+1}) \qquad (\sigma \to \infty).$$

There is also a counterpart of Thm. 2, but for simplicity we will state only a result analogous to Cor. 2b):

THEOREM 4. *If* $f \in C^{(r+s)}(\mathbb{R})$, $r,s \in \mathbb{N}_0$, *satisfies* (2.1) *and if* $f^{(r+s)} \in L^p(\mathbb{R}) \cap \mathrm{Lip}_2(\alpha; C(\mathbb{R})$ *for some* $1 \leq p < \infty$ and $0 < \alpha \leq 1$, *then* $f^\sim \in C^{(r)}(\mathbb{R})$ *and*

$$(4.5) \qquad \|\tilde{S}_\sigma^{(r)} f - (f^\sim)^{(r)}\| = O(\sigma^{-s-\alpha} \log \sigma) \qquad (\sigma \to \infty).$$

Proof. The fact that $f^\sim \in C^{(r)}(\mathbb{R})$ follows from La. 3. As to (4.5), one obtains from La. 2, (1.3) and the equality $((J_\sigma f)^\sim)^{(r)}(t) = (J_\sigma (f^\sim)^{(r)})(t)$, $t \in \mathbb{R}$, (cf. [6, p. 318]) that

$$\|\tilde{S}_\sigma^{(r)} f - (f^\sim)^{(r)}\| \leq \|\tilde{S}_\sigma^{(r)} f - \tilde{S}_\sigma^{(r)}(J_\sigma f)\| + \|J_\sigma (f^\sim)^{(r)} - (f^\sim)^{(r)}\|.$$

Now one can proceed as in the proof of Thm 2. to deduce

$$\|\tilde{S}_\sigma^{(r)} f - \tilde{S}_\sigma^{(r)}(J_\sigma f)\| \leq c\{(1 + \log \sigma)\sigma^{-s}\, \omega_2(\sigma^{-1}; f^{(r+s)}) + \sigma^{-s-1}\},$$

$$\|J_\sigma (f^\sim)^{(r)} - (f^\sim)^{(r)}\| \leq c\, \sigma^{-s}\, \omega_2(\sigma^{-1}; (f^\sim)^{(r+s)})$$

and so (4.5) follows with the aid of La. 3.

The case $r = 0$ can also be found in [15] where, however, it was proved under slightly stronger assumptions upon f and $f^\sim$. For analytic functions see [17;19].

5. INVERSE RESULTS

The aim of this section is to show that the error bounds for $\|S_\sigma f - f\|$ given in (3.1), (3.2) and (3.5) are best possible in the sense that in each case there exists a particular function f for which the right hand side of the corresponding inequality cannot be diminished.

THEOREM 5. a) For each $\sigma>0$ there exists a function $f_1 \in L^2(\mathbb{R}) \cap C(\mathbb{R})$ with $0 < m_\sigma(f_1;0) < \infty$ such that

$$\|S_\sigma f_1 - f_1\| = 2\, m_\sigma(f_1;0).$$

b) For each $s \in \mathbb{N}$ and $0 < \alpha \leq 1$ there exists a function $f_2 \in L^1(\mathbb{R}) \cap C^{(s-1)}(\mathbb{R})$ with $f_2^{(s)} \in \mathrm{Lip}_2(\alpha; L^1(\mathbb{R}))$ such that

$$\|S_\sigma f_2 - f_2\| \neq o(\sigma^{-s-\alpha+1}) \qquad (\sigma \to \infty).$$

c) For each $s \in \mathbb{N}_0$ and $0 < \alpha \leq 1$ there exists a function $f_3 \in C^{(s)}(\mathbb{R})$ with compact support and $f_3^{(s)} \in \mathrm{Lip}_2(\alpha; C(\mathbb{R}))$ such that

$$\|S_\sigma f_3 - f_3\| \neq o(\sigma^{-s-\alpha} \log \sigma) \qquad (\sigma \to \infty).$$

Proof. For part a) see [3]. Concerning b), there holds (cf. [2;9])

$$(S_\sigma f)(t) - f(t) = \frac{1}{\sqrt{2\pi}} \sum_{k=-\infty}^{\infty} (e^{-i2k\sigma t} - 1) \int_{(2k-1)\sigma}^{(2k+1)\sigma} f^\wedge(v) e^{ivt}\, dv,$$

and one obtains

$$\begin{aligned} \|S_\sigma f - f\| &\geq \left| (S_\sigma f)\left(\tfrac{\pi}{2\sigma}\right) - f\left(\tfrac{\pi}{2\sigma}\right)\right| \\ &= \frac{\sigma}{\sqrt{2\pi}} \left| \sum_{k=-\infty}^{\infty} ((-1)^k - 1) \int_{(2k-1)}^{(2k+1)} f^\wedge(v\sigma) e^{i\pi v/2}\, dv \right|. \end{aligned}$$

If the Fourier transform $f^\wedge$ is an even function and nonnegative, then it follows that

$$\|S_\sigma f - f\| \geq \frac{4\sigma}{\sqrt{2\pi}} \sum_{k=0}^{\infty} \int_{(4k+1)}^{(4k+3)} f^\wedge(v\sigma)(-\cos \pi v/2)\, dv. \tag{5.1}$$

Consider now the kernel of the Bessel potential G_κ, $\kappa > 0$, which can be defined via its Fourier transform

$$G_\kappa^\wedge(v) = (1+v^2)^{-\kappa/2} \qquad (v \in \mathbb{R}).$$

Choosing $\kappa = s+\alpha$, it follows that $G_\kappa \in L^1(\mathbb{R}) \cap C^{(s-1)}(\mathbb{R})$ and $f^{(s)} \in \mathrm{Lip}_2(\alpha; L^1(\mathbb{R}))$, i.e., G_κ satisfies the assumptions of the theorem (cf. [16, p. 158]). On the other hand, inserting G_κ into (5.1) yields

$$\| S_\sigma G_\kappa - G_\kappa \| \geq \frac{4\sigma}{\sqrt{2\pi}} \sum_{k=o}^{\infty} \int_{(4k+1)}^{(4k+3)} (2(v\sigma)^2)^{-\kappa/2}(-\cos \pi v/2)\, dv$$

$$= c\ \sigma^{1-\kappa} = c\ \sigma^{-s-\alpha+1} \qquad (\sigma \geq 1)$$

where c is a positive constant. This proves part b) with $f_2 = G_\kappa$ for $\kappa = s+\alpha$. Finally part c) can be deduced from [5, Prop. 4b] (cf. also [11;12]).

Let us conclude with an example showing that Cor. 1 may happen to give a better estimate than does Cor. 2b). Taking $f(t) = e^{-|t|}$, one has that $f \in L^1(\mathbb{R}) \cap C(\mathbb{R})$ and $f' \in \mathrm{Lip}(1; L^1(\mathbb{R}))$, giving the order $\mathcal{O}(\sigma^{-1})$ by Cor. 1. This can be shown to be the exact order of approximation. On the other hand, since $f \in \mathrm{Lip}(1; C(\mathbb{R}))$ but $f' \notin C(\mathbb{R})$, the best possible order which can be deduced from Cor. 2 is $\mathcal{O}(\sigma^{-1} \log \sigma)$.

REFERENCES

[1] N.I. Achieser, Theory of Approximation, Frederick Ungar Publishing Co., New York, 1956.

[2] R.P. Boas, Summation formulas and band-limited signals. Tôhoku Math. J. 24 (1972), 121-125.

[3] J.L. Brown, Jr., On the error in reconstructing a non-band-limited function by means of the bandpass sampling theorem. J. Math. Anal. Appl. 18 (1967), 75-84.

[4] P.L. Butzer, A survey of the Whittaker-Shannon sampling theorem and some of its extensions. J. Math. Res. Exposition 3 (1983), 185-212.

[5] P.L. Butzer, Some recent applications of functional analysis

to approximation theory. In: Proc. of the Euler-Kolloquium, Berlin, Germany, May 1983, in print.

[6] P.L. Butzer and R.J. Nessel, Fourier Analysis and Approximation. Birkhäuser Verlag, Basel; Academic Press, New York, 1971.

[7] P.L. Butzer and W. Splettstösser, A sampling theorem for duration limited functions with error estimates. Inform. and Control 34 (1977), 55-65.

[8] P.L. Butzer and W. Splettstösser, Approximation und Interpolation durch verallgemeinerte Abtastsummen. Forschungsber. des Landes Nordrhein-Westfalen Nr. 2708; Westdeutscher Verlag, Opladen, 1977.

[9] P.L. Butzer and R.L. Stens, The Poisson summation formula, Whittaker's cardinal series and approximate integration. In: Proc. Second Edmonton Conf. on Approximation Theory, Edmonton, Canada, June 1982; Amer. Math. Soc., Providence, R.I., 1983, in print.

[10] P.L. Butzer and W. Trebels, Hilbertransformation, gebrochene Integration und Differentiation. Forschungsber. des Landes Nordrhein-Westfalen Nr 1889; Westdeutscher Verlag, Opladen, 1968.

[11] W. Dickmeis and R.J. Nessel, A unified approach to certain counterexamples in approximation theory in connection with a uniform boundedness principle with rates. J. Approx. Theory 31 (1981), 161-174.

[12] W. Dickmeis and R.J. Nessel, On uniform boundedness principles and Banach-Steinhaus theorems with rates. Numer. Funct. Anal. Optim. 3 (1981), 19-52.

[13] H. Johnen, Inequalities connected with the moduli of smoothness. Mat. Vesnik 9 (24) (1972), 289-303.

[14] I.I. Ogieveckiĭ and L.G. Boičun, On a theorem of Titchmarsh (Russian). Izv. Vysš. Učebn. Zaved. Matematika, 1965, no. 4 (47), 100-103.

[15] W. Splettstösser, R.L. Stens, and G. Wilmes, On approximation by the interpolating series of G. Valiron. Funct. Approx. Comment. Math. 11 (1981), 39-56.

[16] E.M. Stein, Singular Integrals and Differentiability Properties of Functions. Princeton Univ. Press, Princeton, N.J., 1970.

[17] F. Stenger, Numerical methods based on Whittaker cardinal, or sinc functions. SIAM Rev. 23 (1981), 165-224.

[18] R.L. Stens, Approximation to duration-limited functions by sampling sums. Signal Process. 2 (1980), 173-176.

[19] R.L. Stens, A unified approach to sampling theorems for derivatives and Hilbert transforms. Signal Process. 5 (1983), 139-151.

[20] E.C. Titchmarsh, Introduction to the Theory of Fourier Integrals. Oxford Univ. Press, Oxford, 2nd Edition, 1948.

Rudolf Stens, Lehrstuhl A für Mathematik, Rheinisch-Westfälische Technische Hochschule Aachen, Templergraben 55, D - 5100 Aachen, West Germany

International Series of
Numerical Mathematics, Vol. 71

SEQUENTIAL SEARCH FOR ZEROES OF DERIVATIVES

Roger J. Wallace

Abstract. A systematic method of searching for real zeroes of real valued continuous k-th derivatives is investigated. The technique involves successively choosing points to be the abscissae for sequences of k-th divided differences whose signs are used to locate the zeroes. In this paper, various rules for selecting these points are discussed, and several new results concerning error estimates of some of these rules are given.

1. INTRODUCTION

Solutions of boundary value problems that arise from Laplace's equation or from Helmholtz's equation often lead to ordinary differential equations and to the study of Sturm-Liouville systems and their associated eigenvalues and eigenfunctions; see Moon and Spencer [9].

In some situations the fitting of boundary conditions depends upon the convergence of Fourier-type series based on the eigenfunctions, and an important part of that analysis is the estimation of the zeroes of the eigenfunctions or of their derivatives. Often these zeroes are simple and can be localized to finite open intervals by elementary theorems; see Swanson [11]. For example, applying the Sturm comparison theorem to the differential equation $y''(x) + [1 + (1/4x^2)]y(x) = 0$ leads to the conclusion that the difference of two consecutive zeroes $\lambda_n < \lambda_{n+1}$ of J_0 satisfies $\alpha(\lambda_n) < \lambda_{n+1} - \lambda_n < \alpha(\lambda_{n+1})$, where $\alpha(\lambda) = \pi/(1 + 1/4\lambda^2)^{1/2} < \pi$. One of the aims here is to discuss a systematic search method whereby zeroes of this type can be approximated further.

The author wishes to thank the management of Siddons Industries of Melbourne, Australia, for the generous provision of a Travelling Scholarship. Without this Award, attendance at this Conference would not have been possible. Also acknowledged are the many helpful and erudite comments of Dieter K. Ross during the preparation and writing of this paper.

It is to be assumed throughout that a *zero* of a function f is a number ξ which satisfies the inequality $f(\xi-\epsilon)f(\xi+\epsilon) < 0$ for all sufficiently small $\epsilon > 0$, *in addition to* satisfying the equation $f(\xi) = 0$.

A feature of the aforementioned eigenfunctions is that although they can be evaluated as sums of series, their derivatives are not always easy to find; see Sneddon [10]. Accordingly, the search method referred to above is extended so as to enable estimates of isolated real zeroes of real valued continuous k-th derivatives of functions to be found to within prescribed accuracies. There is no recourse to any further properties of these functions or of any of their derivatives. Instead, use is made of the signs of successive values of those k-th divided differences that are defined, for any function f, by the initial condition

$$D^{(0)}[f;x_0] = f(x_0)$$

and by the recurrence

$$D^{(k+1)}[f;x_0,x_1,\ldots,x_{k+1}]=\frac{D^{(k)}[f;x_0,x_1,\ldots,x_k]-D^{(k)}[f;x_1,x_2,\ldots,x_{k+1}]}{x_0-x_{k+1}};$$

$$k = 0,1,2,\ldots .$$

The rate of convergence to a zero depends upon the choice of these distinct abscissae $x_0,x_1,x_2,\ldots,x_k,x_{k+1}, x_{k+2},\ldots$.

This paper extends the work of Booth [2],[3],[4] and [5], Kiefer [7] and [8] and Johnson (cited in Bellman [1]) on the cases $k = 0,1,2,3,4,5$ and 6. An explicit description is given of the abscissae which constitute, in the present context, the *most efficient* scheme of approximation for the case $k=4$. The catalyst for this result is an inequality of Booth [2].

Finally, there is exhibited a lower bound on the maximum possible error that results from n applications of the scheme which, for the case $k=8$, is the most efficient. Also given is a parallel result for $k=10$.

2. PRELIMINARIES AND NOTATION

The problem of estimating an isolated real zero of a real valued continuous function defined on a known open interval can be reduced to that of estimating the unique real zero ξ_0 of a real valued continuous function f_0, where f_0 is defined on the unit interval (0,1) and satisfies $f_0(x) < 0$ if $x \in (0, \xi_0)$ and $f_0(x) > 0$ if $x \in (\xi_0, 1)$.

When no use is to be made of any further information about f_0 or about any of its derivatives, one technique for approximating ξ_0 is the *Bisection strategy*; see Isaacson and Keller [6]. The function is evaluated at the abscissa $\frac{1}{2}$ and the sign of $f_0(\frac{1}{2})$ noted. According as this is non-negative or negative, it is then concluded that $\xi_0 \in (0,\frac{1}{2}]$ or $\xi_0 \in (\frac{1}{2},1)$, whereupon the sign of either $f_0(\frac{1}{4})$ or $f_0(\frac{3}{4})$ is examined. Eventually, this procedure confines ξ_0 to an interval of length $(1-\frac{1}{2})^n$ after n evaluations. By way of comparison, a *Trisection strategy* results in a corresponding interval of length $(1-\frac{1}{3})^n$. In other words, after n evaluations the *maximum possible error* is $(\frac{1}{2})^n$ for Bisection, but is $(\frac{2}{3})^n$ $(\geqslant (\frac{1}{2})^n)$ for Trisection.

Selection of abscissae need not be based solely on the ratios $\frac{1}{2}$ or $\frac{1}{3}$. Other strategies exist for approximating ξ_0, notably stochastic ones that employ a given probability density function Φ to select the abscissae in a random fashion. Booth [3] has shown that, for any such Φ, the maximum expected error after n evaluations exceeds $(\frac{1}{2})^n$.

A problem of greater subtlety is that of estimating the unique real zero ξ_1 of the real valued continuous first derivative $f_1^{(1)}$ of a function f_1, where f_1 is defined on (0,1) and where $f_1^{(1)}$ satisfies $f_1^{(1)}(x) < 0$ if $x \in (0,\xi_1)$ and $f_1^{(1)}(x)>0$ if $x \in (\xi_1,1)$. When no other information is to be assumed, it is usual to select 2 points x_0 and x_1 $(0 < x_0 < x_1 < 1)$, to compute the sign of $D^{(1)}[f_1;x_0,x_1] = (f_1(x_0)-f_1(x_1))/(x_0-x_1)$ and then, according as this is non-negative or negative, to conclude that $\xi_1 \in (0,x_1]$ or $\xi_1 \in (x_0,1)$. Thereupon a third point x_2 is chosen from the appropriate sub-interval, and the process continues.

As with the previous problem concerning ξ_0, many strategies exist for selecting these abscissae $x_0, x_1, x_2, \ldots$ that are to be used to locate ξ_1. One such strategy, formulated by Kiefer [7] and reported in Bellman [1], is based on the classic Fibonacci sequence $\{F(n)\}$, $n=0,1,2,\ldots$ that is defined by the initial conditions $F(0) = F(1) = 1$ and by the difference equation $F(n+2) = F(n+1)+F(n)$, $n=0,1,2,\ldots$. This *Fibonacci strategy* is initiated by choosing

$$(1) \qquad x_0 = \lambda^2 \text{ and } x_1 = \lambda, \text{ where } \lambda = 1/\left(\frac{1+\sqrt{5}}{2}\right) = \lim_{n\to\infty} \frac{F(n)}{F(n+1)} .$$

Next, the quotient $D^{(1)}[f_1;\lambda^2,\lambda]$ is evaluated and its sign examined. If this is non-negative, then $\xi_1 \in (0,\lambda]$, otherwise $\xi_1 \in (\lambda^2,1)$. In the former case $x_2 = \lambda^3 < \lambda^2$ is chosen to be the other abscissa in $(0,\lambda]$. Scaling this new *configuration* $0,\lambda^3,\lambda^2$ and λ by $1/\lambda$ now gives the initial configuration $0,\lambda^2,\lambda$ and 1. In the latter case $x_2 = 2\lambda^2 > \lambda$ is selected to be in $(\lambda^2,1)$, whereupon a translation of this third configuration $\lambda^2,\lambda,2\lambda^2$ and 1 by -1 followed by a reversal in direction and a scaling by $1/\lambda$ yields $0,\lambda^2,\lambda$ and 1 again. In either case, the original ratios are recovered, and so the process can be repeated to locate ξ_1 to any required accuracy.

The present paper concerns itself with the general problem of estimating the unique real zero ξ_k of the real valued continuous k-th derivative $f_k^{(k)}$ of a function f_k, where k is a given non-negative integer, where f_k is defined on $(0,1)$, where $f_k^{(k)}$ satisfies $f_k^{(k)}(x) < 0$ if $x \in (0,\xi_k)$ and $f_k^{(k)}(x) > 0$ if $x \in (\xi_k,1)$ and where no use is to be made of any other properties of f_k or of any of its derivatives.

The *algorithm* by which ξ_k is located involves forming successive k-th divided differences and using them, in the absence of further information about $f_k^{(k)}$, as estimates of $f_k^{(k)}(\xi_k)$. In other words, k+1 abscissae $x_0, x_1, x_2, \ldots$ and x_k $(0 < x_0 < x_1 < x_2 < \ldots < x_k < 1)$ are selected and the ordinates $f_k(x_0), f_k(x_1), f_k(x_2), \ldots$ and $f_k(x_k)$ evaluated. According as the sign of $D^{(k)}[f_k; x_0, x_1, x_2, \ldots, x_k]$ is non-negative or negative, it is concluded that $\xi_k \in (0, x_k]$ or $\xi_k \in (x_0, 1)$, whereupon a k+2-th abscissa x_{k+1} is chosen from the appropriate sub-interval, and the process continues.

A *strategy* is the particular rule by which these abscissae $x_0, x_1, x_2, \ldots, x_k, x_{k+1}, x_{k+2}, \ldots$ are selected. For a given k, one strategy is said to be more *efficient* than another if, for any ξ_k and for any f_k, the maximum possible error that results from using n steps of the former strategy to locate ξ_k is less than that which results from using n steps of the latter. The broad aim here is to discuss efficient strategies for locating ξ_k when $k=4$, $k=8$ and $k=10$.

For $k=0$ and $k=1$, the *most efficient* strategies are Bisection and Fibonacci, respectively. Each of these is said to have *period* 1, for reasons that become apparent shortly. The cases $k=2,3,5$ and 6 need not be explored here; this has already been done, by Kiefer [8] ($k=2$) and by Booth [2] ($k=3$), [5] ($k=5$) and [4] ($k=6$).

The specific purpose of this paper is to extend the work of Booth [4] by giving an explicit description of $\bar{S}_4$, the strategy which is, for the case $k=4$, the most efficient. This result is a consequence of being able to exhibit a closed form expression for the n-th term of the integer sequence $\{U_4(n)\}$, $n=0,1,2,\ldots$ that is defined by the initial conditions

(2) $$U_4(0) = U_4(1) = U_4(2) = U_4(3) = U_4(4) = 1$$

and by the rule

(3) $$U_4(n+5) = \min_{i=0,1,2} (U_4(n+i) + U_4(n+4-i)), n=0,1,2,\ldots .$$

Here, $U_4(n)$ denotes the reciprocal of the maximum possible error which results from n evaluations of $\bar{S}_4$.

REMARK 1. A feature which is common to Bisection and Fibonacci is that it requires only 1 new abscissa in order to recover the original ratios. With Bisection this point is either $\frac{1}{4}$ or $\frac{3}{4}$ whereas with Fibonacci it is λ^3 or $2\lambda^2$. With $\bar{S}_4$, however, it will be found that the requisite number of new abscissae is not 1, but 3. In other words, $\bar{S}_4$ has period 3.

3. SCALING THE INTERVAL

It will prove advantageous to introduce a new way of describing strategies. This is best illustrated by considering first the case $k=0$.

By scaling the unit interval $(0,1)$ by a given factor L, it is seen that locating ξ_0 of f_0 by n evaluations of a particular strategy is equivalent to using the same number of steps of the same strategy to locate the unique zero $L\xi_0$ of the function f_L^* that is defined by $f_L^*(\theta)=f_0(L^{-1}\theta)$ for all $\theta \in (0,L)$.

It is recalled that in n evaluations the Bisection strategy confines ξ_0 to an interval of length 2^{-n}. This means that $(0,2^n)$ is the largest of all the intervals $(0,L)$ upon which n applications of the Bisection strategy B confine $L\xi_0$ to a unit interval with certainty. In a similar fashion it can be concluded that $(0,(\frac{3}{2})^n)$ is the largest of all the intervals $(0,L)$ upon which n evaluations of Trisection confine $L\xi_0$ to a unit interval. The preceding definitions and these two examples suggest another way of stating the fact that, for the case $k=0$, Bisection is the most efficient strategy.

DEFINITION 2. Define $(0,L_{S_0})$ to be the largest interval upon which n evaluations of a particular strategy S_0 confine $L_{S_0}\xi_0$ to a unit interval; and define $L_{\bar{S}_0} = \sup_{S_0} L_{S_0}$.

REMARK 3. Then $L_{\bar{S}_0} = L_B$. Furthermore, it is noted that $L_{\bar{S}_0}$ is the n-th term of the sequence $\{L_0(n)\}$, $n=0,1,2,\ldots$ that is defined by the initial condition

$$L_0(0) = 1$$

and by the rule

$$L_0(n+1) = 2L_0(n), \quad n=0,1,2,\ldots .$$

REMARK 4. Remark 3 indicates how the *structure* of the Bisection strategy can be exhibited in a different manner.

"Denote by f_0^* any $f_{L_0(n)}^*$ defined on $(0,L_0(n))$. Initiate the Bisection strategy by evaluating f_0^* at the abscissa $L_0(n-1)$ and note the sign of $f_0^*(L_0(n-1))$. According as this is non-negative or negative, conclude that $L_0(n)\xi_0 \in (0,L_0(n-1)]$ or $L_0(n)\xi_0 \in (L_0(n-1),L_0(n))$ etc.. Dividing each of these new scaled abscissae 0, $L_0(n-1)$ and $L_0(n)$ by $L_0(n)$ and taking the limit as $n\to\infty$ of each ratio now gives the familiar normalized initial Bisection strategy abscissae 0, $\frac{1}{2}$ and 1."

The reason for the limit statement becomes evident shortly.

In a similar fashion, the locating of ξ_1 of $f_1^{(1)}$ by n evaluations of a particular strategy is equivalent to using the same number of steps of the same strategy to locate the unique zero $L\xi_1$ of the first derivative $f_L^{*(1)}$ of the function f_L^* that is defined, for a given scale L, by $f_L^*(\theta) = f_1(L^{-1}\theta)$ for all $\theta \in (0,L)$. It can now be stated in an alternate way that, for the case k=1, the Fibonacci strategy F is the most efficient.

DEFINITION 5. Define $(0,L_{S_1})$ to be the largest interval upon which n evaluations of a particular strategy S_1 confine $L_{S_1}\xi_1$ to a unit interval; and define $L_{\bar{S}_1} = \sup_{S_1} L_{S_1}$.

REMARK 6. Then $L_{\bar{S}_1} = L_F$. That is to say, $L_{\bar{S}_1}$ equals the n-th term of the sequence $\{L_1(n)\}, n=0,1,2,\ldots$ that is defined by the initial conditions

$$L_1(0) = L_1(1) = 1$$

and by the rule

$$L_1(n+2) = L_1(n+1) + L_1(n), n=0,1,2,\ldots;$$

see Kiefer [7].

REMARK 7. Remark 6 indicates how the structure of the Fibonacci strategy can be exhibited in another way.

"Denote by f_1^* any $f_{L_1(n)}^*$ defined on $(0,L_1(n))$. Initiate the Fibonacci strategy by choosing $x_0 = L_1(n-2)$ and $x_1 = L_1(n-1)$ and note the sign of the first divided difference $D^{(1)}[f_1^*; L_1(n-2), L_1(n-1)]$. According as this is non-negative or negative, conclude that $L_1(n)\xi_1 \in (0, L_1(n-1)]$ or $L_1(n)\xi_1 \in (L_1(n-2), L_1(n))$ etc.. Dividing each of these new scaled abscissae 0, $L_1(n-2)$, $L_1(n-1)$ and $L_1(n)$ by $L_1(n)$ and taking the limit as $n \to \infty$ of each ratio now gives the familiar normalized initial Fibonacci strategy abscissae $0, \lambda^2, \lambda$ and 1; see (1)".

In general, for any non-negative integer k, the locating of ξ_k of $f_k^{(k)}$ by n evaluations of a particular strategy is equivalent to using the same number of steps of the same strategy to locate the unique zero $L\xi_k$ of the k-th derivative $f_L^{*(k)}$ of the function f_L^* that is defined, for a given scale L, by $f_L^*(\theta) = f_k(L^{-1}\theta)$ for all $\theta \in (0,L)$.

DEFINITION 8. For a given non-negative integer k, define $(0, L_{S_k})$ to be the largest interval upon which n evaluations of a particular strategy S_k confine $L_{S_k}\xi_k$ to a unit interval. Define $L_{\bar{S}_k} = \sup_{S_k} L_{S_k}$ and denote $L_{\bar{S}_k}$ by $L_k(n)$ and $f^*_{L_k(n)}$ by f_k^*.

For example, $L_0(n) = L_B = 2^n$ whereas $L_1(n) = L_F = F(n)$.

REMARK 9. The maximum possible error which results from n applications of $\bar{S}_k$ is $1/L_k(n)$.

REMARK 10. Remarks 4 and 7 illustrate the importance, for a general k, of the scaled interval $(0, L_k(n))$ and of the associated scaled abscissae that lie within it. Each of the *normalized* abscissae $x_0, x_1, x_2, \ldots, x_k, x_{k+1}, x_{k+2}, \ldots$ that are to be used to locate ξ_k is obtained from its corresponding *scaled* abscissa by first dividing the latter by $L_k(n)$ and by then taking the limit as $n \to \infty$ of the resultant ratio.

For any given $k > 1$, it is difficult to find a closed form expression for $L_k(n)$. Booth [2], however, has proved the following important inequality.

For any given non-negative integer k,

$$L_k(0) = L_k(1) = L_k(2) = \ldots = L_k(k) = 1;$$

$$(4)\qquad L_k(n+k+1) \leqslant \begin{cases} L_k(n) + L_k(n+k) \\ L_k(n+1) + L_k(n+k-1) \\ \vdots \qquad\qquad \vdots \\ L_k(n+[\frac{k}{2}]) + L_k(n+k-[\frac{k}{2}]) \end{cases},\quad n=0,1,2,\ldots .$$

Booth's proof of (4) is an adaption of an unpublished technique of Johnson (cited in Bellman [1]) concerning the case k=1. As (4) is to reappear later in this paper when the cases k=4, k=8 and k=10 are discussed, it is pertinent to elaborate upon its proof. The salient features of such are best illustrated by considering the case k=2.

LEMMA 11. (Booth [2]). Let $L_2(n)$ be as specified in Definition 8. Then

$$L_2(0) = L_2(1) = L_2(2) = 1;$$

$$L_2(n+3) \leqslant \begin{cases} L_2(n) + L_2(n+2) \\ 2L_2(n+1) \end{cases},\quad n=0,1,2,\ldots .$$

Proof. The initial conditions are due to the fact that evaluating f_2^* at less than 3 abscissae gives no information as to the location of any zero of $f_2^{*(2)}$; hence each of $L_2(0)$, $L_2(1)$ and $L_2(2)$ cannot be greater than 1. The required equalities now follow.

Next, let $(0,L_{S_2}(n))$ be the largest interval upon which n evaluations of a strategy S_2 confine any $L_{S_2}(n)\xi_2 \in (0,L_{S_2}(n))$ to a unit interval. Further, let x_0, x_1 and x_2 $(0 < x_0 < x_1 < x_2 < L_{S_2}(n))$ be the first 3 abscissae selected by S_2, let y_0 denote any point which satisfies $0 < y_0 < x_0$ and consider the situation when $L_{S_2}(n)\xi_2 \in (0,y_0)$. Then, the fact that $L_{S_2}(n)\xi_2$ is less than y_0, which in turn is less than each of the 3 points x_0,x_1 and x_2, implies that $y_0 \leqslant L_2(n-3)$; whereupon taking the supremum over y_0 gives

$$(5)\qquad x_0 \leqslant L_2(n-3).$$

Now, suppose that $L_{S_2}(n)\xi_2 \in (x_0, L_{S_2}(n))$. Since $L_{S_2}(n)\xi_2$ is greater than x_0, but less than $L_{S_2}(n)$, it follows that $L_{S_2}(n) - x_0 \leqslant L_2(n-1)$. By taking the supremum over S_2, and by recalling Definition 8, it transpires that

(6) $$L_2(n) - x_0 \leqslant L_2(n-1) \;;$$

whereupon addition of (5) and (6) gives

(7) $$L_2(n) \leqslant L_2(n-3) + L_2(n-1) \;.$$

In a similar fashion, let y_1 denote any point which satisfies $x_0 < y_1 < x_1$ and consider when $L_{S_2}(n)\xi_2 \in (0,y_1)$. Because $L_{S_2}(n)\xi_2$ is less than y_1, which is less than the 2 points x_1 and x_2, it is seen that $y_1 \leqslant L_2(n-2)$, and so

(8) $$x_1 \leqslant L_2(n-2) \;.$$

Next, suppose that $L_{S_2}(n)\xi_2 \in (x_1, L_{S_2}(n))$. Since $L_{S_2}(n)\xi_2$ is greater than the 2 points x_0 and x_1, but less than $L_{S_2}(n)$, it follows that $L_{S_2}(n)-x_1 \leqslant L_2(n-2)$. Hence,

(9) $$L_2(n)-x_1 \leqslant L_2(n-2) \;;$$

whereupon addition of (8) and (9) gives

(10) $$L_2(n) \leqslant 2\, L_2(n-2).$$

Finally, letting y_2 be any point which satisfies $x_1 < y_2 < x_2$, and considering when $L_{S_2}(n)\xi_2 \in (0,y_2)$, leads to

(11) $$x_2 \leqslant L_2(n-1);$$

whereupon, by supposing that $L_{S_2}(n)\xi_2 \in (x_2, L_{S_2}(n))$, it transpires that

(12) $$L_2(n)-x_2 \leqslant L_2(n-3).$$

Therefore, it is seen that (11) and (12) are superfluous, as their addition merely duplicates (7); they are included here solely for completeness.

Together, (7) and (10) state that

$$L_2(n) \leqslant \begin{cases} L_2(n-3)+L_2(n-1) \\ 2L_2(n-2) \end{cases}$$

and the proof is complete.

REMARK 12. The importance of (4) lies in the fact that it suggests a "natural" upper bound on the sequence $\{L_k(n)\}$, $n=0,1,2,\ldots$, namely the sequence $\{U_k(n)\}$ that is defined, for any given non-negative integer k and for all $n=0,1,2,\ldots$, by the initial conditions

$$U_k(0) = U_k(1) = U_k(2) = \ldots = U_k(k) = 1$$

and by the rule

$$U_k(n+k+1) = \min_{i=0,1,2,\ldots[\frac{k}{2}]} (U_k(n+i)+U_k(n+k-i)),\ n=0,1,2,\ldots .$$

For each of the cases $k=0,1,2,3$, and 6 it is not difficult to obtain a closed form expression for $U_k(n)$. Upon comparison with the corresponding expressions for $L_k(n)$, it is observed that

$$(13)\qquad L_k(n) = U_k(n),\quad n=0,1,2,\ldots$$

for $k=0,1,2,3$, and 6.

Booth [5] has shown that (13) is false when $k=5$. Booth [4], however, has established that (13) is *true* when $k=4$ and that

$$U_4(3t+1) = 2^{t-1},\quad t=2,3,4,\ldots .$$

A complete closed form expression for $U_4(n)$ will now be derived. First of all, some preliminary results are required.

LEMMA 13. Let $\{U_4(n)\}$, $n=0,1,2,\ldots$ be the sequence that is defined by the initial conditions

$$U_4(0) = U_4(1) = U_4(2) = U_4(3) = U_4(4) = 1$$

and by the rule

$$U_4(n+5) = \min_{i=0,1,2} (U_4(n+i)+U_4(n+4-i)),\ n=0,1,2,\ldots .$$

Then, for all $n=0,1,2,\ldots$,

$$(14)\quad \begin{cases} U_4(n+5) = U_4(n+1) + U_4(n+3), & \text{if } n \equiv 0 \text{ or } 1 \pmod 3; \\ U_4(n+5) = 2U_4(n+2), & \text{if } n \equiv 2 \pmod 3. \end{cases}$$

Proof. These conclusions follows from results of Booth in [4], and by performing a small number of evaluations.

LEMMA 14. *Let the sequence* $\{U_4(n)\}$, $n=0,1,2,\ldots$ *be as defined in Lemma* 13 *and let the translation operator* E_m^a *be defined by*

(15) $$E_m^a U_k(mt+j) = U_k(mt+ma+j)$$

for any non-negative integers a, k, t *and* j, *and for any positive integer* m. *Then, for each* $j=0,1$ *and* 2,

(16) $$E_3^2(E_3-2)(E_3^2-1)U_4(3t+j) = 0, \quad t = 0,1,2,\ldots.$$

Proof. Use (15) with $k=4$ and $m=3$ to write (14) as the matrix equation

(17) $$A\mathbf{U}_4(t) = \mathbf{0}, \quad t=0,1,2,\ldots ,$$

where $$A = \begin{pmatrix} -E_3 & -1 & E_3 \\ E_3^2 & -E_3 & -1 \\ 0 & E_3(E_3-2) & 0 \end{pmatrix}; \quad \mathbf{U}_4(t) = \begin{pmatrix} U_4(3t) \\ U_4(3t+1) \\ U_4(3t+2) \end{pmatrix} \text{ and } \mathbf{0} = \begin{pmatrix} 0 \\ 0 \\ 0 \end{pmatrix}.$$

Now, apply Cramer's rule to (17) and conclude that, for each $j=0,1$ and 2,

$$\det(A)\, U_4\, (3t+j) = \det(A_j), \quad t = 0,1,2,\ldots ,$$

where A_j is the matrix A with the j-th column replaced by $\mathbf{0}$. Thus $\det(A_j) = 0$ and so the proof is complete because $\det(A) = E_3^2(E_3-2)(E_3^2-1)$.

THEOREM 15. *Let* $\{U_4(n)\}, n=0,1,2,\ldots$ *be the sequence that is defined by the initial conditions*

$$U_4(0) = U_4(1) = U_4(2) = U_4(3) = U_4(4) = 1$$

and by the rule

$$U_4(n+5) = \min_{i=0,1,2} (U_4(n+i)+U_4(n+4-i)), n=0,1,2,\ldots .$$

Then $U_4(5) = 2$ *whereas*

(18) $$\begin{cases} U_4(3t) = \frac{5}{12}\cdot 2^t - \frac{1}{6}(-1)^t + \frac{1}{2} \\ U_4(3t+1) = \frac{1}{2}\cdot 2^t \\ U_4(3t+2) = \frac{2}{3}\cdot 2^t - \frac{1}{6}(-1)^t + \frac{1}{2} \end{cases} \quad \textit{for all } t = 2,3,4,\ldots .$$

Proof. The first result is trivial. The other follows by using Lemma 14 to write each $U_4(3t+j)$, $j=0,1$ or 2, as

$$U_4(3t+j) = \alpha_j 2^t + \beta_j(-1)^t + \gamma_j, \quad t=2,3,4,\ldots .$$

where α_0, β_0, γ_0,... and γ_2 are 9 real unknowns whose values are found through computing $U_4(6)$, $U_4(7)$, $U_4(8)$,... and $U_4(14)$.

Note that the index t in (18) commences at 2 because of the power of the factor E_3 in (16). Section 4 below will make it apparent that the periodicity of the $U_4(n)$ of (18) implies the aforementioned periodicity of $\bar{S}_4$; see Remark 1.

COROLLARY 16.

$$U_4(6t) = \frac{5}{12}\cdot 4^t + \frac{1}{3}\ ; \qquad U_4(6t+1) = \frac{1}{2}\cdot 4^t\ ;$$
$$U_4(6t+2) = \frac{2}{3}\cdot 4^t + \frac{1}{3}\ ; \qquad U_4(6t+3) = \frac{5}{6}\cdot 4^t + \frac{2}{3}\ ;$$
$$U_4(6t+4) = 4^t\ ; \qquad U_4(6t+5) = \frac{4}{3}\cdot 4^t + \frac{2}{3}\ ;$$
$$\underline{\text{for all}}\quad t = 1,2,3,\ldots\ .$$

Booth [4] has given an implicit description of $\bar{S}_4$. He begins by selecting 5 points on the scaled interval $(0, U_4(n))$; some of the subsequent abscissae then are merely translations of linear combinations of values of $U_4(n)$ at smaller n. This parallels the behaviour of the Bisection and Fibonacci strategies described earlier; see Remarks 3, 4, 6 and 7. However, some of the later points are selected according to other criteria. For example, one of these scaled abscissae in $(0,U_4(n))$ is the point which is the minimum of $U_4(n-6)+U_4(n-12)$ and $U_4(n-5)$ when $n\equiv 2 \pmod 3$. The explicit formulae for $U_4(n)$ exhibited in Theorem 15 and Corollary 16 now permit a numerical value to be assigned to all of these abscissae of [4], particularly ones such as these. These *scaled* abscissae of $\bar{S}_4$ will not be now exhibited. This is in order to avoid repetition, as the *normalized* abscissae of $\bar{S}_4$ are about to be given below. These latter abscissae have been obtained from the former by dividing by the appropriate endpoints, and by taking limits as $n\to\infty$ of ratios; see Remark 10.

4. AN EXPLICIT DESCRIPTION OF $\bar{S}_4$.

On the interval (0,1) select the 5 points $x_0 = \frac{1}{4}$, $x_1 = \frac{3}{8}$, $x_2 = \frac{1}{2}$, $x_3 = \frac{5}{8}$ and $x_4 = \frac{3}{4}$. This symmetric configuration is written as

$$C_0 = [0;\ \tfrac{1}{4},\ \tfrac{3}{8},\ \tfrac{1}{2},\ \tfrac{5}{8},\ \tfrac{3}{4};\ 1].$$

The notation that is to be used here to denote configurations is that of Booth [4].

Each of the 5 chosen fractions is identified as a limiting ratio of certain values of $U_4(n)$; see Theorem 15 or Corollary 16. Namely,

$$x_0 = \frac{1}{4} = \lim_{t\to\infty} \frac{U_4(3t-4)}{U_4(3t+2)} ; \qquad x_1 = \frac{3}{8} = \lim_{t\to\infty} \frac{U_4(3t-2)}{U_4(3t+2)} ;$$

$$x_2 = \frac{1}{2} = \lim_{t\to\infty} \frac{U_4(3t-1)}{U_4(3t+2} ; \qquad x_3 = \frac{5}{8} = \lim_{t\to\infty} \frac{U_4(3t)}{U_4(3t+2)} ;$$

$$x_4 = \frac{3}{4} = \lim_{t\to\infty} \frac{U_4(3t+1)}{U_4(3t+2)} .$$

Attention is drawn to the fact that the first fraction $x_0 = \frac{1}{4}$ is *not* $\lim_{t\to\infty} U_4(3t-3)/U_4(3t+2)$.

According as the sign of $D^{(4)}[f_4; \frac{1}{4}, \frac{3}{8}, \frac{1}{2}, \frac{5}{8}, \frac{3}{4}]$ is non-negative or negative, it is now concluded that $\xi_4 \in (0, \frac{3}{4}]$ or $\xi_4 \in (\frac{1}{4},1)$. The configuration that pertains to $(0, \frac{3}{4}]$ is

$$C_1 = [0; \frac{1}{4}, \mathbf{\frac{5}{16}}, \frac{3}{8}, \frac{1}{2}, \frac{5}{8}; \frac{3}{4}].$$

Note that extra configuration points are to be written in boldface type. The other configuration is $[\frac{1}{4}; \frac{3}{8}, \frac{1}{2}, \frac{5}{8}, \mathbf{\frac{11}{16}}, \frac{3}{4}; 1]$ but this need not be considered because translating it by -1 and then reversing direction merely gives C_1 again. The new fraction $\frac{5}{16}$ stems from the fact that $\frac{5}{16}:\frac{3}{4} = \lim_{t\to\infty} U_4(3t-3)/U_4(3t+1)$, where the $\frac{3}{4}$ is connected with $U_4(3t+1)$ because of x_4. One of the choices that arises from C_1 is

$$C_2 = [0; \mathbf{\frac{3}{16}}, \frac{1}{4}, \frac{5}{16}, \frac{3}{8}, \frac{1}{2}; \frac{5}{8}] .$$

Here, $\frac{3}{16}$ results from the fact that $\frac{3}{16} : \frac{5}{8} = \lim_{t\to\infty} U_4(3t-5)/U_4(3t)$ the correspondence between $\frac{5}{8}$ and $U_4(3t)$ is due to x_3. The other choice that emanates from C_1 pertains to the interval $(\frac{1}{4}, \frac{3}{4})$. Instead of now using $\frac{5}{16}, \frac{3}{8}, \frac{1}{2}, \frac{5}{8}$ and one other point as abscissae, the points $\frac{3}{8}, \frac{7}{16}, \frac{1}{2}, \frac{9}{16}$ and $\frac{5}{8}$ are used. In other words, $\frac{5}{16}$ is discarded, and *two* new abscissae $\frac{7}{16}$ and $\frac{9}{16}$ chosen; the configuration is, therefore,

$$C_2' = [\frac{1}{4}; \frac{3}{8}, \mathbf{\frac{7}{16}}, \frac{1}{2}, \mathbf{\frac{9}{16}}, \frac{5}{8}; \frac{3}{4}] .$$

A translation of C_2' by $-\frac{1}{4}$, followed by a scaling by 2, gives the initial configuration C_0; thus the original ratios are recovered. Note that it has required the selection of 3 new abscissae $\frac{5}{16}, \frac{7}{16}$ and $\frac{9}{16}$ in order to do this. A similar conclusion applies to one of the choices from C_2, namely

$$C_3 = [0; \frac{1}{8}, \frac{3}{16}, \frac{1}{4}, \frac{5}{16}, \frac{3}{8}; \frac{1}{2}].$$

Here, C_3 reduces to C_0 by scaling by 2; and the 3 new abscissae are $\frac{5}{16}$, $\frac{3}{16}$ and $\frac{1}{8}$. The last of these 3 fractions ($\frac{1}{8}$) results from the fact that $\frac{1}{8}:\frac{1}{2} = \lim_{t\to\infty} U_4(3t-7)/U_4(3t-1)$, where the $\frac{1}{2}$ is related to $U_4(3t-1)$ through x_2.

The other choice that arises from C_2 pertains to the interval $(\frac{3}{16}, \frac{5}{8})$; the relevant configuration is

$$C_3' = [\frac{3}{16}; \frac{1}{4}, \frac{5}{16}, \frac{3}{8}, \frac{7}{16}, \frac{1}{2}; \frac{5}{8}] .$$

As f_4 is not evaluated at endpoints of configurations, C_3' can be extended to

$$C_3'' = [\frac{1}{8}; \frac{1}{4}, \frac{5}{16}, \frac{3}{8}, \frac{7}{16}, \frac{1}{2}; \frac{5}{8}] .$$

A translation by $-\frac{5}{8}$, a scaling by 2 and a reversal in direction now reduces C_3'' to C_0. The new abscissae in this instance are $\frac{5}{16}$, $\frac{3}{16}$ and $\frac{7}{16}$.

It is seen, therefore, that *wherever* ξ_4 *might be*, 3 additional abscissae are required in order to recover the original ratios; see Remark 1. This periodic process can now be repeated to achieve any prescribed accuracy.

5. A CLOSED FORM EXPRESSION FOR $U_8(n)$.

By using induction on n, it is found that, for all $n=0,1,2,\ldots$,

$$U_8(n+9) = U_8(n+2) + U_8(n+6), \text{ if } n\equiv 1 \text{ or } 2 \pmod 5;$$
$$U_8(n+9) = U_8(n+3) + U_8(n+5), \text{ if } n\equiv 0 \text{ or } 3 \pmod 5;$$
$$U_8(n+9) = 2U_8(n+4) \quad , \text{ if } n\equiv 4 \pmod 5),$$

from which it follows that

$$E_5^4(E_5-2)(E_5^4-1)U_8(5t+\ell) = 0$$

for all $t=0,1,2,\ldots$ and for each $\ell = 0,1,2,3$ and 4.

In a similar fashion to Corollary 16, it is now concluded that $U_8(9) = U_8(10) = U_8(11) = U_8(12) = U_8(13) = 2$, $U_8(14) = U_8(15) = 3$, $U_8(16) = U_8(17) = U_8(18) = 4$ and $U_8(19) = 5$; and that, for all $t = 1,2,3,\ldots$,

$$(19)\begin{cases} U_8(20t) = \frac{7}{20}\cdot 16^t+\frac{2}{5};\; U_8(20t+1) = \frac{2}{5}\cdot 16^t+\frac{3}{5};\; U_8(20t+2) = \frac{9}{20}\cdot 16^t+\frac{4}{5};\\ U_8(20t+3) = \frac{1}{2}\cdot 16^t;\; U_8(20t+4) = \frac{3}{5}\cdot 16^t+\frac{2}{5};\; U_8(20t+5) = \frac{7}{10}\cdot 16^t+\frac{4}{5};\\ U_8(20t+6) = \frac{4}{5}\cdot 16^t+\frac{1}{5};\; U_8(20t+7) = \frac{9}{10}\cdot 16^t+\frac{3}{5};\; U_8(20t+8) = 16^t;\\ U_8(20t+9) = \frac{6}{5}\cdot 16^t+\frac{4}{5};\; U_8(20t+10) = \frac{7}{5}\cdot 16^t+\frac{3}{5};\; U_8(20t+11) = \frac{8}{5}\cdot 16^t+\frac{2}{5};\\ U_8(20t+12) = \frac{9}{5}\cdot 16^t+\frac{1}{5};\; U_8(20t+13) = 2\cdot 16^t;\; U_8(20t+14) = \frac{12}{5}\cdot 16^t+\frac{3}{5};\\ U_8(20t+15) = \frac{14}{5}\cdot 16^t+\frac{1}{5};\; U_8(20t+16) = \frac{16}{5}\cdot 16^t+\frac{4}{5};\; U_8(20t+17) = \frac{18}{5}\cdot 16^t+\frac{2}{5};\\ U_8(20t+18) = 4\cdot 16^t;\; U_8(20t+19) = \frac{24}{5}\cdot 16^t+\frac{1}{5}. \end{cases}$$

6. A CLOSED FORM EXPRESSION FOR $U_{10}(n)$

By methods that parallel those of Section 5 above, it is found that, for all $n=0,1,2,\ldots$,

$$U_{10}(n+11)=U_{10}(n+2)+U_{10}(n+8), \text{ if } n\equiv 2 \pmod 6;$$

$$U_{10}(n+11)=U_{10}(n+3)+U_{10}(n+7), \text{ if } n\equiv 1 \text{ or } 3 \pmod 6;$$

$$U_{10}(n+11)=U_{10}(n+4)+U_{10}(n+6), \text{ if } n\equiv 0 \text{ or } 4 \pmod 6;$$

$$U_{10}(n+11)=2U_{10}(n+5), \text{ if } n\equiv 5 \pmod 6;$$

which in turn implies that

$$E_6^4(E_6-2)(E_6^2-1)U_{10}(6t+p) = 0$$

for all $t=0,1,2,\ldots$ and for each $p=0,1,2,3,4$ and 5.

Accordingly, there results a closed form expression for $U_{10}(n)$, given by $U_{10}(11) = U_{10}(12) = U_{10}(13) = U_{10}(14) = U_{10}(15) = U_{10}(16) = 2$, $U_{10}(17) = U_{10}(18) = U_{10}(19) = 3$, $U_{10}(20) = U_{10}(21) = U_{10}(22) = 4$ and $U_{10}(23) = 5$; and by

$$\begin{aligned} &U_{10}(12t) = \tfrac{1}{3}\cdot 4^t+\tfrac{2}{3};\; U_{10}(12t+1) = \tfrac{3}{8}\cdot 4^t;\; U_{10}(12t+2) = \tfrac{5}{12}\cdot 4^t+\tfrac{1}{3};\\ &U_{10}(12t+3) = \tfrac{11}{24}\cdot 4^t+\tfrac{2}{3};\; U_{10}(12t+4) = \tfrac{1}{2}\cdot 4^t;\; U_{10}(12t+5) = \tfrac{7}{12}\cdot 4^t+\tfrac{2}{3};\\ &U_{10}(12t+6) = \tfrac{2}{3}\cdot 4^t+\tfrac{1}{3};\; U_{10}(12t+7) = \tfrac{3}{4}\cdot 4^t;\; U_{10}(12t+8) = \tfrac{5}{6}\cdot 4^t+\tfrac{2}{3};\\ &U_{10}(12t+9) = \tfrac{11}{12}\cdot 4^t+\tfrac{1}{3};\; U_{10}(12t+10) = 4^t;\; U_{10}(12t+11) = \tfrac{7}{6}\cdot 4^t+\tfrac{1}{3}; \end{aligned}$$

for all $t=2,3,4,\ldots$.

7. CONCLUSION

It is recalled, from Remark 12 and from Remark 9, respectively, that $U_k(n)$ is an upper bound on $L_k(n)$ and that $1/L_k(n)$ is the maximum possible error which results from n applications of $\bar{S}_k$. Hence the closed form expression for $1/U_8(n)$ that is obtained from (19) is a lower bound on the maximum possible error which results from n applications of $\bar{S}_8$. Analogously, $1/U_{10}(n)$ is a lower bound on the maximum possible error which results from n evaluations of $\bar{S}_{10}$.

REFERENCES

1. R. Bellman, Dynamic Programming. Princeton University Press, Princeton, 4th Edition, 1965.

2. R.S. Booth, Location of Zeros of Derivatives. SIAM J. Appl. Math. 15 (1967), 1496-1501.

3. R.S. Booth, Random Search for Zeroes. J. Math. Anal. Appl. 20 (1967), 239-257.

4. R.S. Booth, Location of Zeros of Derivatives.II. SIAM. J. Appl. Math. 17 (1969), 409-415.

5. R.S. Booth, An Odd Order Search Problem. SIAM. J. Alg. Disc. Meth. 3 (1982), 135-143.

6. E. Isaacson and H.B. Keller, Analysis of Numerical Methods. John Wiley and Sons, Inc., New York, London, Sydney, 1966.

7. J. Kiefer, Sequential Minimax Search for a Maximum. Proc. Amer. Math. Soc. 4 (1953), 502-506.

8. J. Kiefer, Optimal Sequential Search and Approximation Methods under Minimum Regularity Assumptions. SIAM. J. Appl. Math. 5 (1957), 105-136.

9. P. Moon and D.E. Spencer, Field Theory Handbook. Springer-Verlag, Berlin, 1961.

10. I.N. Sneddon, Elements of Partial Differential Equations. McGraw-Hill Kogakusha Ltd., Tokyo, 1957.

11. C.A. Swanson, Comparison and Oscillation Theory of Linear Differential Equations. Academic Press, New York and London, 1968.

Roger J. Wallace, Department of Applied Mathematics, La Trobe University, Victoria, 3083, Australia.

Inequalities of Functional Analysis

Wolf Valley

International Series of
Numerical Mathematics, Vol. 71

A WEAK-TYPE INEQUALITY AND ALMOST EVERYWHERE CONVERGENCE

Franziska Fehér

Abstract. This paper is concerned with the problem of finding conditions such that $T_k f(u)$ converges a.e., where T_k ($k \in \mathbb{N}$) are linear, bounded operators on a Banach function space X and $f \in X$. It is shown that an essential rôle is played by a weak-type inequality, thus generalizing the respective theorems of E.M.Stein and S.Sawyer to the abstract setting of function spaces. Finally, an application is given to Fourier series on Lorentz spaces L_{1q} with $0 < q < 1$.

1. INTRODUCTION

Let (Ω, Σ, μ) denote a measure space with $\mu(\Omega) = 1$, and X,Y rearrangement invariant Banach function spaces of realvalued μ-measurable functions on Ω. Moreover, for $k = 1,2,\cdots$ let $T_k : X \to Y$ be bounded linear operators from X into Y. The problem of almost everywhere convergence consists in finding necessary and sufficient conditions such that for each $f \in X$ the sequence $\{T_k f(u)\}$ converges a.e. on Ω. A famous example is the problem of a.e. convergence of Fourier series in L_p-spaces, which was solved by A. Kolmogorov [10], L. Carleson [1], and R.A. Hunt [8].

A basic theorem inbehind is the following [4, p.2]

BANACH PRINCIPLE. With notations as above, let T denote the maximal operator of the sequence $\{T_k\}_{k=1}^{\infty}$ defined by

$$(Tf)(u) := \sup_{k \in \mathbb{N}} |(T_k f)(u)| \qquad (u \in \Omega,\ f \in X)$$

<u>and let</u> $X_0 \subset X$ <u>be a dense subspace of</u> X. <u>Then the following are equivalent</u> :

(1) $\forall f \in X: \ |(T_k f)(u) - (T_l f)(u)| \to 0 \quad$ a.e. $\quad (k,l \to \infty)$

(2) a) $\forall f \in X_0: \ |(T_k f)(u) - (T_l f)(u)| \to 0 \quad$ a.e. $\quad (k,l \to \infty)$

b) <u>There exists a decreasing function</u> $C: (0,\infty) \to (0,\infty)$ <u>with</u> $C(\lambda) \downarrow 0 \quad (\lambda \uparrow \infty)$ <u>such that</u>

(1.1) $$\mu\{u \in \Omega;\ (Tf)(u) > \lambda\} \leq C(\lambda/\|f\|) \qquad (\lambda > 0,\ f \in X).$$

This principle is quite similar to the Banach-Steinhaus theorem on norm convergence of sequences of operators. The crucial point is that the condition of uniform boundedness of the operators is here replaced by the <u>weak-type inequality</u> (1.1) for the maximal operator of the operator sequence.

The main tool in the proof of Banach's principle,namely the Baire category theorem, is valid for any complete metrizable space. In particular, it holds for complete p-normed spaces $(0 < p < \infty)$, that is to say, for complete spaces $(X, \| \ \|_X)$ with the properties

$$\text{(i)} \ \|f\|_X = 0 \ \text{ iff } f = 0; \qquad \text{(ii)} \ \|\alpha f\|_X = |\alpha| \ \|f\|_X$$

$$\text{(iii)} \ \|f+g\|_X^p \leq \|f\|_X^p + \|g\|_X^p \ .$$

Hence we have the following

REMARK. Banach's principle remains valid, if X and Y are complete p-normed function spaces.

With respect to applications of the Banach principle it would be very important to know more details about the function C which figures at the right hand side of (1.1). For the particular case that $X = Y = L_p$ is a Lebesgue space $(1 \leq p < \infty)$, E.M. Stein [17] showed that $C(\lambda) =: \lambda^{-p}$, if $1 \leq p \leq 2$ and the operators T_k commute with translations. In [16] S. Sawyer extended this result to $p > 2$, but with the additional assumption that the operators T_k are positive. A complete theory for $X = L_p$, $Y = L_q$ was given by E.M. Nikišin [14] and B. Maurey [13]; see also [5] and [6].

2. THE WEAK-TYPE INEQUALITY

The problem, which arises when passing to abstract Banach function spaces, is, how to find the function C in this general setting. We begin by recalling some notations:

The _fundamental function_ τ_X of a function space X (in the sense of [12]) is defined as $\tau_X(t) := \|\chi_{E_t}\|$ for any set $E_t \in \Sigma$ with $0 \leq \mu(E_t) = t \leq 1$. Here χ_A denotes the characteristic function of the set A. The _nonincreasing rearrangement_ f^* of a μ-measurable function f on Ω is given by $f^*(t)=\inf\{\lambda>0;\ D_\lambda f \leq t\}$ for $t > 0$, where $D_\lambda f := \mu\{u \in \Omega;\ |f(u)| > \lambda\}$ is the _distribution function_ of f.

The _Lorentz space_ $\Lambda(X)$ associated with X is the space of (classes of) μ-measurable functions on Ω equipped with the norm

$$\|f\|_{\Lambda(X)} := \int_0^1 f^*(s)\,d\tau_X(s) \qquad (f \in \Lambda(X)) .$$

If X,Y are Banach function spaces of μ-measurable functions on Ω, then a linear operator $S: X \to Y$ is called _of weak-type_ (X,Y) if and only if

$$(Sf)^*(t)\ \tau_Y(t) \leq \text{const.}\ \|f\|_{\Lambda(X)} \qquad (f \in \Lambda(X)) .$$

LEMMA 1. _A linear operator_ $S: X \to L_1(\Omega,\mu)$ _is of weak-type_ (X,Y) _if and only if_

$$(2.1)\qquad D_\lambda(Sf) \equiv \mu\{u \in \Omega;\ |(Sf)(u)| > \lambda\} \leq \tau_Y^{-1}\Big(\frac{\text{const}}{\lambda}\|f\|_{\Lambda(X)}\Big),$$

$\lambda > 0$, _where_ τ_X _is assumed to be strictly increasing._

Proof. If S is of weak-type (X,Y), then for each $\lambda > 0$

$$D_\lambda(Sf) \leq t \quad \text{implies} \quad \lambda \leq \frac{\text{const.}}{\tau_Y(t)} \|f\|_{\Lambda(X)} .$$

Since the latter is equivalent to $\tau_Y(t) \leq (\text{const.}/\lambda)\|f\|_{\Lambda(X)}$ that is to say,

$$t \leq \tau_Y^{-1}\left[\frac{\text{const.}}{\lambda}\|f\|_{\Lambda(X)}\right],$$

one obtains (2.1). Conversely, suppose (2.1) holds for each $\lambda > 0$. In particular, for $\lambda_0 = \text{const.}\|f\|_{\Lambda(X)}/\tau_Y(t)$ one has

$$D_{\lambda_0}(Sf) \leq \tau_Y^{-1}\left[\frac{\text{const.}}{\lambda_0}\|f\|_{\Lambda(X)}\right] = t,$$

and hence $(Sf)^*(t) = \inf\{\lambda > 0;\ D_\lambda(Sf) \leq t\} \leq \lambda_0$, that is to say, S is of weak-type (X,Y).

A Banach function space X is said to be <u>of Rademacher type</u> p, $1 \leq p \leq 2$, if and only if there exists a constant $K_p > 0$, such that

$$\text{(2.2)} \qquad \int_0^1 \|\textstyle\sum_{k=1}^n r_k(t) f_k\|_X dt \leq K_p\left(\sum_{k=1}^n \|f_k\|_X^p\right)^{1/p}$$

for any finite set $\{f_k\}_{k=1}^n$ of elements $f_k \in X$; here $r_k(t) := \text{sign}\sin(2^{k+1}\pi t)$ are the Rademacher functions. Note that X is of Rademacher type p if and only if there exist $r \in (1,\infty)$ and $K_r > 0$ such that

$$\text{(2.3)} \qquad \left[\int_0^1 \|\textstyle\sum_{k=1}^n r_k(t) f_k\|_X^r dt\right]^{1/r} \leq K_r\left[\sum_{k=1}^n \|f_k\|_X^p\right]^{1/p}$$

for any finite set $\{f_k\}_{k=1}^\infty$ in X, see J.P. Kahane [9].

Finally, a family $\{\varphi_\alpha\}_{\alpha \in I}$ of measure preserving mappings $\varphi_\alpha : \Omega \to \Omega$ is called a <u>mixing family</u> if and only if for each $\rho > 1$ and all $A, B \in \Sigma$ there exists an $\alpha \in I$ such that

$$\mu(A \cap \varphi_\alpha^{-1}(B)) \leq \mu(A)\mu(B) .$$

With these notations the theorems of E.M. Stein [17] and of S. Sawyer [16] can be generalized in the following manner:

THEOREM 1. *Let* (Ω,Σ,μ) *be a finite measure space with* $\mu(\Omega) = 1$, *and* X,Y *rearrangement invariant Banach function spaces of* μ*-measurable functions on* Ω; *assume that*

(2.4) X *is of Rademacher type* p $(1 \leq p \leq 2)$,

(2.5) *there exists a number* $a > 0$ *such that for all* $n \in \mathbb{N}$

$$\frac{1}{\tau_Y(2/n)\, n^{1/p}} > a \quad .$$

Moreover, let $T_k : X \to L_1(\Omega,\mu)$ *be linear, bounded operators from* X *into* $L_1(\Omega,\mu)$, $k \in \mathbb{N}$, *and* $\varphi_\alpha : \Omega \to \Omega$ *a mixing family of norm- and measure preserving transformations*, $\alpha \in I$, *commuting with the* T_k, i.e. $T_k\varphi_\alpha = \varphi_\alpha T_k$, $\alpha \in I$ *and* $k \in \mathbb{N}$. *Finally, let* T *denote the maximal operator, defined by*

$$(Tf)(u) := \sup_{k\in\mathbb{N}} |(T_kf)(u)| \qquad (u \in \Omega, f \in X).$$

If

(2.6) $$|(T_kf)(u) - (T_lf)(u)| \to 0 \quad \text{a.e.} \qquad (k,l \to \infty)$$

for all $f \in X$, *then* T *is of weak-type* (X,Y) *in the sense of* (2.1).

Note that for $X = Y = L_q$, $1 \leq q \leq 2$, one has $\tau_Y^{-1}(s) = s^q$. Hence the generalized weak-type inequality (2.1) reduces to (1.1) with $C(s) = s^{-p}$, and (2.5) is just the condition that $q \leq p$, yielding the theorem of [17] as a corollary, since L_q is of Rademacher type $p := \min\{2,q\}$; see e.g. [11], [15]. Beyond the Lebesgue case, Theorem 1 can also be applied e.g. to Lorentz and Orlicz spaces. Indeed, the Lorentz space L_{rs} has the fundamental function $t^{1/r}$ and (see [3]) is of Rademacher type

$$p = \begin{cases} \min\{2,s\} & (1 \leq s < r) \\ \min\{2,r\} & (1 < r < s,\ r \neq 2) \\ 2-\varepsilon \quad \text{for any } \varepsilon > 0 & (2 = r < s) \end{cases} .$$

Concerning Orlicz spaces, it is well known that $\tau_{L_{M\Psi}}(t) = 1/\Psi^{-1}(1/t)$, and $L_{M\Psi}$ is of type p if $\Phi(t^{1/q})$ is convex; here $1/q + 1/p = 1$, and Φ denotes the conjugate Young function of Ψ.

REMARK. Since any Banach space is obviously of Rademacher type 1, condition (2.4) in Theorem 1 can be dropped if (2.5) is replaced by the stronger condition $1/\tau_Y(2/n)n > a$, $n \in \mathbb{N}$.

Essentially, the proof of Theorem 1 follows the pattern as elaborated in [6], but of course mutatis mutandis: In [6] the theorem is first proved for the Hilbert case $X = Y = L_2$, using the Parseval formula. Then the other L_p spaces are included by means of some technical tricks which do work only for $1 \leq p \leq 2$. In order to show that the theorem is false for $p > 2$, Stein gives a conterexample. A condition like (2.4) seems to be welcome, since it explains the restriction to $p \leq 2$. In fact, the Lebesgue space L_q is of Rademacher type $p = \min\{2,q\}$, and there does not exist any Banach space of Rademacher type p with $p > 2$. With this respect compare also the version of J.E.Gilbert [5] of Stein's theorem. The second essential point of Theorem 1 is that the weak-type inequality (1.1) is now replaced by (2.1).

The Proof of Theorem 1 is indirect: Assume that for each $c > 0$ there exists a function $f_c \in X$ with $f_c > 0$ and there is a number $\lambda > 0$ such that

$$D_\lambda(Tf_c) > \tau_Y^{-1}\left[\frac{c}{\lambda}\|f_c\|_{\Lambda(X)}\right] .$$

Let $A := \{u \in \Omega;\ (Tf_c)(u) > \lambda\}$, $A_k := \varphi_k^{-1}(A)$ for some $\varphi_k \in \{\varphi_\alpha\}_{\alpha \in I}$, $k = 1,2,\cdots,n$, and $f_k(u) := f_c(\varphi_k(u))$ for $u \in \Omega$, $k = 1,2,\cdots,n$. Note that for $u \in A_k$ one has $\varphi_k(u) \in A$ and hence

$$(2.7) \qquad (Tf_k)(u) = Tf_c(\varphi_k(u)) > \lambda \qquad (k = 1,2,\cdots,n)$$

by definition of the set A. Now consider the finite linear combination of Rademacher functions, namely

$$(2.8) \qquad F(u,t) := \frac{1}{M} \sum_{k=1}^{n} r_k(t) f_k(u) \qquad (u \in \Omega, t \in [0,1))$$

with a constant $M > 0$ which will be chosen later. Obviously, $F(\cdot,t) \in X$ for all $t \in [0,1)$. Moreover, since X is of Rademacher type p (see (2.2)) and $\| f_k \|_X = \| f_c \circ \varphi_k \|_X = \| f_c \|_X$, one has

$$\int_0^1 \| F(\cdot,t) \|_X \, dt = \frac{1}{M} \int_0^1 \| \sum_{k=1}^{n} r_k(t) f_k \|_X \, dt \leq$$

$$\leq \frac{K_p}{M} \left[\sum_{k=1}^{n} \| f_k \|_X^p \right]^{1/p} = \frac{K_p}{M} \left[\sum_{k=1}^{n} \| f_c \|_X^p \right]^{1/p}$$

$$= \frac{K_p \, n^{1/p}}{M} \| f_c \|_X \leq \frac{K_p \, n^{1/p}}{M} d \, \| f_c \|_{\Lambda(X)} \quad .$$

For the last estimate we used that $\Lambda(X) \subset X$ with a continuous embedding, i.e. $\| f_c \|_X \leq d \, \| f_c \|_{\Lambda(X)}$ with some constant $d > 0$. Chosing M and n in such a way that

$$(2.9) \qquad \frac{K_p \, n^{1/p}}{M} d \, \| f_c \|_{\Lambda(X)} = \frac{1}{4} \quad , \qquad 1 < n\mu(A) \leq 2 \quad ,$$

the above estimate yields ($m :=$ Lebesgue measure)

$$m\{t \in [0,1); \ \| F(\cdot,t) \|_X > 1\} \leq \frac{1}{4} \quad .$$

Therefore, there exists a set $B \subset [0,1)$ with $m(B) \geq 3/4$ and such that $\|F(\cdot,t)\|_X \leq 1$ for all $t \in B$.

On the other hand, since $(Tf_k)(u) > \lambda$, $u \in A_k$, by (2.7), there exists for each $u \in A^* := \bigcup_{k=1}^{n} A_k$ a set $L_u \subset [0,1)$ with $m(L_u) \geq 1/2$ and

$$TF(u,t) = \frac{1}{M} \sum_{k=1}^{n} r_k(t)(Tf_k)(u) > \frac{\lambda}{M} \qquad (t \in L_u)$$

on account of a basic property of Rademacher functions.

Collecting the results so far obtained, one has for all $u \in A^*$ and $I_u := B \cap L_u$ that

$$(2.10) \qquad I_u \subset [0,1) \quad , \quad m(I_u) \geq 1/4$$

$$\|F(\cdot,t)\|_X \leq 1 \quad , \quad TF(u,t) > \lambda/M \qquad (t \in I_u) .$$

The next step is, how to estimate the measure of A^*. Since $A^* = \bigcup_{k=1}^{n} A_k = \Omega \setminus \bigcap_{k=1}^{n} (\Omega \setminus A_k)$, it follows that $\mu(A^*) = 1 - \mu\left(\bigcap_{k=1}^{n} (\Omega \setminus A_k)\right)$. Now choose the mappings φ_k $(k = 1,2,\cdots,n)$ such that

$$\mu\left(\bigcap_{k=1}^{n} (\Omega \setminus A_k)\right) \leq \frac{e}{2} (\mu(\Omega \setminus A))^n \quad .$$

This is possible, because the family $\{\varphi_\alpha\}_{\alpha \in I}$ is supposed to be mixing. Therefore

$$\mu(A^*) \geq 1 - \frac{e}{2} (\mu(\Omega \setminus A))^n = 1 - \frac{e}{2} (1 - \mu(A))^n$$

$$\geq 1 - \frac{e}{2} e^{-n\mu(A)} > 1 - \frac{e}{2} e^{-1} = \frac{1}{2} \quad ,$$

that is to say, $\mu(A^*) > 1/2$. Passing to the product measure $\mu \times m$ shows that

$$(\mu \times m)\{(u,t) \in \Omega \times [0,1);\ u \in A^* \text{ and } t \in I_u\} > \frac{1}{2} \cdot \frac{1}{4} = \frac{1}{8}\ .$$

Hence, there exists a $t^* \in [0,1)$ such that $\mu\{u \in A\ ;\ t^* \in I_u\} > 1/8$, and for this t^* the following holds

$$\|F(\cdot, t^*)\|_X \leq 1\ , \tag{2.11}$$

$$\mu\{u \in \Omega;\ TF(u,t^*) > \frac{\lambda}{M}\} \geq \mu\{u \in A^*;\ t^* \in I_u\} > \frac{1}{8}\ .$$

Finally, one has to consider the relation between the different constants, namely (2.9) and $\mu(A) = D_\lambda(Tf_c) > \tau_Y^{-1}((c/\lambda)\|f_c\|_{\Lambda(X)})$. From these conditions it readily follows that

$$\frac{\lambda}{M} \geq \frac{c}{4K_p d} \quad \frac{1}{\tau_Y(2/n)n^{1/p}} > \text{const. } a > 0 \qquad (n \in \mathbb{N})$$

by assumption of the theorem. From this inequality one concludes that for all $c' > 0$

$$\mu\{u \in \Omega;\ TF(u,t^*) > c'a\} \geq \mu\{u \in \Omega;\ TF^*(u,t) > \frac{\lambda}{M}\} > \frac{1}{8}$$

which is bounded away from zero, in contradiction to Banach's principle.

If the operators T_k $(k \in \mathbb{N})$ are positive, the assumptions (2.4) and (2.5) can be dropped, as the following generalization of S. Sawyer's theorem [16] shows.

THEOREM 2. *With the notations as above let $T_k: X \to Y$, $k \in \mathbb{N}$, be positive linear operators between Banach function spaces X,Y and let $\varphi_\alpha: \Omega \to \Omega$, $\alpha \in I$, be a family of norm- and measure preserving transformations that commute with the T_k, $k \in \mathbb{N}$. If $|(T_k f)(u) - (T_l f)(u)| \to 0$ a.e. as $k,l \to \infty$, then the maximal* *operator* T *is of weak-type* (X,Y) *in the sense of* (2.1).

The proof of this theorem is readily obtained from Sawyer's proof by suitable generalizations, as shown above; therefore it is omitted.

3. APPLICATION TO FOURIER SERIES

A detailed study of the proof of Theorem 1 shows that it remains valid if X is not a Banach function space but only a complete p-normed space for some $0 < p < \infty$. Since Banach's principle holds also for this case (see above), the only part of the proof that has to be altered is the part where the Rademacher type p of X is involved. More precisely, if X is p-normed, the estimate

$$\int_0^1 \|F(\cdot,t)\|_X \, dt \leq \frac{K_p n^{1/p}}{M} d\|f_c\|_{\Lambda(X)}$$

has to be replaced in the following manner:

$$\int_0^1 \|F(\cdot,t)\|_X^p \, dt = \frac{1}{M} \int_0^1 \|\textstyle\sum_{k=1}^n r_k(t) f_k\|_X^p \, dt \leq$$

$$\leq \frac{1}{M} \int_0^1 \textstyle\sum_{k=1}^n \|r_k(t) f_k\|_X^p \, dt = \frac{1}{M} \int_0^1 \textstyle\sum_{k=1}^n \|f_k\|^p \, dt$$

$$= \frac{1}{M} n \|f_c\|_X^p \leq \frac{1}{M} n d^p \|f_c\|_{\Lambda(X)}^p .$$

Instead of (2.9) one now has to choose the constants M,n such that

$$\frac{d^p n}{M} \|f_c\|_{\Lambda(X)}^p = \frac{1}{4} , \qquad 1 < n\mu(A) \leq 2 . \tag{3.1}$$

Since $\|F(\cdot,t)\|_X^p < 1$ iff $\|F(\cdot,t)\|_X < 1$, one has as before

$$m\{t \in [0,1);\ \|F(\cdot,t)\|_X > 1\} \leq \frac{1}{4} .$$

On account of (3.1), the relation between the constants is now

$$\frac{\lambda}{M} \geq \frac{c\|f_c\|^{1-p}_{\Lambda(X)}}{4\,d^p} \cdot \frac{1}{\tau_Y(2/n)n} .$$

Therefore the following holds:

THEOREM 3. _Let_ (Ω,Σ,μ) _be a finite measure space with_ $\mu(\Omega) = 1$, _and_ X,Y _complete_ p-_normed function spaces of_ μ-_measurable functions on_ Ω _for some_ $0 < p < \infty$. _Assume that there exists a number_ $a > 0$ _such that for all_ $n \in \mathbb{N}$

$$\frac{1}{\tau_Y(2/n)n} > a \quad , \tag{3.2}$$

and let T_k, $k \in \mathbb{N}$, T, _and_ φ_α, $\alpha \in I$, _be as in Theorem_ 1. _If_ $|(T_kf)(u) - (T_lf)(u)| \to 0$ a.e. $(k,l \to \infty)$, _then the maximal operator_ T _is of weak-type_ (X,Y) _in the sense of_ (2.1).

We now whish to apply this theorem to Fourier series on Lorentz spaces L_{pq}. These Lorentz spaces are Banach spaces for $p,q \geq 1$, whereas the spaces L_{1q}, $0 < q < 1$, are complete q-normed spaces (see [7]). The Lorentz spaces are related to the Lebesgue spaces as follows (see e.g. [2]):

$$\left.\begin{array}{c} L_{22} \simeq L_2 \subset \underset{(1<p<2)}{L_{pp}} \simeq L_p \\[2em] \underset{(0<q<1)}{L_{1q}} \end{array}\right\} \subset L_{11} \simeq L_1 \subset \underset{(p,q>1)}{L_{pq}}$$

On L_p with $1 < p \leq 2$ the Fourier series converges a.e., by the Carleson - Hunt theorem. In L_1 there exists a function f such that the Fourier series of f diverges, by Kolmogorov's theorem. The question now is, whether the Fourier series converges a.e. on L_{1q} $(0 < q < 1)$ or not.

COROLLARY. *There exists a function* $f \in L_{1q}$, $0 < q < 1$, *such that the Fourier series of f diverges on a set of positive measure*.

For the *Proof* apply Theorem 3 to Ω := Torus, $d\mu := dm/2\Pi$, $X := Y := L_{1q}$, $T_k := S_k :=$ Fourier partial sum, $\varphi_\alpha(t) := t - \alpha$, and note that $\tau_{L_{1q}}(t) = t = \tau_{L_1}(t)$. Therefore, $1/\tau_Y(2/n)n = 1/(2/n)n = 1/2$ and $\Lambda(L_{1q}) = L_{11}$; moreover, the maximal operator

$$(Sf)(u) := \sup_{k \in \mathbb{N}} |(S_k f)(u)|$$

satisfies the weak-type inequality (2.1) if and only if S is of weak-type (1.1). But this cannot be true, since otherwise (again by Theorem 3) the Fourier series of any L_1-function would converge a.e. in contradiction to Kolmogorov's divergence result. Hence the corollary follows from Theorem 3.

ACKNOWLEDGEMENT. The author would like to thank Professor P.L. Butzer for his encouragement to write this paper and for many fruitful discussions.

REFERENCES

1. L. Carleson, On convergence and growth of partial sums of Fourier series. Acta Math. 116 (1966), 135-157.

2. M. Cotlar and R. Cignoli, An Introduction to Functional Analysis. North Holland, Amsterdam and London, 1974

3. J. Creekmoore, Type and cotype in Lorentz L_{pq} spaces. Indag. Math. 43 (1981), 145-152.

4. A. Garsia, Topics in Almost Everywhere Convergence. Markham Publishing Company, Chicago, 1970.

5. J.E. Gilbert, Nikišin - Stein theory and factorization with applications. Proc. Symposia in Pure Mathematics XXXV, Part 2, American Mathematical Society (1979), 233-267.

6. M. de Guzman, Real Variable Methods in Fourier Analysis. Mathematics Studies 46, North Holland, Amsterdam, New York and Oxford, 1981.

7. R.A. Hunt, On L(p,q) spaces. L'Enseignement mathém. 12 (1967), 249-276.

8. R.A. Hunt, On the convergence of Fourier series. Orthogonal Expansions and their Continuous Analogues. Proc. Conf. Edwardsville, III (1967), 235-255. Southern Illinois Univ. Press, Carbondale, III (1968).

9. J.P. Kahane, Series of Random Functions. Heath Math. Monographs. Heath and Co., Lexington, Mass. 1968.

10. A. Kolmogorov, Sur les fonctions harmoniques conjugées et les séries de Fourier. Fundamentals Math. 7 (1925), 23-28.

11. J. Lindenstrauss and L. Tzafriri, Classical Banach Spaces II, Function Spaces. Springer Verlag, Berlin and New York, 1979.

12. W.A.J. Luxemburg, Banach function spaces. Thesis. Delft 1955.

13. B. Maurey, Espaces L^p, applications radonifiantes et géometrie des espaces de Banach. Séminaire Maurey - Schwartz, Exposé V, 1973/74.

14. E.M. Nikišin, Resonance theorems and superlinear operators Russ. Math. Surveys 25 (1970), 125-187.

15. G. Pisier, "Type" des espaces normés. Séminaire Maurey - Schwartz, Exposé III, 1973/74.

16. S. Sawyer, Maximal inequalities of weak type. Ann. of Math. (2) 84 (1966), 157-174.

17. E.M. Stein, On limits of sequences of operators. Ann. of Math. (2) 74 (1961), 140-170.

Franziska Fehér, Lehrstuhl A für Mathematik, RWTH Aachen, D-5100 Aachen, West Germany

International Series of
Numerical Mathematics, Vol. 71
© 1984 Birkhäuser Verlag Basel

SOME INEQUALITIES FOR ℓ_p NORMS OF MATRICES

Moshe Goldberg[1)]

Abstract. The purpose of this note is to prove several inequalities which describe certain submultiplicativity properties of ℓ_p norms for matrices.

Let $1 \leq p \leq \infty$, and consider the ℓ_p norm of an $m \times n$ complex matrix $A = (\alpha_{ij}) \in \underset{\sim}{C}_{m\times n}$:

$$|A|_p = \left(\sum_{i=1}^{m} \sum_{j=1}^{n} |\alpha_{ij}|^p\right)^{1/p}.$$

Investigating this norm, Ostrowski [3] proved the following result:

THEOREM 1 ([3], Theorem 7). *If A and B are rectangular matrices such that AB exists, and if* $1 \leq p \leq 2$, *then*

(1) $$|AB|_p \leq |A|_p |B|_p.$$

While Ostrowski showed that this theorem may fail for $p > 2$, the following extension was given in [1]:

THEOREM 2 ([1], Theorem 1.3). *If* $A \in \underset{\sim}{C}_{m\times k}$, $B \in \underset{\sim}{C}_{k\times n}$, *and* $p \geq 2$, *then*

(2) $$|AB|_p \leq k^{1-2/p} |A|_p |B|_p.$$

In this note we generalize the above theorems as follows:

1) Research sponsored in part by the Air Force Office of Scientific Research, Air Force System Command, Grants AFOSR-83-0150.

THEOREM 3. *Let* $A \in \underset{\sim}{C}_{m\times k}$ *and* $B \in \underset{\sim}{C}_{k\times n}$. *Then, for* $1 \leq q \leq p \leq 2$,

(3) $$|AB|_p \leq |A|_p|B|_q,$$

(4) $$|AB|_p \leq |A|_q|B|_p;$$

for $1 \leq p \leq q \leq 2$,

(5) $$|AB|_p \leq n^{1/p-1/q}|A|_p|B|_q,$$

(6) $$|AB|_p \leq m^{1/p-1/q}|A|_q|B|_p;$$

for $p \geq q \geq 2$,

(7) $$|AB|_p \leq k^{1-1/p-1/q}|A|_p|B|_q,$$

(8) $$|AB|_p \leq k^{1-1/p-1/q}|A|_q|B|_p;$$

and for $q \geq p \geq 2$,

(9) $$|AB|_p \leq k^{1-1/p-1/q}\ n^{1/p-1/q}|A|_p|B|_q,$$

(10) $$|AB|_p \leq k^{1-1/p-1/q}\ m^{1/p-1/q}|A|_q|B|_p.$$

Obviously, for $p = q$ each of the inequalities in (3)-(6) reduces to (1), and each of (7)-(10) gives (2). Thus, Theorem 3 generalizes Theorems 1 and 2.

The inequalities in (3)-(10) become equalities if A and B are chosen to be certain incidence (i.e., (0,1)) matrices. Hence, none of these inequalities can be improved.

In order to prove Theorem 3 we start by quoting:

LEMMA 1 (see [2], Corollary I.4.5). *If* $x \in \underset{\sim}{C}^n$ *is an n-vector and* $p \geq q \geq 1$, *then*

$$|x|_p \leq |x|_q \leq n^{1/q-1/p}|x|_p.$$

Now, for convenience only we define

$$\mu_{pq}(n) = \begin{cases} 1, & p \geq q \geq 1, \\ n^{1/p-1/q}, & q \geq p \geq 1. \end{cases}$$

With this we can evidently restate Lemma 1 as follows:

LEMMA 2. *If* $x \in \mathbb{C}^n$ *and* $p,q \geq 1$, *then*

(11) $$|x|_p \leq \mu_{pq}(n)|x|_q.$$

Next, for $p,q \geq 1$ we consider the mixed $\ell_{p,q}$ norm of a matrix $A = (\alpha_{ij}) \in \mathbb{C}_{m\times n}$, [3],

$$|A|_{pq} = \left(\sum_{j=1}^{n}\left(\sum_{i=1}^{m}|\alpha_{ij}|^p\right)^{q/p}\right)^{1/q},$$

and prove:

LEMMA 3. *For* $p,q \geq 1$ *and* $A \in \mathbb{C}_{m\times n}$:

(12) $$|A|_{pq} \leq \mu_{pq}(m)|A|_q,$$

(13) $$|A|_{pq} \leq \mu_{qp}(n)|A|_p.$$

Proof. Denote the columns of A by $a_1,\ldots,a_n$. By (11),

$$\begin{aligned} |A|_{pq} &= |(|a_1|_p,\ldots,|a_n|_p)|_q \\ &\leq |(\mu_{pq}(m)|a_1|_q,\ldots,\mu_{pq}(m)|a_n|_q)|_q \\ &= \mu_{pq}(m)|(|a_1|_q,\ldots,|a_n|_q)|_q = \mu_{pq}(m)|A|_q, \end{aligned}$$

so (12) is established. For (13) we use (11) again, to obtain

$$\begin{aligned} |A|_{pq} &= |(|a_1|_p,\ldots,|a_n|_p)|_q \\ &\leq \mu_{qp}(n)|(|a_1|_p,\ldots,|a_n|_p)|_p = \mu_{qp}(n)|A|_p, \end{aligned}$$

and the proof is complete. □

We shall also need:

LEMMA 4. *Let* $A \in \underset{\sim}{C}_{m\times k}$, $B \in \underset{\sim}{C}_{k\times n}$. *Let* $p,q \geq 1$, *and let* q' *be the conjugate of* q, *so that*

$$\frac{1}{q} + \frac{1}{q'} = 1.$$

Then,

(14) $$|AB|_p \leq |A^T|_{q'p}|B|_{qp} ,$$

(15) $$|AB|_p \leq |A^T|_{qp}|B|_{q'p} ,$$

where T *denotes the transpose.*

Proof. Put $C = AB$, $C = (\gamma_{ij})$. Then, by Hölder's Inequality,

$$|\gamma_{ij}| = |\sum_{\ell=1}^{k}\alpha_{i\ell}\beta_{\ell j}| \leq \left(\sum_{\ell=1}^{k}|\alpha_{i\ell}|^{q'}\right)^{1/q'}\left(\sum_{\ell=1}^{k}|\beta_{\ell j}|^{q}\right)^{1/q}.$$

So

$$|AB|_p^p = \sum_{i=1}^{m}\sum_{j=1}^{n}|\gamma_{ij}|^p = \sum_{i=1}^{m}\sum_{j=1}^{n}\left[\left(\sum_{\ell=1}^{k}|\alpha_{i\ell}|^{q'}\right)^{1/q'}\left(\sum_{\ell=1}^{k}|\beta_{\ell j}|^{q}\right)^{1/q}\right]^p$$

$$= \left[\sum_{i=1}^{m}\left(\sum_{\ell=1}^{k}|\alpha_{i\ell}|^{q'}\right)^{p/q'}\right]\left[\sum_{j=1}^{n}\left(\sum_{\ell=1}^{k}|\beta_{\ell j}|^{q}\right)^{p/q}\right] = |A^T|_{q'p}|B|_{qp} ,$$

and (14) holds. For (15) we repeat the proof with q and q' interchanged. □

We are finally ready for

Proof of Theorem 3. Use (14), (12) and (13) (in this order) to obtain

$$|AB|_p \leq |A^T|_{q'p}|B|_{qp} \leq \mu_{q'p}(k)|A^T|_p\cdot\mu_{pq}(n)|B|_q$$

(16)

$$= \mu_{q'p}(k)\mu_{pq}(n)|A|_p|B|_q.$$

Hence, if $2 \geq p \geq q$, then $q' \geq p$ and (3) follows; if $2 \geq q \geq p$ then $q' \geq p$ and (5) follows; if $p \geq q \geq 2$, then $p \geq q'$, so

$$\mu_{q'p}(k) = k^{1/q'-1/p} = k^{1-1/q-1/p},$$

and (7) follows; and if $q \geq p \geq 2$, then $p \geq q'$ and (16) yields (9).

Using (15), (13), and (12), we get

$$|AB|_p \leq |A^T|_{qp}|B|_{q'p} \leq \mu_{pq}(m)|A^T|_q \cdot \mu_{q'p}(k)|B|_p$$

$$= \mu_{pq}(m)\mu_{q'p}(k)|A|_q|B|_p.$$

Thus, considering as before the cases $2 \geq p \geq q$, $2 \geq q \geq p$, $p \geq q \geq 2$ and $q \geq p \geq 2$, we obtain, respectively, (4), (6), (8), and (10). □

Inequalities analogous to (3)-(10), for the cases $p \geq 2 \geq q$ and $q \geq 2 \geq p$, will appear in a forthcoming paper.

REFERENCES

1. M. Goldberg and E.G. Straus, Multiplicity of ℓ_p norms for matrices. Linear algebra Appl. 52/53 (1983), 351-360.

2. A.M. Krull, Linear Methods of Applied Analysis. Addison-Wesley Pub. Co., Reading, Massachusetts, 1973.

3. A. Ostrowski, Über Normen von Matrizen. Math. Z. 63 (1955), 2-18.

Moshe Goldberg, Mathematics Department, Technion--Israel Institute of Technology, Haifa 32000, Israel, *and* Institute for the Interdisciplinary Applications of Algebra and Combinatorics, University of California, Santa Barbara, California 93106, U.S.A.

International Series of
Numerical Mathematics, Vol. 71

ON INEQUALITIES RELATED TO SEQUENCE SPACES $ces[p,q]$

P.D. Johnson Jr. and R.N. Mohapatra

Abstract. Inclusion between two sequence spaces gives rise to inequalities. For instance $\ell_p \subset ces_p$ $(1 < p \leq \infty)$ is essentially Hardy's inequality for sequences apart from the constant. This paper contains inclusion relations between diagonal copies of ℓ_p and $ces[p,q]$ which interalia yield inequalities related to $ces[p,q]$.

1. PRELIMINARIES

Let w be the vector space of all real or complex sequences and let Φ be the subspace of all eventually zero sequences, i.e. sequences with finitely many non-zero entires. We regard a sequence space to be a subspace of w containing Φ. If $S \subset w$, we write $aS = \{ab : b \in S\}$.

If $a \in w$, the normal hull of a is given by

$$N(a) = \{b \in w : |b_n| \leq |a_n| , n = 0,1,\ldots\} .$$

If $S \subset w$, the normal hull of S is given by

$$N(S) = \bigcup_{a \in S} N(a) .$$

A subset S of w is said to be normal if $S = N(S)$.

Let $A = (a_{nk})$ $(n,k \geq 0)$ be an infinite matrix of real numbers. The domain of A, denoted by $D(A)$, is given by

$$D(A) = \{x \in w : \sum_{k=0}^{\infty} a_{nk} x_k \text{ converges for } n = 0,1,\ldots\}$$

For $x \in D(A)$, $Ax = ((Ax)_n)$ is given by

$$(Ax)_n = \Sigma_{k=0}^{\infty} a_{nk} x_k .$$

For a sequence space λ , we write

(1.1) $$A^{-1}\lambda = \{x \in D(A) : Ax \in \lambda\} ,$$

(1.2) $$\text{nor} - A^{-1}\lambda = \{x \in w : A|x| \in \lambda\} .$$

For $\lambda,\mu \subset w$, let $D(\mu,\lambda) = \{x \in w : x\mu \subset \lambda\}$.

If (p_n) is a sequence of positive real numbers, we write

(1.3) $\ell(p_n) = \{x \in w : \Sigma_{n=0}^{\infty} |x_n|^{p_n} < \infty\}$;

(1.4) $\sigma(x) = (\frac{1}{n+1} \Sigma_{k=0}^{n} x_k)_n \qquad (x \in w)$;

(1.5) $\alpha - ces(p_n) = \{x \in w : \sigma(x) \in \ell(p_n)\}$

(1.6) $ces(p_n) = \{x \in w : \sigma(|x|) \in \ell(p_n)\} \qquad (|x| = (|x|_n))$.

For sequences of positive real numbers (p_n), (q_n), we write

$$Q_n = q_0 + q_1 + \ldots + q_n ,$$

(1.7) $\alpha - ces[(p_n),(q_n)] = \{x \in w : \{Q_n^{-1} \Sigma_{k=0}^{n} q_k x_k\}_n \in \ell(p_n)\}$;

(1.8) $ces[(p_n),(q_n)] = \{x \in w : \{Q_n^{-1} \Sigma_{k=0}^{n} q_k |x_k|\}_n \in \ell(p_n)\}$.

Note that if $q_n = 1$ for all n , then (1.7) and (1.8) reduce to (1.5) and (1.6) respectively. Further, if in (1.6), $p_n = p$ for all n , then $ces(p_n)$ reduces to ces_p which was studied by Shiue [13] and Leibowitz [8]. If $p_n = p$ for all n , (1.7) and (1.8) reduce to $\alpha - ces[p,q]$ and $ces[p,q]$ respectively.

Let $C = (C_{nk})$ and $C_q = (C^q_{nk})$ be respectively the matrices associated with $(C,1)$ and $(\overline{N},q_n)$ methods of summation given by

$$C_{nk} = \begin{cases} (n+1)^{-1} & (0 \leq k \leq n), \\ 0 & (k > n); \end{cases} \quad \text{and} \quad C^q_{nk} = \begin{cases} q_k / Q_n & (0 \leq k \leq n), \\ 0 & (k > n). \end{cases}$$

Then we find that (1.5), (1.6), (1.7) and (1.8) are respectively $C^{-1}(\ell(p_n))$, $\text{nor} - C^{-1}(\ell(p_n))$, $C_q^{-1}(\ell(p_n))$ and $\text{nor} - C_q^{-1}(\ell(p_n))$. Thus, these vector spaces of sequences are inverse images of normal sequence spaces $\ell(p_n)$ under matrix transformations. $\ell(p_n)$ was first studied by Simons [14] and some related work was done by Maddox [11].

Topological vector space properties of $\text{nor} - A^{-1}(\lambda)$ have been studied by Johnson and Mohapatra [4,6,7]. In [4], it is shown that under suitable conditions $\text{nor} - A^{-1}(\lambda)$ shares many of the properties of λ. Leibowitz [9] has considered Hausdorff sequence spaces along the same line. In [5] we have considered conditions for the existence of best possible HPD inequalities.

The following facts are known:

PROPOSITION [6]. The following are equivalent:

(a) $\Phi \subset \text{ces}[(p_n),(q_n)]$;

(b) $\Phi \subset \alpha - \text{ces}[(p_n),(q_n)]$

(c) $\text{ces}[(p_n),(q_n)] \neq 0$

(d) $\frac{1}{Q} \in \ell(p_n)$.

It can be shown that $\text{ces}[(p_n),(q_n)]$ is a complete topological vector space of sequences.

2. INEQUALITIES AND INCLUSION AMONG SEQUENCE SPACES

Hardy and Landau (see [3; p.239]) proved

THEOREM A. If $p > 1$ and $x_n \geq 0$ $(n = 0,1,\dots)$, then

$$\Sigma((n+1)^{-1} \Sigma_{k=0}^{n} x_k)^p \leq \left(\frac{p}{p-1}\right)^p \Sigma\, x_n^p \tag{2.1}$$

unless the x_n's are zero, or $x \notin \ell p$.

Petersen [12], Davies and Petersen [2] produced sufficient conditions on a matrix A and an auxiliary sequence $f \in w$ for the existence of an inequality of the form

$$\| A|x| \|_p \leq K \| f \cdot a \cdot x \|_p \qquad (1 \leq p < \infty) \tag{2.2}$$

where K is a positive constant depending on p and $a = (a_{mm})$ is the main diagonal sequence of A . Love [10] has considered inequalities of the form

$$\| Ax \|_p \leq C \| x \|_q \qquad (p \neq q) \tag{2.3}$$

$\| \cdot \|_p$ being ℓ_p norm. In [5], we have considered inequalities of the form

$$\| A|x|_p \|_p \leq K \| bx \|_q \quad \text{for} \quad 0 < p , q < \infty , \tag{2.4}$$

or indeed of the form

$$\| A|x| \|_\lambda \leq K \| bx \|_\mu \tag{2.5}$$

where λ , μ are normed or quasinormed Fk spaces satisfying some conditions.

Recalling ℓ_p and ces_p , we find that Hardy's inequality amounts to the inclusion $\ell_p \subset ces_p$. We also observe that the inclusion $\ell_p \subset ces_p$ gives rise to an inequality equivalent to that of Hardy apart from a constant. Thus there is a close connection between inclusions among sequence spaces and inequalities.

In this paper we shall consider inclusions between diagonal copies of ℓ_p and $ces[p,q]$.

In [4] we studied the spaces $ces[p,q]$ as instances of the spaces nor - $A^{-1}(\lambda)$. In that study we supposed that the inclusion $q^{-1} \ell_p \subset ces[p,q]$ or equivalently $q^{-1} \in D(\ell_p , ces[p,q])$ is a natural generalization of Hardy's inequality. We wondered under what circumstances this inclusion would hold. In [5], we have proved

THEOREM B ([5]). If $p \in (1,\infty)$ and

(2.6) $$\{Q_n^{p-1} \Sigma_{m=n}^{\infty} Q_m^{-p}\} \in \ell_\infty$$

then

(2.7) $$q^{-1} \in D(\ell_p, ces[p,q]) .$$

Let $J = (j_{nk})$, $j_{nk} = 1$ $(n \leq k)$, $j_{nk} = 0$ $(n > k)$. Let J^t be the transpose of the matrix J and J^{-1} be the inverse of J .

Before proceeding to special results on the question stated above, we remark the following:

REMARKS 1. When $p = 1$, we have from (2.7) and Corollary 4.9 of [5],

$$D(\ell_1, ces[p,q]) = q^{-1}(J^t(Q^{-1}))^{-1}\ell_\infty$$

and consequently

$$q^{-1}\ell_1 \subset ces[1,q]$$

if and only if $Q^{-1} \in \ell_1$.

2. For any p , $0 < p \leq \infty$, $q^{-1} \in D(\ell_p, ces[p,q])$ if and only if $C_q D_{q^{-1}} = D_{Q^{-1}} J$ maps ℓ_p into ℓ_p where D_m is the diagonal matrix whose diagonal elements are m . For $p \neq 1$ or ∞ we do not know if there are easy numerical conditions on Q for $D_{Q^{-1}} J$ to map ℓ_p into ℓ_p .

3. If $p = \infty$ then the above remark implies that $q^{-1} \in D(\ell_\infty, ces[\infty,q])$ iff $\{nQ_n^{-1}\} \in \ell_\infty$, for $\{nQ_n^{-1}\}$ is the sequence of row sums of $D_{Q^{-1}} J$.

4. It can be shown that $ces[\infty,q] \subset q^{-1}\ell_\infty$ if and only if $q \in \ell_1$.

3. COPSON'S INEQUALITY AND SEQUENCE SPACE $ces[p,q]$

Generalizing Theorem A, Copson [1] proved.

THEOREM C ([1]). Suppose $1 < p < \infty$, q is a positive sequence and $Q_n = q_0 + q_1 + \ldots + q_n$. For $x_n \geq 0$,

$$(3.1) \qquad \left(\Sigma\, q_n\, Q_n^{-p} (\Sigma_{k=0}^{n}\, q_k\, x_k)^p\right)^{1/p} \leq \left(\frac{p}{p-1}\right) \left(\Sigma\, q_n\, x_n^p\right)^{1/p} .$$

From this we get

COROLLARY 1. For $1 < p < \infty$, $q^{-1/p}\, \ell_p \subset \text{nor} - C_q^{-1}(q^{-1/p}\, \ell_p)$.

We first prove

THEOREM 1. <u>If</u> (q_n) <u>is bounded away from zero and</u> $1 < p < \infty$, <u>then</u>

$$(3.2) \qquad q^{-1}\, \ell_p \subset q^{-1/p}\, \ell_p \subset \text{nor} - C_q^{-1}\, (q^{-1/p}\, \ell_p) \subset \text{ces}[p,q] .$$

<u>Proof.</u> By hypothesis on (q_n) , $q^{-1/p}$ is bounded and hence $q^{-1/p}\, \ell_p \subset \ell_p$. Thus

$$\text{nor} - C_q^{-1}\, (q^{-1/p}\, \ell_p) \subset \text{nor} - C_q^{-1}\, (\ell_p) = \text{ces}[p,q] .$$

Also note that $q^{-1/p'} = q^{-1}\, q^{1/p}\, (\frac{1}{p} + \frac{1}{p'} = 1)$ and hence $q^{-1}\, \ell_p \subseteq q^{-1/p}\, \ell_p$.

Finally from Theorem C, we obtain the inclusion

$$q^{-1/p}\, \ell_p \subset \text{nor} - C_q^{-1}\, (q^{-1/p}\, \ell_p) .$$

On collecting all these we have Theorem 1.

REMARK. Under the hypothesis of Theorem 1, the inclusion

$$(3.3) \qquad q^{-1/p}\, \ell_p \supseteq q^{-1}\, \ell_p$$

is an equality if and only if q is also bounded; because in that case both spaces turn out to be ℓ_p .

In what follows we are going to investigate the strictness of the inclusion

$$(3.4) \qquad \text{nor} - C_q^{-1}(q^{-1/p}\, \ell_p) \subset \text{ces}[p,q]$$

when $p \in (1,\infty)$ and q is bounded away from zero.

We observe that if q is bounded above and is also bounded below by a positive number then surely

$$\text{nor} - C_q^{-1}(q^{-1/p}\,\ell_p) = \text{ces}[p,q]\ . \tag{3.5}$$

Hence, it is natural to ask if for the equality (3.5) boundedness of q is necessary. In this connection we prove

THEOREM 2. <u>There exists (q_n) such that q_n is bounded away from zero and (q_n) is unbounded for which the equality (3.5) holds.</u>

<u>Proof</u>. Set

$$q_n = \begin{cases} k & (\text{if } n = 2^k\ ,\ k = 0,1,\ldots)\ ; \\ 1 & (\text{if } n \neq 2^k)\ . \end{cases} \tag{3.6}$$

Clearly q is bounded away from zero and is not bounded. In view of Theorem 1, it is sufficient to show that

$$\text{nor} - C_q^{-1}(q^{-1/p}\,\ell_p) \supseteq \text{ces}[p,q]\ . \tag{3.7}$$

Thus one has to show that

$$Q^{-1}\,J(q|x|) \in \ell_p \quad \text{implies} \quad q^{1/p}\,Q^{-1}\,J(q|x|) \in \ell_p\ . \tag{3.8}$$

Observing that $Q_n = nC_n$ where (C_n) is a bounded sequence and replacing $\{J(q|x|)^p\}$ by an arbitrary non-decreasing, non-negative sequence $\alpha \equiv (\alpha_n)$, it will be enough to prove that

$$\Sigma\, n^{-p}\,\alpha_n < \infty \quad \text{implies} \quad \Sigma\, q_n\, n^{-p}\,\alpha_n < \infty \tag{3.9}$$

for a sequence (α_n) which is non-negative and non-decreasing.

Since,

$$\Sigma_{n\geq 2}\, q_n\, n^{-p}\,\alpha_n = \Sigma_{k=1}^{\infty}\left[k2^{-kp}\,\alpha_{2^k} + \Sigma_{2^k<n<2^{k+1}}\, n^{-p}\,\alpha_n\right]\ ,$$

and since,

$$\Sigma_{k=1}^{\infty} \Sigma_{2^k<n<2^{k+1}} n^{-p} \alpha_n \leq \Sigma_{n=1}^{\infty} n^{-p} \alpha_n < \infty ,$$

we have to prove

$$(3.10) \qquad \Sigma_{k=1}^{\infty} k2^{-kp}\alpha_{2^k} < \infty$$

to establish (3.9).

We have

$$\Sigma_{k=1}^{\infty} k2^{-kp} \alpha_{2^k} < \Sigma_{k=1}^{\infty} 2^{-kp} \Sigma_{2^k<n<2^{k+1}} \alpha_n$$

$$= 2^p \Sigma_{k=1}^{\infty} \Sigma_{2^k<n<2^{k+1}} 2^{-(k+1)p} \alpha_n$$

$$\leq 2^p \Sigma_{k=1}^{\infty} \Sigma_{2^k<n<2^{k+1}} n^{-p} \alpha_n$$

$$\leq 2^p \Sigma_{n=1}^{\infty} n^{-p} \alpha_n < \infty$$

in view of the hypothesis in (3.9).

Our next result is concerned with the strictness of the inclusion (3.4). We prove

THEOREM 3. *If* $1 < p < \infty$ *and* (Q_n) *diverges to* ∞ *then the inclusion* (3.4) *is strict*.

The proof depends upon the following

LEMMA. *Let* $y \in \ell_1 \setminus \Phi$, *and* x *be a positive sequence diverging properly to* ∞ (i.e. $x^{-1} \in c_0$) . *Then there exists a positive non-decreasing sequence* α *such* that $\alpha y \in \ell_1$ *but* $\alpha y x \notin \ell_1$.

Proof. We may assume y to be non-negative. Inductively determine a sequence of positive integers (n_k) such that the following hold:

(i) $n_1 < n_2 < \ldots$;

(ii) $n \geqslant n_k$ implies $x_n \geqslant k^2$;

(iii) if $\beta_k = k^{-2} (\Sigma_{n \geq n_k} y_n)^{-1}$, then

$$\Sigma_{n \geq n_{k+1}} y_n \leq (k+1)^{-2} \beta_k^{-1} .$$

Note that

$$\beta_1 = (\Sigma_{n \geq n_1} y_n)^{-1} \geq 1$$

and for $k \geq 1$

$$\beta_{k+1} = (k+1)^{-2} (\Sigma_{n \geq n_{k+1}} y_n)^{-1}$$

$$\geq (k+1)^{-2} (k+1)^2 \beta_k = \beta_k .$$

Define $\alpha_n = 1(n < n_1)$, and $\alpha_n = \beta_k(n_k \leq n < n_{k+1})$. Then α is non-decreasing. Also, we have

$$\Sigma_{n \geq n_1} \alpha_n y_n = \Sigma_k \beta_k \Sigma_{n_k \leq n < n_{k+1}} y_n$$

$$\leq \Sigma_{k=1}^{\infty} k^{-2} < \infty .$$

Hence $\alpha y \in \ell_1$. However,

$$\Sigma_{n > n_1} \alpha_n x_n y_n \geq \sum_1^{\infty} k^2 \beta_k \Sigma_{n_k \leq n < n_{k+1}} y_n \geq \Sigma_{k=1}^{\infty} k^2 \beta_k (\Sigma_{n \geq n_k} y_n - (k+1)^{-2} \beta_k^{-1})$$

$$= \Sigma_{k=1}^{\infty} (2k+1)(k+1)^{-2} = \infty .$$

Thus $\alpha xy \notin \ell_1$.

<u>Proof of Theorem 3</u>. By Theorem 1,

$$\text{nor} - C_q^{-1}(q^{-1/p} \ell_p) \subset \text{ces}[p,q] .$$

Since $\text{ces}[p,q]$ is a non-trivial space of sequences, $Q^{-1} \in \ell_p$ or $Q^{-p} \in \ell_1$. By the lemma there exists a non-negative, non-decreasing sequence α such that $\alpha Q^{-p} \in \ell_1$ but $q \alpha Q^{-p} \notin \ell_1$. Since $\alpha^{1/p}$ is non-decreasing, $J^{-1}(\alpha^{1/p})$ is non-negative. Hence we can show that

$$q^{-1} J^{-1}(\alpha^{1/p}) \in \text{ces}[p,q] \setminus \text{nor} - C_q^{-1}(q^{-1/p} \ell_p) .$$

This completes the proof.

COROLLARY. If $1 < p < \infty$ and q is bounded away from zero, then $q^{-1} \ell_p \subset ces[p,q]$.

4. REMARK

One might wonder if for $1 < p < \infty$ it is always the case that

(4.1) $$q^{-1} \ell_p \subset ces[p,q] .$$

Our example will show that (4.1) is not always true.

EXAMPLE. Let $1 < p < \infty$ and $Q_n = n^{1/p} \log(n + 1)$. Hence $q = J^{-1}(Q)$ is positive. We claim that $q^{-1} \ell_p \not\subset ces[p,q]$. It suffices to show that $D_{Q^{-1}} J$ does not map ℓ_p into ℓ_p . Take $x \equiv x_n$ to be such that $x_n = n^{-1}$, $x \in \ell_p$. But

$$D_{Q^{-1}} J(n^{-1}) = (n^{-1/p}(\log(n + 1))^{-1} \Sigma_{k=1}^{n} k^{-1})_n = d_n n^{-1/p}$$

where (d_n) is a bounded sequence. Since $(n^{-1/p})$ does not belong to ℓ_p our assertion is verified.

This raises the following problem:

PROBLEM. Characterize the positive sequences q such that $ces[p,q]$ is non-trivial for all $p > 1$ but

$$q^{-1} \ell_p \not\subset ces[p,q] .$$

REFERENCES

1. E.T. Copson, Note on series of positive terms, J. London Math. Soc. 2 (1927), 9-12.

2. G.S. Davies and G.M. Petersen, On an inequality of Hardy's (II), Quart. J. Math. (Oxford) (2) 15 (1964), 35-40.

3. G.H. Hardy, J.E. Littlewood and G. Polya, Inequalities (Cambridge), 1934.

4. P.D. Johnson Jr. and R.N. Mohapatra, The maximal normal subspace of the inverse image of a normal space of sequences by a non-negative matrix

transformations, to appear.

5. P.D. Johnson Jr. and R.N. Mohapatra, Inequalities involving lower-triangular matrices, Proc. London Math. Soc. (3) 41 (1980), 83-137.

6. P.D. Johnson Jr. and R.N. Mohapatra, Density of finitely non-zero sequences in some sequence spaces, Math. Japonica 24 (1979), 253-262.

7. P.D. Johnson Jr. and R.N. Mohapatra, Sectional convergence in spaces obtained as inverse images of sequence spaces under matrix transformations, Math. Japonica 24 (1979), 179-185.

8. G.M. Leibowitz, A note on cesáro sequence spaces, Tamkang J. Math. 2 (1971), 151-157.

9. G.M. Leibowitz, Some Hausdorff sequence spaces, Math. Japonica 26 (1981), 91-103.

10. E.R. Love, Inequalities between norms in sequence spaces, Inequalities III (Birkhauser Verlag), 1983.

11. I.J. Maddox, Spaces of strongly summable sequences, Quarterly J. Math. (Oxford) 18 (1967), 345-355.

12. G.M. Petersen, An inequality of Hardy's, Quart. J. Math. (Oxford) (2) 13 (1962), 237-240.

13. J.S. Shiue, On the cesáro sequence spaces, Tamkang J. Math. 1 (1970), 19-25.

14. S. Simons, The sequence spaces $\ell(p_n)$ and $m(p_n)$, Proc. London Math. Soc. (3) 15 (1965), 422-436.

P.D. Johnson Jr., Department of Mathematics, Auburn University, Auburn, Alabama, U.S.A.

and

Ram N. Mohapatra, Department of Mathematics, York University, Downsview, Ontario M3J 1P3, Canada

International Series of
Numerical Mathematics, Vol. 71

TWO THEOREMS IN SUPERCONVEX ANALYSIS

Heinz König

Abstract. The present paper deals with the superconvex theory developed in the Thesis 1977 and in two subsequent papers 1980/81 of Rodé. A central result of it is a Dini type theorem: If on the superconvex space X an increasing sequence of superconvex functions $f_n: X \to [-\infty, \infty[$ with limit function $f: X \to [-\infty, \infty[$ is such that at each $x \in X$ one has $f_n(x) = f(x)$ for some $n \in \mathbb{N}$, then $\text{Inf}\, f_n \to \text{Inf}\, f$. The condition that the limit function be attained at each point seems to be severe but is fulfilled quite often. However, it will be shown below that it is not essential: It suffices that there exists a sequence of positive numbers $c_n \downarrow 0$ such that $f(x) - f_n(x) = O(c_n)$ for $n \to \infty$ at each $x \in X$. Rodé proved his result via a certain intersection theorem which resembles the classical Baire theorem. We retain this procedure and first extend the intersection theorem, to its ultimate limit as a counterexample reveals.

1. SUPERCONVEX SPACES

Let Q consist of the sequences $t = (t_l)_l$ of real numbers $t_l \geqq 0$ $\forall l \in \mathbb{N}$ with $\sum_{l=1}^{\infty} t_l = 1$, and P of the $t = (t_l)_l \in Q$ with $t_l = 0$ for almost all $l \in \mathbb{N}$. For $r \in \mathbb{N}$ let P^r denote the set of the $t = (t_1, \ldots, t_r) \in \mathbb{R}^r$ with $t_1, \ldots, t_r \geqq 0$ and $\sum_{l=1}^{r} t_l = 1$. And for a set X let X^∞ consist of the sequences $x = (x_l)_l$ of elements $x_l \in X$ $\forall l \in \mathbb{N}$.

Our definition of a superconvex space is different from but equivalent to the one in Rodé [2]. Let X be a set $\neq \emptyset$. A superconvex structure on X is intended to imitate and formalize the formation of countable convex combinations

$$\sum_{l=1}^{\infty} t_l x_l \text{ with } t=(t_l)_l \in Q \text{ and } x=(x_l)_l \in X^{\infty}.$$

The precise definition is as follows: A superconvex structure on X is defined to be a map

$$I: Q \times X^{\infty} \to X, \text{ written } I(t,x) = \mathop{I}_{l=1}^{\infty} (t_l, x_l) = I\begin{pmatrix} t_1 \ldots t_l \ldots \\ x_1 \ldots x_l \ldots \end{pmatrix},$$

with the properties

1) $\mathop{I}_{l=1}^{\infty} (\delta_l^p, x_l) = x_p \ \forall p \in \mathbb{N}$, where as usual $\delta_l^p = \begin{cases} 1 \text{ for } l=p \\ 0 \text{ for } l \neq p \end{cases}$;

2) $\mathop{I}_{p=1}^{\infty} \left(t_p, \mathop{I}_{l=1}^{\infty} (t_l^p, x_l) \right) = \mathop{I}_{l=1}^{\infty} \left(\sum_{p=1}^{\infty} t_p t_l^p, x_l \right) \ \forall t \in Q$ and $t^p \in Q \ \forall p \in \mathbb{N}$;

observe that $\left(\sum_{p=1}^{\infty} t_p t_l^p \right)_l \in Q$ as well. In the sequel we fix a superconvex structure I on X, in short a superconvex space. The axioms 1) and 2) permit to a wide extent to handle the operation I like conventional countable convex combinations. We illustrate this with one example, and then shall make rather free use of computational rules.

1.1 PROPOSITION. Assume that $t \in Q$ and $u,v \in X^{\infty}$ are such that $u_l = v_l$ for all $l \in \mathbb{N}$ with $t_l > 0$. Then $I(t,u) = I(t,v)$.

Proof. We fix some $k \in \mathbb{N}$ with $t_k > 0$ and define $\varphi: \mathbb{N} \to \mathbb{N}$ to be $\varphi(l) = l$ if $t_l > 0$ and $\varphi(l) = k$ if $t_l = 0$. Then

$$\sum_{p=1}^{\infty} t_p \delta_l^{\varphi(p)} = \sum_{p=1}^{\infty} t_p \delta_l^p = t_l \quad \forall l \in \mathbb{N}.$$

It follows that

$$\begin{aligned} \mathop{I}_{l=1}^{\infty} (t_l, u_l) &= \mathop{I}_{l=1}^{\infty} \left(\sum_{p=1}^{\infty} t_p \delta_l^{\varphi(p)}, u_l \right) \\ &= \mathop{I}_{p=1}^{\infty} \left(t_p, \mathop{I}_{l=1}^{\infty} (\delta_l^{\varphi(p)}, u_l) \right) = \mathop{I}_{p=1}^{\infty} (t_p, u_{\varphi(p)}). \end{aligned}$$

The same holds true for v, and $u_{\varphi(p)} = v_{\varphi(p)} \ \forall p \in \mathbb{N}$. We are done.

The above proposition permits to define in the obvious manner the combinations

$$\mathop{I}_{l=1}^{r}(t_l,x_l) = I\begin{pmatrix}t_1 \dots t_r\\ x_1 \dots x_r\end{pmatrix}$$

for $t=(t_1,\dots,t_r)\in P^r$ and $x=(x_1,\dots,x_r)\in X^r$ $\forall r\in\mathbb{N}$.

A subset $A\subset X$ is defined to be <u>convex</u> iff $I(t,x)\in A$ for all $t\in P$ and $x\in A^\infty$, and to be <u>superconvex</u> iff $I(t,x)\in A$ for all $t\in Q$ and $x\in A^\infty$. And a function $f:X\to[-\infty,\infty[$ is defined to be <u>convex</u> iff for all $t\in P$ and $x\in X^\infty$ one has

(*) $$f(I(t,x)) \leqq \sum_{l=1}^{\infty} t_l f(x_l), \quad \text{with the convention } 0(-\infty):=0,$$

and to be <u>superconvex</u> iff (*) holds true for all $t\in Q$ and $x\in X^\infty$ such that the sequence $(f(x_l))_l$ is bounded above and hence the infinite series involved has a sum $\in[-\infty,\infty[$. A superconvex function is convex in view of 1.1. Let us recall from [2] two rapid consequences of the definitions.

1.2 REMARK. i) <u>A convex function</u> $f:X\to[-\infty,\infty[$ <u>which is bounded above is superconvex</u>. ii) <u>A superconvex function</u> $f:X\to\mathbb{R}$ <u>is bounded below</u>.

2. THE INTERSECTION THEOREM

2.1 THEOREM. <u>Assume that for the sequence of superconvex</u> $A_n\subset X$ <u>there exist</u> $a_n\in A_n$ <u>and</u> $0\leqq\alpha_n<1$ <u>with</u>

$$A_{n+1}\subset I\begin{pmatrix}\alpha_n & 1-\alpha_n\\ a_n & A_n\end{pmatrix} := \left\{I\begin{pmatrix}\alpha_n & 1-\alpha_n\\ a_n & x\end{pmatrix} : x\in A_n\right\} \quad \forall n\in\mathbb{N}.$$

<u>If</u> $\sum_{n=1}^{\infty}\alpha_n=\infty$ <u>then</u> $\bigcap_{n=1}^{\infty}A_n\neq\emptyset$.

<u>Proof</u>. The proof resembles that of the special case $\alpha_n=\frac{1}{2}$ $\forall n\in\mathbb{N}$ in [2]. i) We define inductively $a_l^n\in A_n$ $\forall n,l\in\mathbb{N}$ as follows: For $l=1$ put $a_1^n:=a_n$ $\forall n\in\mathbb{N}$. In the induction step $1\leqq l\Rightarrow l+1$ take

$a_{l+1}^n \in A_n$ such that

$$a_l^{n+1} = I\begin{pmatrix}\alpha_n & 1-\alpha_n \\ a_n & a_{l+1}^n\end{pmatrix} \quad \forall n \in \mathbb{N}.$$

ii) We form $\alpha_p^q := \prod_{l=p}^{q-1} (1-\alpha_l)$ for $1 \leqq p < q$ and $\alpha_p^p := 1$ for $p \geqq 1$, and then $t_l^n := \alpha_n^{n+l-1} - \alpha_n^{n+l} = \alpha_n^{n+l-1}\alpha_{n+l-1} \geqq 0$ $\forall n,l \in \mathbb{N}$. It follows that $t_1^n = \alpha_n$ and $t_{l+1}^n = \alpha_n^{n+l}\alpha_{n+l} = (1-\alpha_n)\alpha_{n+1}^{n+l}\alpha_{n+l} = (1-\alpha_n)t_l^{n+1}$. Furthermore $\sum_{l=1}^{r} t_l^n = 1-\alpha_n^{n+r}$ $\to 1$ for $r \to \infty$ in view of the hypothesis, so that $(t_l^n)_l \in Q$ $\forall n \in \mathbb{N}$. iii) We form $z_n := \mathop{I}_{l=1}^{\infty} (t_l^n, a_l^n) \in A_n$ $\forall n \in \mathbb{N}$. For fixed $n \in \mathbb{N}$ then

$$z_{n+1} = \mathop{I}_{p=1}^{\infty} (t_p^{n+1}, a_p^{n+1}) = \mathop{I}_{p=1}^{\infty}\left(t_p^{n+1}, I\begin{pmatrix}\alpha_n & 1-\alpha_n \\ a_n & a_{p+1}^n\end{pmatrix}\right)$$
$$= \mathop{I}_{p=1}^{\infty}\left(t_p^{n+1}, \mathop{I}_{l=1}^{\infty} (s_l^p, x_l)\right) = \mathop{I}_{l=1}^{\infty}\left(\sum_{p=1}^{\infty} t_p^{n+1} s_l^p, x_l\right),$$

where $s_1^p = \alpha_n$, $s_{p+1}^p = 1-\alpha_n$ and $s_l^p = 0$ for $l \neq 1, p+1$; and where $x_1 = a_n = a_1^n$ and $x_l = a_l^n$ for $l \geqq 2$. It follows that

$$\sum_{p=1}^{\infty} t_p^{n+1} s_l^p = \left\{\begin{matrix} \alpha_n = t_1^n & \text{for } l=1 \\ (1-\alpha_n)t_{l-1}^{n+1} = t_l^n & \text{for } l \geqq 2 \end{matrix}\right\} = t_l^n \quad \forall l \in \mathbb{N},$$

so that $z_{n+1} = \mathop{I}_{l=1}^{\infty} (t_l^n, a_l^n) = z_n$. Hence $z_1 = z_n \in A_n$ $\forall n \in \mathbb{N}$ which proves the assertion.

We continue with an example which shows that in the intersection theorem the assumption $\sum_{n=1}^{\infty} \alpha_n = \infty$ is sharp.

In a real Banach space the closed bounded convex subsets $X \neq \emptyset$ are superconvex spaces under the natural operation $I: I(t,x) = \sum_{l=1}^{\infty} t_l x_l$ for $t \in Q$ and $x \in X^{\infty}$. In particular this applies to $X = Q \subset l^1$.

2.2 THEOREM. _Consider a sequence of real numbers_ $0 \leqq \alpha_n < 1$ $\forall n \in \mathbb{N}$ _with_ $\sum_{n=1}^{\infty} \alpha_n < \infty$. _Then there exist sequences of superconvex subsets_ $A_n \subset Q$ _and points_ $a_n \in A_n$ _with_ $A_{n+1} \subset \alpha_n a_n + (1-\alpha_n) A_n$ $\forall n \in \mathbb{N}$ _such that_ $\bigcap_{n=1}^{\infty} A_n = \emptyset$.

Proof. Let $e^n := (\delta_l^n)_l \in Q$ and $Q^n := \{t \in Q : t_l = 0 \text{ for } l \leqq n\}$ $\forall n \in \mathbb{N}$. Define $\beta_o := 1$ and $\beta_n := \prod_{l=1}^{n} (1-\alpha_l)$ $\forall n \in \mathbb{N}$. In view of the hypothesis we have $\beta_n \to \beta > 0$ for $n \to \infty$. Put $u^n := \sum_{l=1}^{n} (\beta_{l-1} - \beta_l) e^l$, so that $u^n \in l^1$ with nonnegative components and $||u^n|| = 1 - \beta_n$ $\forall n \in \mathbb{N}$. Then define

$$A_1 := Q \quad \text{and} \quad A_n := u^{n-1} + \beta_{n-1} Q^{n-1} \quad \text{for } n \geqq 2,$$

$$a_1 := e^1 \quad \text{and} \quad a_n := u^{n-1} + \beta_{n-1} e^n \quad \text{for } n \geqq 2,$$

so that $A_n \subset Q$ is superconvex and $a_n \in A_n$. For $n=1$ we have

$$A_2 = u^1 + \beta_1 Q^1 \subset u^1 + \beta_1 Q = \alpha_1 e^1 + (1-\alpha_1) Q = \alpha_1 a_1 + (1-\alpha_1) A_1,$$

and for $n \geqq 2$ we have

$$\begin{aligned} A_{n+1} &= u^n + \beta_n Q^n \subset u^n + \beta_n Q^{n-1} = u^{n-1} + \alpha_n \beta_{n-1} e^n + (1-\alpha_n) \beta_{n-1} Q^{n-1} \\ &= \alpha_n (u^{n-1} + \beta_{n-1} e^n) + (1-\alpha_n)(u^{n-1} + \beta_{n-1} Q^{n-1}) = \alpha_n a_n + (1-\alpha_n) A_n. \end{aligned}$$

In order to prove $\bigcap_{n=1}^{\infty} A_n = \emptyset$ assume that $t \in A_n$ $\forall n \in \mathbb{N}$. For $l \in \mathbb{N}$ then $t \in A_n$ for $n-1 \geqq l$ implies that $t_l = \beta_{l-1} - \beta_l$. It follows that $\sum_{l=1}^{r} t_l = 1 - \beta_r \to 1 - \beta < 1$ for $r \to \infty$, which contradicts $t \in Q$.

3. THE DINI TYPE THEOREM

3.1 THEOREM. _Consider an increasing sequence of superconvex functions_ $f_n : X \to [-\infty, \infty[$ _and a function_ $f : X \to \mathbb{R}$ _with_ $f_n \leqq f$ $\forall n \in \mathbb{N}$, _also a_ *sequence of real numbers* $1 \leqq t_n \uparrow \infty$. *For each nonvoid superconvex* $A \subset X$ *then*

$$\operatorname*{Inf}_{x\in A} f(x) \leqq \lim_{n\to\infty} \operatorname*{Inf}_{x\in A} f_n(x) + \operatorname*{Sup}_{x\in A} \operatorname*{Inf}_{n\in\mathbb{N}} t_n(f(x)-f_n(x)),$$

with the convention $\infty+(-\infty):=\infty$.

Proof. For nonvoid superconvex $A \subset X$ we put

$$\alpha(A) := \lim_{n\to\infty} \operatorname*{Inf}_{x\in A} f_n(x) \quad \text{and} \quad \beta(A) := \operatorname*{Inf}_{x\in A} f(x),$$

so that $-\infty \leqq \alpha(A) \leqq \beta(A) < \infty$. i) Fix a nonvoid superconvex $A \subset X$, real numbers $a > \alpha(A)$ and $b > \beta(A)$, and $0 \leqq t < 1$ as well as $n \in \mathbb{N}$. Then there exist a point $u \in A$ and a nonvoid superconvex $B \subset I\binom{t\ \ 1-t}{u\ \ A} \subset A$ with

$$\alpha(B) \leqq tb+(1-t)a \quad \text{and} \quad f_n(x) \leqq tb+(1-t)a \ \forall x \in B.$$

In fact, fix some $u \in A$ with $f(u) < b$; then $f_l(u) < b$ $\forall l \in \mathbb{N}$ as well. And $[f_n \leqq a]$ is superconvex with $A \cap [f_n \leqq a] \neq \emptyset$, so that

$$B := I\begin{pmatrix} t & 1-t \\ u & A\cap[f_n \leqq a] \end{pmatrix} \subset I\begin{pmatrix} t & 1-t \\ u & A \end{pmatrix}$$

is superconvex and $\neq \emptyset$. Now for $l \geqq n$ there is a point $v \in A$ with $f_l(v) < a$, and in view of $f_n(v) \leqq f_l(v)$ we have $v \in A \cap [f_n \leqq a]$ and hence $x := I\binom{t\ \ 1-t}{u\ \ v} \in B$. We have $f_l(x) \leqq t f_l(u) + (1-t) f_l(v) \leqq tb+(1-t)a$. It follows that $\alpha(B) \leqq tb+(1-t)a$. On the other hand each $x \in B$ is of the form $x = I\binom{t\ \ 1-t}{u\ \ v}$ with $v \in A \cap [f_n \leqq a]$, so that we obtain $f_n(x) \leqq t f_n(u) + (1-t) f_n(v) \leqq tb+(1-t)a$. Thus i) is proved.

ii) From the $1 \leqq t_n \uparrow \infty$ $\forall n \in \mathbb{N}$ and $t_o := 1$ we form the numbers $s_n := 1-(t_{n-1}/t_n)$ $\forall n \in \mathbb{N}$, so that $0 \leqq s_n < 1$. We have

$$\prod_{l=1}^{n} (1-s_l) = \prod_{l=1}^{n} (t_{l-1}/t_l) = \frac{1}{t_n} \to 0 \text{ for } n\to\infty \text{ and hence } \sum_{n=1}^{\infty} s_n = \infty.$$

iii) We fix a nonvoid superconvex $A \subset X$. The assertion is clear if $\beta(A) \leqq \alpha(A)$, so that we can assume $\beta(A) > \alpha(A)$. We fix a number $c > 0$ with $\alpha(A) < \beta(A) - c$ and prove the existence of an $x \in A$ with $t_n(f(x) - f_n(x)) \geqq c$ $\forall n \in \mathbb{N}$. Then

$$\sup_{x \in A} \inf_{n \in \mathbb{N}} t_n(f(x) - f_n(x)) \geqq c,$$

and this implies the assertion in both the cases $\alpha(A) > -\infty$ and $\alpha(A) = -\infty$. iv) To achieve this we start from $A_0 := A$ and form inductively nonvoid superconvex $A_n \subset X$ and $a_n \in A_{n-1}$ $\forall n \in \mathbb{N}$ with

$$\text{o)} \qquad A_n \subset I\begin{pmatrix} (1/3)s_n & 1-(1/3)s_n \\ a_n & A_{n-1} \end{pmatrix} \subset A_{n-1} \text{ for } n \geqq 1,$$

$$\text{1)} \quad \alpha(A_n) < \beta(A_n) - \frac{c}{t_n} \text{ for } n \geqq 0,$$

$$\text{2)} \quad f_n(x) \leqq \beta(A_n) - \frac{c}{t_n} \ \forall x \in A_n \text{ for } n \geqq 1.$$

The case $n=0$ is clear. For the induction step $0 \leqq n-1 \Rightarrow n$ we apply i) to A_{n-1}, to

$$a := \beta(A_{n-1}) - \frac{c}{t_{n-1}} \quad \text{and} \quad b := \beta(A_{n-1}) + \frac{c}{t_{n-1}},$$

and to $t := (1/3)s_n$. Then

$$\begin{aligned} tb + (1-t)a &= \beta(A_{n-1}) - (1 - \tfrac{2}{3}s_n)\frac{c}{t_{n-1}} \\ &< \beta(A_{n-1}) - (1 - s_n)\frac{c}{t_{n-1}} = \beta(A_{n-1}) - \frac{c}{t_n}. \end{aligned}$$

Hence we obtain a point $u =: a_n \in A_{n-1}$ and a nonvoid superconvex $B =: A_n \subset X$ with o), and 1)2) follow since $A_n \subset A_{n-1}$ and hence $\beta(A_{n-1}) \leqq \beta(A_n)$. The inductive choice is thus completed. Now from the intersection theorem we obtain a point $x \in A_n$ $\forall n \in \mathbb{N}$. In particular $x \in A$, and from 2) combined with $\beta(A_n) \leqq f(x)$ we see that $t_n(f(x) - f_n(x)) \geqq c$ $\forall n \in \mathbb{N}$. The proof is complete.

3.2 COROLLARY. <u>Consider an increasing sequence of superconvex functions</u> $f_n : X \to [-\infty, \infty[$ $\forall n \in \mathbb{N}$ <u>with limit function</u> $f : X \to \mathbb{R}$. <u>Assume</u>

that there exists a sequence of positive numbers $c_n \downarrow 0$ *such that* $f(x)-f_n(x)=O(c_n)$ *for* $n\to\infty$ *at each* $x\in X$. *On each nonvoid superconvex* $A\subset X$ *then*

$$\inf_{x\in A} f_n(x) \to \inf_{x\in A} f(x) \quad \textit{for } n\to\infty.$$

We conclude with two examples which can serve to understand the situation.

3.3 EXAMPLE. *An increasing sequence of superconvex functions* $f_n:X\to[0,1]$ *with limit function* $f=1$ *such that* $\operatorname{Inf} f_n=0\ \forall n\in\mathbb{N}$. The condition imposed in 3.2 will be violated in an obvious manner.

Let X consist of the continuous functions $u:[0,1]\to[0,1]$ with $u(1)=1$, with the natural operation $I:I(t,u)=\sum_{l=1}^{\infty} t_l u_l$ for $t\in Q$ and $u\in X^{\infty}$. Fix a sequence of numbers $0<\alpha_n<1$ with $\alpha_n\uparrow 1$, and define $f_n:X\to[0,1]$ to be $f_n(u)=\operatorname{Max}(u|[0,\alpha_n])\ \forall u\in X\ \forall n\in\mathbb{N}$. The f_n are convex and hence superconvex in view of 1.2.i), and we have $f_n\uparrow f=1$ on X. It is clear that $\operatorname{Inf} f_n=0\ \forall n\in\mathbb{N}$.

3.4 EXAMPLE. *An increasing sequence of convex* (but not superconvex) *functions* $f_n:X\to\mathbb{R}$ *with limit function* $f:X\to\mathbb{R}$ *such that*

$$\operatorname{Inf} f=0,\ \operatorname{Inf} f_n=-\infty\ \forall n\in\mathbb{N},\ \textit{and}\ \sup_{x\in X}\inf_{n\in\mathbb{N}} n(f(x)-f_n(x))=0.$$

We start with a convex function $F:X\to\mathbb{R}$ on a superconvex space X which is not bounded below. Then F is not superconvex in view of 1.2.ii) and hence not bounded above either in view of 1.2.i). For example, let X be the closed unit ball of a real Banach space E and F the restriction to X of a discontinuous linear functional on E. The functions $f_n:=\operatorname{Max}(F,2^{-n}F) = F^+-2^{-n}F^-\ \forall n\in\mathbb{N}$ are convex and $f_n\uparrow F^+=:f$ on X. The assertions are all obvious.

REFERENCES

1. Gerd Rodé, Ein Grenzwertsatz für stützende Funktionen auf superkonvexen Räumen.
Dissertation, Universität des Saarlandes, Saarbrücken 1977.

2. Gerd Rodé, Superkonvexe Analysis.
Arch.Math. 34(1980),452-462.

3. Gerd Rodé, Superkonvexität und schwache Kompaktheit.
Arch.Math. 36(1981),62-72.

Heinz König, Fachbereich Mathematik, Universität des Saarlandes,
D-6600 Saarbrücken, Fed.Rep.Germany

International Series of
Numerical Mathematics, Vol. 71

SOME INEQUALITIES FOR THE EIGENVALUES OF COMPACT OPERATORS

Hermann König

Abstract. Some inequalities between the eigenvalues of compact operators in Banach spaces and their approximation- and Weyl-numbers as well as their p-summing norms are given. Good constants in these estimates are of interest.

1. SUMMABILITY OF s-NUMBERS AND EIGENVALUES.

If $T : X \to Y$ is a continuous linear operator between (complex) Banach spaces X and Y, its approximation numbers are given by

$$a_n(T) := \inf \{ \|T - T_n\| \mid \text{rank } T_n < n \}$$

and its Weyl-numbers by

$$x_n(T) := \sup \{ a_n(TA) \mid A : l_2 \to X \text{ with } \|A\| = 1 \}$$

for all $n \in \mathbb{N}$. Clearly $x_n(T) \leq a_n(T)$. If $X = Y$ and T is compact, $T \in K(X)$, we denote its eigenvalues by $\lambda_n(T)$ - counted according to their multiplicity and ordered non-increasingly in absolute value. In Hilbert spaces

$$a_n(T) = x_n(T) = \lambda_n(T^*T)^{1/2},$$

cf. Pietsch [6], ch. 11. Moreover, a classical inequality of Weyl [9]relates the $a_n(T)$ and $\lambda_n(T)$:

THEOREM 1. _Let_ X _be_ _a_ _Hilbert_ _space_, $T \in K(X)$ _and_ $n \in \mathbb{N}$, $0 < p < \infty$. _Then_

$$\prod_{j=1}^{n} |\lambda_j(T)| \leq \prod_{j=1}^{n} a_j(T) \tag{1}$$

$$\sum_{j=1}^{n} |\lambda_j(T)|^p \leq \sum_{j=1}^{n} a_j(T)^p . \tag{2}$$

Formula (2) follows from (1) by purely analytic arguments. The question whether formulas (1), (2) also hold in Banach spaces X, has to be answered negatively. Let

$$T = \begin{pmatrix} 1 & 1 \\ 0 & 2 \end{pmatrix} : l_\infty^2 \to l_\infty^2 , \quad T_1 = \begin{pmatrix} 1/3 & 1 \\ 2/3 & 2 \end{pmatrix} : l_\infty^2 \to l_\infty^2$$

where l_∞^2 is $\mathbb{C}^2$ with the max-norm. Then $\lambda_1(T) = 2$, $\lambda_2(T) = 1$, but $a_1(T) = 2$, $a_2(T) \leq \|T - T_1\| = 2/3 < 1$. Thus (1), (2) could possibly hold in Banach spaces only with additional constants. This is true for (2), even for the smaller Weyl numbers.

THEOREM 2. *Let* X *be a Banach space,* $T \in K(X)$ *and* $n \in \mathbb{N}$, $0 < p < \infty$. *Then with* $c_p := 2e/\sqrt{p}$ *for* $p \leq 1$ *and* $c_p := \sqrt{2e}\, 2^{1/p}$ for $p \geq 1$

$$(3) \qquad \sum_{j=1}^{n} |\lambda_j(T)|^p \leq c_p^p \sum_{j=1}^{n} x_j(T)^p .$$

This important result is due to Pietsch [7] with $c_p := 2^{1/p}e$. We give a modification of his proof which yields the above better value of c_p as $p \to 0$. By a factorization of T over Hilbert spaces, theorem 2 is reduced to theorem 1. We need the following notion.

For $1 \leq p < \infty$, an operator $T : X \to Y$ is p-summing if there is $c > 0$ such that for any $x_1, \ldots x_m \in X$

$$\sum_{i=1}^{m} \|Tx_i\|^p \leq c^p \sup \{ \sum_{i=1}^{m} | < x^*, x_i > |^p : \|x^*\|_{X^*} = 1 \} .$$

The infimum over all such values c is denoted by $\pi_p(T)$.

Proof. Let $X_n \subseteq X$ be n-dimensional with $T(X_n) \subseteq X_n$ such that $T_n := T|_{X_n} : X_n \to X_n$ has the same first n eigenvalues as T, i.e. X_n is a space of (generalized) eigenvectors. By Garling-Gordon [2] or Pietsch [6], 17.3.7 and 28.2.4. there are operators $A_n : X \to l_2^n$, $B_n : l_2^n \to X_n$ with $B_n A_n|_{X_n} = Id$ and $\pi_2(A_n) \leq \sqrt{n}$, $\|B_n\| \leq 1$. Here l_2^n denotes the n-dimensional Hilbert space. (This basic fact on 2-summing operators relies on $\pi_2(Id_{X_n}) = \sqrt{n}$.) Let $J_n : X_n \to X$ denote the canonical injection. Since T has the same first n eigenvalues as $T_n = (B_n A_n) T J_n = B_n (A_n T J_n)$ which in turn has the same eigenvalues as the Hilbert-space operator $\tilde{T}_n := (A_n T J_n) B_n$, we find using theorem 1

$$\prod_{j=1}^{n} |\lambda_j(T)| = \prod_{j=1}^{n} |\lambda_j(\tilde{T}_n)| \leq \prod_{j=1}^{n} a_j(\tilde{T}_n) = \prod_{j=1}^{n} x_j(\tilde{T}_n) .$$

The Weyl-numbers are submultiplicative,

$$x_{r+s-1}(A_n(TJ_nB_n)) \leq x_r(A_n) \cdot x_s(TJ_nB_n) \quad \text{for } r,s \in \mathbb{N} ,$$

cf. Pietsch [7], proposition 2 - the corresponding fact for the approximation numbers is evident. We also use lemma 8 of [7],

$$r^{1/2} x_r(A_n) \leq \pi_2(A_n) .$$

Thus

$$\begin{aligned} x_{r+s-1}(\tilde{T}_n) &\leq x_r(A_n)\, x_s(T)\, \|J_n\|\, \|B_n\| \\ &\leq \pi_2(A_n)/\sqrt{r}\, x_s(T) \leq \sqrt{n/r}\, x_s(T). \end{aligned}$$

Let $k \in \mathbb{N}$ be fixed. Let $m := [n/k]$, $n = mk + l$, $0 \leq l < k$. Then we get for $j \in \{1,\dots m\}$

$$\prod_{i=1}^{k} x_{(j-1)k+i}(\tilde{T}_n) \leq \sqrt{n/j}^{\,k} \prod_{i=1}^{k} x_{(j-1)(k-1)+i}(T) ,$$

letting $r := j$, $s := (j-1)(k-1)+i$ and

$$\prod_{i=1}^{l} x_{mk+i}(\tilde{T}_n) \leq \sqrt{n/(m+1)} \prod_{i=1}^{l} x_{m(k-1)+i}(T) .$$

Forming the product of these products, we find

$$\prod_{j=1}^{n} x_j(\tilde{T}_n) \leq \left(\frac{n^{n/2}}{m!^{k/2}(m+1)^{l/2}}\right) \prod_{j=1}^{n-m} x_j(T) \prod_{i=1}^{m} x_{i(k-1)+1}(T) ;$$

the last m terms occuring twice. If we denote the sequence of Weyl numbers where every $(k-1)^{st}$ number occurs twice (starting at index k) by $\dot{x}_k(T)$, this means

$$\prod_{j=1}^{n} |\lambda_j(T)| \leq \prod_{j=1}^{n} x_j(\tilde{T}_n) \leq \left(\frac{n^n}{m!^k(m+1)^l}\right)^{1/2} \prod_{j=1}^{n} \dot{x}_j(T) .$$

Elementary estimates using Stirlings's formula show

$$m!^k (m+1)^l \geq n^n \cdot (ek)^n .$$

Thus

$$\prod_{j=1}^{n} |\lambda_j(T)| \leq \prod_{j=1}^{n} (\sqrt{ek}\, \dot{x}_j(T)).$$

These inequalities for $n \in \mathbb{N}$ imply in the same way as (1) implies (2) that for all $0 < p < \infty$

$$\sum_{j=1}^{n} |\lambda_j(T)|^p \leqslant \sum_{j=1}^{n} (\sqrt{ek}\, \dot{x}_j(T))^p .$$

Since every $(k-1)^{st}$ Weyl-number starting with k occurs twice, the monotonicity of the Weyl-numbers yields

$$(4) \qquad (\sum_{j=1}^{n} |\lambda_j(T)|^p)^{1/p} \leqslant \sqrt{ek}\, (1 + \frac{1}{k-1})^{1/p} (\sum_{j=1}^{n} x_j(T)^p)^{1/p}$$

If $p \leqslant 1$, we now choose $k := 2 + [2/p]$. Then

$$\sqrt{ek}\, (1 + \frac{1}{k-1})^{1/p} \leqslant \sqrt{2e(1+1/p)}\, (1 + p/2)^{1/p} \leqslant 2e/\sqrt{p} .$$

For $p \geq 1$ let $k = 2$. Then

$$\sqrt{ek}\, (1 + \frac{1}{k-1})^{1/p} = 2^{1/p} \sqrt{2e} .$$

This and (4) proves theorem 2.

Problem. Does an inequality corresponding to (1) hold for compact operators $T \in K(X)$ in Banach spaces, i.e. does

$$\prod_{j=1}^{n} |\lambda_j(T)| \leqslant c^n \prod_{j=1}^{n} x_j(T)$$

hold with some absolute $c > 1$? This would follow if theorem 2 could be shown to hold even with $\sup \{c_p | 0 < p \leqslant 1\} < \infty$ since the arithmetic means $(\sum_{j=1}^{n} |\alpha_j|^p/n)^{1/p}$ tend to the geometric mean $|\alpha_1 \cdots \alpha_n|^{1/n}$ for $p \to 0$.

2. FURTHER REMARKS AND APPLICATIONS

Estimates of the form $|\lambda_n(T)| \leq c\, a_n(T)$ do not hold, even in Hilbert spaces: Let $0 < \alpha < 1$ and consider

$$T = \begin{pmatrix} 0 & 1 & & \\ & \ddots & \ddots & \\ & & \ddots & 1 \\ \alpha & & & 0 \end{pmatrix} : l_2^n \to l_2^n .$$

Then $T^n = \alpha \,\mathrm{Id}$, thus $\lambda_n(T) = \alpha^{1/n}$, $a_n(T) = \alpha$. Thus in Hilbert spaces $|\lambda_n(T)| \leq a_n(T)^{1/n}$ for $\|T\| \leq 1$ is best possible; in Banach spaces one has

<u>Proposition 1.</u> <u>Let</u> X <u>be</u> <u>a</u> <u>Banach</u> <u>space</u>, $n \in \mathbb{N}$ <u>and</u> $T \in K(X)$ <u>with</u> $\|T\| \leq 1$. <u>Then</u>

$$|\lambda_n(T)| \leq 2.5\, x_n(T)^{1/n} .$$

A proof, at least for approximation numbers with 16 instead of 2.5, can be found in Carl-Triebel [1]. The p-summing operators introduced in the last section coincide in Hilbert spaces with the Hilbert-Schmidt operators, i.e. those maps T with $\sum_{n \in \mathbb{N}} a_n(T)^2 < \infty$, and thus have square-summable eigenvalues $(1 \leq p < \infty)$. In general Banach spaces their eigenvalues behave slightly differently:

<u>Proposition 2.</u> <u>Let</u> X <u>be</u> <u>a</u> <u>Banach</u> <u>space</u>, $1 \leq p < \infty$ <u>and</u> $T \in K(X)$ <u>be</u> p-<u>summing</u>. <u>Then</u> <u>the</u> <u>eigenvalues</u> <u>of</u> T <u>are</u> $q := \max(2,p)$-<u>summable</u> <u>with</u>

$$\left(\sum_{n=1}^{\infty} |\lambda_n(T)|^q\right)^{1/q} \leq \pi_p(T) .$$

This result is best possible, in general, and can be found in [3]. It has an equivalent reformulation in terms of integral operators which states

<u>Proposition 3.</u> <u>Let</u> (Ω,μ) <u>be</u> <u>a</u> <u>measure</u> <u>space</u>, $k : \Omega^2 \to \mathbb{C}$ <u>be</u> μ^2-<u>measurable</u> <u>with</u>

$$c_p := (\int_\Omega (\int_\Omega |k(x,y)|^{p'} d\mu(y))^{p/p'} d\mu(x))^{1/p} < \infty$$

where $2 \leq p < \infty$, $1/p + 1/p' = 1$. *Then* k *defines an integral operator* T_k *in* $L_p(\Omega,\mu)$ *with* p-*th power summable eigenvalues*,

$$(\sum_{n=1}^{\infty} |\lambda_n(T_k)|^p)^{1/p} \leq c_p .$$

This follows from proposition 2 since $\pi_p(T_k) \leq c_p$ is seen to hold by applying Hölder's inequality, but actually proposition 2 is equivalent to proposition 3, cf. [3], [5]. The Weyl-numbers of integral operators can be estimated asymptotically from above under summability or differentiability assumptions on the kernel, cf. Pietsch [8] or [4]. This gives another possibility to estimate the asymptotic behaviour of eigenvalues of integral operators.

REFERENCES

1. B. Carl and H. Triebel, Inequalities between eigenvalues, entropy numbers and related quantities in Banach spaces. Math. Ann. 251 (1980), 129 - 133.

2. D.J.H. Garling and Y. Gordon, Relations between some constants associated with finite-dimensional Banach spaces. Isr. J. Math. 9 (1971), 346 - 361.

3. W.B. Johnson, H. König, B. Maurey and J.R. Retherford, Eigenvalues of p-summing and l_p-type operators in Banach spaces. J. Funct. Anal. 32 (1979), 353 - 380.

4. H. König, Some remarks on weakly singular integral operators. Int. Eq. Oper. Th. 3 (1980), 397 - 407.

5. H. König, L. Weis, On the eigenvalues of orderbounded integral operators. To appear in Int. Eq. Oper. Th.

6. A. Pietsch, Operator Ideals. Verlag der Wissenschaften, Berlin, 1978.

7. A. Pietsch, Weyl numbers and eigenvalues of operators in Banach spaces. Math. Ann. 247 (1980), 149 - 168.

8. A. Pietsch, Eigenvalues of integral operators. I. Math. Ann. 247 (1980), 169 - 178.

9. H. Weyl, Inequalities between the two kinds of eigenvalues of a linear transformation. Proc. Nat. Acad. Sci. USA 35 (1949), 408 - 411.

H. König, Mathematisches Seminar, Universität Kiel, D-23 Kiel, West-Germany.

International Series of
Numerical Mathematics, Vol. 71

SOME NEGATIVE RESULTS IN CONNECTION WITH MARCHAUD - TYPE INEQUALITIES

Rolf Joachim Nessel and Erich van Wickeren[1)]

Abstract. Continuing our previous investigations on quantitative uniform boundedness and condensation principles, the present paper is concerned with some negative results in connection with Marchaud - type inequalities. The existence of the relevant counter-examples follows by means of a general theorem, given in terms of bounded, sublinear functionals on Banach spaces. The method of proof essentially consists in a quantitative version of the familiar gliding hump method. Emphasis is laid upon the treatment of some significant applications, illustrating the usefulness of the abstract approach.

1. INTRODUCTION

Let $C_{2\pi}$ be the Banach space of 2π - periodic, continuous functions on the real axis $\mathbb{R}$, endowed with the usual norm $\|f\|_C := \sup\{|f(x)|;\ x \in \mathbb{R}\}$. For $f \in C_{2\pi}$ the *rth* modulus of continuity is defined by ($r \in \mathbb{N}$, the set of natural numbers)

$$(1.1) \qquad \omega_r(t,f;C_{2\pi}) := \sup\{\|\Delta_u^r f\|_C;\ |u| \leq t\} ,$$

$$\Delta_u^r f(x) := \sum_{k=o}^{r} \binom{r}{k}(-1)^k f(x+ku) .$$

It trivially follows from the definition that the $(r+q)$ *th* modulus may be estimated by the *rth* one ($r,q \in \mathbb{N}$):

1) Supported by Deutsche Forschungsgemeinschaft Grant No. Ne 171/5-1.

$$(1.2)\qquad \omega_{r+q}(t,f;C_{2\pi}) \leq 2^q\, \omega_r(t,f;C_{2\pi})\,.$$

The converse assertion is given by Marchaud's inequality (in fact, one example of a weak-type inequality in the sense of [4]): It states that the *rth* modulus may also be estimated in terms of the (r+q)*th* one, provided an appropriate averaging is taken into account, namely (cf. [16, p. 104 ff])

$$(1.3)\qquad \omega_r(t,f;C_{2\pi}) \leq C_{r,q}\, t^r \int_t^1 \omega_{r+q}(u,f;C_{2\pi})u^{-(r+1)}\,du\,.$$

Assertions (1.2,3) fit together in as much as they show that

$$(1.4)\qquad \mathrm{Lip}(\beta,r;C_{2\pi}) = \mathrm{Lip}(\beta,r+q;C_{2\pi}) \qquad (0<\beta<r)$$

with the standard definition of the Lipschitz classes ($t\to 0+$)

$$(1.5)\qquad \mathrm{Lip}(\beta,r;C_{2\pi}) := \{f\in C_{2\pi}: \omega_r(t,f;C_{2\pi}) = \mathcal{O}(t^\beta)\}\,,$$

$$\mathrm{lip}(\beta,r;C_{2\pi}) := \{f\in C_{2\pi}: \omega_r(t,f;C_{2\pi}) = o(t^\beta)\}\,.$$

In fact, (1.4) suggests that Marchaud's inequality (1.3) might be strengthened to an estimate of the more direct type (1.2), provided one deals with <u>non</u>smooth (e.g., nondifferentiable) functions. For example, one may pose the question: Does there exist an estimate of type

$$(1.6)\qquad \omega_r(t,f;C_{2\pi}) \leq C_f\, \omega_{r+q}(t,f;C_{2\pi})$$

for <u>each</u> function

$$(1.7)\qquad f\in \mathrm{Lip}(\beta,r;C_{2\pi})\setminus \mathrm{lip}(\beta,r;C_{2\pi}) \qquad (0<\beta<r)\ ?$$

Of course, this question is not meaningful for smooth (e.g., (r+q)-times differentiable) functions. But even for nonsmooth elements the answer is negative. Thus there exists f_β with prescribed (smoothness and) nonsmoothness (1.7) for which

$$\limsup_{t \to 0+} \omega_r(t,f_\beta;C_{2\pi}) / \omega_{r+q}(t,f_\beta;C_{2\pi}) = \infty . \tag{1.8}$$

In other words, in spite of (1.4) there are also bad elements within the class $\mathrm{Lip}(\beta,r;C_{2\pi})$ which distinguish between the bounded, sublinear functionals ω_r and ω_{r+q} on the space $C_{2\pi}$.

Such negative results can in fact be obtained by means of a rather general theorem, formulated in Section 2 for bounded, sublinear functionals on Banach spaces. Indeed, the present paper surveys and extends our previous results on the subject (cf. [6-8]). Let us point out, however, that we do not wish to comment or motivate the theoretical aspects of the conditions and assertions of Section 2. This may be found in [15] on the basis of some suggestions of Favard [10 ; 11] concerning negative results in connection with the problem of the comparison of different approximation processes. On the contrary, in the present note emphasis is laid upon the treatment of some significant applications, illustrating the usefulness of the abstract approach of Section 2. This is carried out in Section 3. Of course, our first application (Section 3.1) is concerned with the problem posed at the beginning, including a quantitative extension of (1.8). Neglecting the more theoretical aspects of Favard's work, it seems that the type of questions we are discussing here originates in a problem, posed by S.B. Steckin in connection with his inverse approximation theorem (cf. Section 3.2). This problem was solved by Boman [2] whose concrete reasoning was the starting point for our general approach (cf. [6]). Some further applications are worked out in Section 3.3 - 4, including the existence of certain counterexamples in the theory of monotone approximation.

The authors would like to express their sincere gratitude to Dr. W. Dickmeis for his critical reading of the manuscript and many valuable suggestions.

2. A GENERAL THEOREM

Let X be a (complex) Banach space (with norm $\|\cdot\|_X$), and X^* be the class of functionals T on X which are sublinear, i.e.,

$$|T(f+g)| \leq |Tf| + |Tg|, \quad |T(af)| = a|Tf|$$

for all $f,g \in X$ and $a \geq 0$, and which are bounded, i.e.,

$$\|T\|_{X^*} := \sup\{|Tf|; \|f\|_X = 1\} < \infty .$$

Let ω denote a function, continuous on $[0,\infty)$ such that

$$(2.1) \qquad 0 = \omega(0) < \omega(s) \leq \omega(t), \ \omega(t)/t \leq \omega(s)/s \qquad (0 < s \leq t)$$

(abstract modulus of continuity, cf. [16, p. 96 ff]). Additionally we assume that $\omega(t) \neq \mathcal{O}(t)$, i.e.,

$$(2.2) \qquad \lim_{t\to 0+} \omega(t)/t = \infty .$$

THEOREM 2.1. _Suppose that for_ $S_t, T_n, R_n, V_n \in X^*$ _there are a sequence_ $\{\varphi_n\} \subset (0,\infty)$, _tending to zero, a positive function_ $\sigma(t)$ _on_ $(0,\infty)$, _and elements_ $h_n \in X$ _satisfying_

$$(2.3) \qquad \|h_n\|_X \leq C_1 \qquad (n \in \mathbb{N}),$$

$$(2.4) \qquad |S_t h_n| \leq C_2 \min\{1, \sigma(t)/\varphi_n\} \qquad (n \in \mathbb{N}, t > 0),$$

$$(2.5) \qquad \limsup_{n\to\infty} |T_n h_n| \geq C_3 > 0 ,$$

$$(2.6) \qquad \limsup_{n\to\infty} |R_n h_n| \geq C_4 > 0 ,$$

$$(2.7) \qquad |V_n h_n| \leq C_5 \qquad (n \in \mathbb{N}),$$

$$\text{(2.8)} \qquad \limsup_{m\to\infty} |V_m h_n| / \varphi_m \leq C_{6,n} \qquad (n \in \mathbb{N}) .$$

<u>Then for each</u> ω <u>with</u> (2.1,2) <u>there exists</u> $f_\omega \in X$ <u>such that simultaneously</u> $(t \to 0+,\ n \to \infty)$

$$\text{(2.9)} \qquad |S_t f_\omega| = O(\omega(\sigma(t))) ,$$

$$\text{(2.10)} \qquad |T_n f_\omega| \neq o(\omega(\varphi_n)) ,$$

$$\text{(2.11)} \qquad |R_n f_\omega| \neq o(|V_n f_\omega|) .$$

<u>Proof.</u> We proceed essentially as in [5] via the gliding hump method. Thus, starting with $n_1 = 1$, $\delta_1 = 1$ one may construct sequences $\{n_k\} \subset \mathbb{N}$, $\{\delta_k\} \subset \{0,1\}$ such that for $k \geq 2$

$$\text{(2.12)} \qquad n_k > n_{k-1}, \quad \varphi_{n_k} < \min\{1/k,\ \varphi_{n_{k-1}}\} ,$$

$$\text{(2.13)} \qquad \omega(\varphi_{n_k}) \max\{1, \|T_{n_{k-1}}\|_{X^*}, \|R_{n_{k-1}}\|_{X^*}, \|V_{n_{k-1}}\|_{X^*}\} \leq \omega(\varphi_{n_{k-1}})/k ,$$

$$\text{(2.14)} \qquad \sum_{j=1}^{k-1} \omega(\varphi_{n_j})/\varphi_{n_j} \leq \omega(\varphi_{n_k})/\varphi_{n_k} ,$$

$$\text{(2.15)} \qquad |V_{n_k} h_{n_j}| \leq \omega(\varphi_{n_k}) \qquad (1 \leq j \leq k-1) ,$$

$$\text{(2.16)} \qquad \omega(\varphi_{n_k}) \leq 10 \begin{cases} |T_{n_k} g_k| / C_3 , & k \text{ even} \\ |R_{n_k} g_k| / C_4 , & k \text{ odd} , \end{cases}$$

$$g_k := \sum_{j=1}^{k} \delta_j\ \omega(\varphi_{n_j}) h_{n_j} .$$

Indeed, if the first $k-1$ elements of the sequences, and thus $g_{k-1} \in X$, are given, consider, for k even,

$$M_{k-1} := \limsup_{n\to\infty} |T_n g_{k-1}| / \omega(\varphi_n) .$$

In case $M_{k-1} \leqslant C_3/5$ there exists $N_k \in \mathbb{N}$ such that for $n \geqslant N_k$ (cf. (2.8))

$$(2.17) \qquad |T_n g_{k-1}| \leqslant (M_{k-1}+C_3/5)\omega(\varphi_n) \leqslant (2/5)C_3\ \omega(\varphi_n) ,$$

$$(2.18) \qquad \max_{1\leqslant j\leqslant k-1} |V_n h_{n_j}| / (C_{6,n_j}+1) \leqslant \varphi_n .$$

Then choose $n_k \geqslant N_k$ large enough to satisfy (2.12 - 14) (cf. 2.1,2)) as well as (cf. (2.2,5))

$$\max_{1\leqslant j\leqslant k-1} (C_{6,n_j}+1)\varphi_{n_k} \leqslant \omega(\varphi_{n_k}) , |T_{n_k} h_{n_k}| \geqslant C_3/2 ,$$

thus (2.15) in view of (2.18). Setting $\delta_k = 1$, the element g_k is well - defined and

$$|T_{n_k} g_k| \geqslant \omega(\varphi_{n_k}) |T_{n_k} h_{n_k}| - |T_{n_k} g_{k-1}| \geqslant (C_3/10)\omega(\varphi_{n_k}) ,$$

i.e., (2.16) holds true. On the other hand, if $M_{k-1} > C_3/5$, take $\delta_k = 0$, thus $g_k = g_{k-1}$. Then, of course, one may find n_k large enough to satisfy (2.12 - 16). If k is odd, one proceeds analogously, replacing T_n by R_n.

Since X is complete and (cf. (2.3, 13))

$$(2.19) \qquad \sum_{j=k}^{\infty} \delta_j\, \omega(\varphi_{n_j}) \| h_{n_j} \|_X \leqslant C_1 \sum_{j=k}^{\infty} \omega(\varphi_{n_j}) \leqslant 2\, C_1\, \omega(\varphi_{n_k}) ,$$

the element

$$f_\omega := \sum_{j=1}^{\infty} \delta_j\, \omega(\varphi_{n_j}) h_{n_j}$$

is well-defined in X. Consider $t>0$ with $\sigma(t)\leqslant\varphi_1$. Then there exists $k\in\mathbb{N}$ such that $\varphi_{n_{k+1}}<\sigma(t)\leqslant\varphi_{n_k}$ (cf. (2.12)), and (2.1,4,14,19) deliver

$$\begin{aligned}|S_t f_\omega| &\leqslant \Big(\sum_{j=1}^{k}+\sum_{j=k+1}^{\infty}\Big)\omega(\varphi_{n_j})|S_t h_{n_j}|\\ &\leqslant C_2\sigma(t)\sum_{j=1}^{k}\omega(\varphi_{n_j})/\varphi_{n_j}+C_2\sum_{j=k+1}^{\infty}\omega(\varphi_{n_j})\\ &\leqslant 2C_2\sigma(t)\omega(\varphi_{n_k})/\varphi_{n_k}+2\,C_2\,\omega(\varphi_{n_{k+1}})\leqslant 4\,C_2\,\omega(\sigma(t)),\end{aligned}$$

thus (2.9) (choose $k=0$ if $\sigma(t)>\varphi_1$). For k even one has by (2.13,16,19)

$$\begin{aligned}|T_{n_k}f_\omega| &\geqslant |T_{n_k}g_k|-\|T_{n_k}\|_{X^*}\|f_\omega-g_k\|_X\\ &\geqslant (C_3/10)\omega(\varphi_{n_k})-2\,C_1\|T_{n_k}\|_{X^*}\omega(\varphi_{n_{k+1}})\\ &\geqslant \omega(\varphi_{n_k})(C_3/10-2\,C_1/(k+1)),\end{aligned}$$

thus (2.10), and analogously for k odd

$$|R_{n_k}f_\omega|\geqslant\omega(\varphi_{n_k})(C_4/10+o(1)). \tag{2.20}$$

Moreover, in view of (2.7,13,15,19)

$$\begin{aligned}|V_{n_k}f_\omega| &\leqslant \omega(\varphi_{n_k})|V_{n_k}h_{n_k}|+\Big(\sum_{j=1}^{k-1}+\sum_{j=k+1}^{\infty}\Big)\omega(\varphi_{n_j})|V_{n_k}h_{n_j}|\\ &\leqslant \omega(\varphi_{n_k})[C_5+\sum_{j=1}^{k-1}\omega(\varphi_{n_j})]+2\,C_1\|V_{n_k}\|_{X^*}\omega(\varphi_{n_{k+1}})\\ &\leqslant \omega(\varphi_{n_k})[C_5+2\,\omega(\varphi_1)+2\,C_1],\end{aligned}$$

thus (2.11) by (2.20). □

REMARK 2.2. The argument is in fact shape preserving in the following sense: Let Λ be a closed cone in X, i.e., a closed subset of X for which $f,g \in \Lambda$ implies $\alpha f + \beta g \in \Lambda$ for all $\alpha,\beta \geq 0$. Then the proof indeed guarantees $f_\omega \in \Lambda$, provided $h_n \in \Lambda$ for all $n \in \mathbb{N}$.

3. APPLICATIONS

This section is devoted to some (old and new) applications. Apart from those already mentioned explicitly in the Introduction (to Marchaud's inequality, to Steckin's problem), it is the shape preserving property of our approach (cf. Remark 2.2) which is used in Section 3.3 to derive the existence of a counterexample in the theory of monotone approximation. Finally, Section 3.4 treats the necessity of a condition posed upon a sequence $\{\delta_n\}$ in connection with a comparison of $\|\Delta_\delta^r f\|$, as $\delta \to 0$, with $\|\Delta_{\delta_n}^r f\|$, as $\delta_n \to 0$. Summarizing, the aim of this section is to indicate the usefulness of Theorem 2.1. Further applications, in particular to numerical analysis, are worked out in [6-8].

3.1 Marchaud's inequality. Our first application is concerned with a proof of (1.8). In fact, for any positive function $\alpha(t)$ with ($q \in \mathbb{N}$)

$$(3.1) \qquad \lim_{t\to 0+} \alpha(t) = \infty, \qquad \lim_{t\to 0+} \alpha(t)t^q = 0$$

one has the following quantitative extension, the rate of divergence in (1.8) being measured by α.

COROLLARY 3.1. *For each ω,α satisfying (2.1,2), (3.1), respectively, there exists $f_{\omega,\alpha} \in C_{2\pi}$ such that* ($t \to 0+$)

$$(3.2) \qquad \omega_r(t,f_{\omega,\alpha};C_{2\pi}) \begin{cases} = O(\omega(t^r)) \\ \neq o(\omega(t^r)), \end{cases}$$

$$\omega_r(t,f_{\omega,\alpha};C_{2\pi}) \neq o(\alpha(t)\omega_{r+q}(t,f_{\omega,\alpha};C_{2\pi})) .$$

<u>Proof.</u> For $X = C_{2\pi}$ consider $h_n(x) = e^{inx}$, $S_t f = \omega_r(t,f;C_{2\pi})$, $T_n = S_{1/n}$. Since $(n \in \mathbb{N})$

$$(3.3)\qquad |\Delta_u^r h_n(x)| = |1-e^{inu}|^r \begin{cases} \leqslant \min\{2^r,(n|u|)^r\} & u \in \mathbb{R} \\ & \text{for} \\ \geqslant (2n|u|/\pi)^r & |u| \leqslant \pi/n , \end{cases}$$

conditions (2.3-5) are satisfied with $\sigma(t) = t^r$, $\varphi_n = n^{-r}$, $C_1 = 1$, $C_2 = 2^r$, $C_3 = (2/\pi)^r$. In order to realize the gliding hump for the ratio of the functionals one has to select suitable subsequences. To this end, in view of (3.1) there exists $m_n \in \mathbb{N}$ with

$$(3.4)\qquad n < m_n, \quad \alpha(1/m_n) \leqslant (m_n/n)^q .$$

Now one may set

$$R_n = (m_n/n)^r S_{1/m_n} , \quad V_n f = (m_n/n)^{r+q}\omega_{r+q}(1/m_n,f;C_{2\pi}) .$$

Again with (3.3) it follows that also (2.6-8) hold true with $C_4 = C_3$, $C_5 = C_1$, $C_{6,n} = 0$. Therefore Theorem 2.1 delivers the existence of a function $f_{\omega,\alpha} \in C_{2\pi}$ with properties (2.9,10) which coincide with (3.2). Moreover, (2.11) reads

$$\omega_r(1/m_n,f_{\omega,\alpha};C_{2\pi}) \neq o((m_n/n)^q\omega_{r+q}(1/m_n,f_{\omega,\alpha};C_{2\pi}))$$

which in view of (3.4) completes the proof. □

Note that for $\omega(t) = t^{\beta/r}$, $0 < \beta < r$, assertion (3.2) is equivalent to (1.7), a choice of interest for $\alpha(t)$ may then be $t^{-\gamma}$, $0 < \gamma < q$. Of course, the present arguments also hold true for the Banach spaces $L^p_{2\pi}$, $1 \leqslant p < \infty$, with the same testfunctions h_n, as well as for the spaces $L^p(\mathbb{R})$ of functions, defined on the noncompact domain $\mathbb{R}$, where, however, a different choice of h_n is needed (cf. Section 3.4). See [6] for a corresponding treatment of Marchaud's inequality for functions, defined on compact intervals $[a,b] \subset \mathbb{R}$.

3.2 Steckin's problem. As already mentioned in Section 1, the whole story seems to go back to the following problem: For Π_n, the set of trigonometrical polynomials $p_n(x) := \sum_{k=-n}^{n} c_k e^{ikx}$ of degree n, consider the bounded, sublinear functional

$$E_n(f;C_{2\pi}) := \inf\{\|f-p_n\|_C ; p_n \in \Pi_n\} \tag{3.5}$$

of best approximation on $C_{2\pi}$. Then the theorem of Jackson states that for any $r \in \mathbb{N}$ (cf. [16, p. 260])

$$E_n(f;C_{2\pi}) \leq C_r\, \omega_r(1/n,f;C_{2\pi}) \qquad (f \in C_{2\pi}). \tag{3.6}$$

Conversely, it was shown by S.B. Steckin in 1951 that (cf. [16, p. 331 ff])

$$\omega_r(1/n,f;C_{2\pi}) \leq C_r\, n^{-r} \sum_{k=o}^{n} (k+1)^{r-1} E_k(f;C_{2\pi}) \qquad (f \in C_{2\pi}).$$

Both results match each other in as much as for $0 < \beta < r$

$$f \in \mathrm{Lip}(\beta,r;C_{2\pi}) \iff E_n(f;C_{2\pi}) = \mathcal{O}_f(n^{-\beta}).$$

In 1977 Steckin (cf. [2]) then posed the problem: Does there even exist an estimate of the more direct type

$$\omega_r(1/n,f;C_{2\pi}) \leq C_f\, E_n(f;C_{2\pi})$$

for each nondifferentiable function f, say? Again the answer is negative and was given by Boman [2]. Let us show how to regain (a quantitative extension of) Boman's result from the general theorem of Section 2.

COROLLARY 3.2. For each ω, satisfying (2.1.2), and each sequence $M := \{M_n\}$, tending to infinity, there exists $f_{\omega,M} \in C_{2\pi}$ with (3.2) and

$$\omega_r(1/n, f_{\omega,M}; C_{2\pi}) \neq o(M_n E_n(f_{\omega,M}; C_{2\pi})).$$

Proof. For $X = C_{2\pi}$ consider the functionals

$$S_t f = \omega_r(t, f; C_{2\pi}) \ , \quad T_n = R_n = S_{1/n} \ , \quad V_n f = M_n \, E_n(f; C_{2\pi}) \ .$$

With $h_n(x) = e^{inx}$ one obtains (2.3 - 6) with $\sigma(t) = t^r$, $\varphi_n = n^{-r}$, again using the elementary estimates (3.3). Since $V_m h_n = 0$ for $m \geq n$ (one of the particular features of the problem at hand), conditions (2.7,8) also hold true (with $C_5 = C_{6,n} = 0$). Hence (2.9 - 11) deliver the assertions. □

Note that, choosing $\omega(t) = t^{\beta/r}$, $0 < \beta < r$, and $M_n = n^\gamma, \gamma > 0$, the parameter β may be arbitrarily small (cf. (1.7)), whereas γ may be arbitrarily large !

3.3 Monotone approximation. Let $C[0,1]$ be the space of functions, continuous on the compact interval $[0,1]$, endowed with the standard sup - norm $\|\circ\|_C$. As usual, $C^{(r)}[0,1]$, $r \in \mathbb{N}$, denotes the subset of r - times continuously differentiable functions. For $n \in \mathbb{N}$ let $E_n(f; C[0,1])$ be the error of best approximation of $f \in C[0,1]$ by elements of P_n, the set of algebraic polynomials of degree at most n (cf. (3.5)). Consider the cone ($m \in \mathbb{N}$)

$$K_m := \{f \in C[0,1]; \ (\Delta_u^m f)(x) \geq 0 \ \text{for} \ 0 \leq x \leq x + mu \leq 1\}$$

and the error of best monotone approximation

$$E_n^{(m)}(f; C[0,1]) := \inf\{\|f - p_n\|_C \ ; \ p_n \in P_n \cap K_m\} \ .$$

In [14I] it is proved that for $f \in K_1$ there holds true the Jackson theorem

$$E_n^{(1)}(f;C[0,1]) \leqslant C \quad \omega_1(1/n,f;C[0,1])$$

which may be compared with the classical one (cf. (3.6))

$$E_n(f;C[0,1]) \leqslant C \quad \omega_1(1/n,f;C[0,1]) .$$

Thus one may expect an estimate of type (cf. [14I])

$$E_n^{(1)}(f;C[0,1]) \leqslant C \quad E_n(f;C[0,1]) \qquad (f \in K_1)$$

(the converse is trivial). But in [14 II] it is shown that there exists a m-times differentiable function $f_o \in K_m$ such that

$$\limsup_{n \to \infty} E_n^{(m)}(f_o;C[0,1])/E_n(f_o;C[0,1]) = \infty .$$

This result can also be reestablished by Theorem 2.1, in fact equipped with rates.

COROLLARY 3.3. <u>Let</u> $m,r \in \mathbb{N}$. <u>Then for each</u> ω, <u>satisfying</u> (2.1,2), <u>and each sequence</u> $M := \{M_n\}$, <u>tending to infinity, there exists</u> $f_{\omega,M} \in K_m$ <u>such that</u> $(t \to 0+,\ n \to \infty)$

$$\omega_r(t,f_{\omega,M};C[0,1]) \begin{cases} = O(\omega(t^r)) \\ \neq o(\omega(t^r)) , \end{cases}$$

$$E_n^{(m)}(f_{\omega,M};C[0,1]) \neq o(M_n\, E_n(f_{\omega,M};C[0,1])) .$$

<u>Proof.</u> Let us first construct the testfunctions h_n (cf. proof of Lemma in [14 II]). For $0 < a < 1$ let $g_a(x) := (x-a)^{2r+m+1} \in P_{2r+m+1}$ and

$$G_a(x) := \begin{cases} g_a(x), & a \leqslant x \leqslant 1 \\ 0 \quad , & 0 \leqslant x < a . \end{cases}$$

Then $G_a \in C^{(2r+m)}[0,1] \cap K_m$ with

$$\text{(3.7)} \qquad \| G_a^{(j)} \|_C \leqslant (2r+m+1)!/(2r+m+1-j)! \qquad (0 \leqslant j \leqslant r+m),$$

$$\text{(3.8)} \qquad E_n(G_a; C[0,1]) \leqslant \| G_a - g_a \|_C = a^{2r+m+1} \qquad (n \geqslant 2r+m+1).$$

Since $g_a^{(m)}(0) = -(2r+m+1)!a^{2r+1}/(2r+1)! < 0$ and $q_n^{(m)}(0) \geqslant 0$ for any $q_n \in \mathcal{P}_n \cap K_m$, Markoff's inequality (cf. [16, p. 218]) and (3.8) deliver for $n \geqslant 2r+m+1$

$$a^{2r+1} \leqslant |g_a^{(m)}(0)| \leqslant |g_a^{(m)}(0) - q_n^{(m)}(0)|$$

$$\leqslant n^{2m} \| g_a - q_n \|_C \leqslant n^{2m} (\| q_n - G_a \|_C + a^{2r+m+1}) ,$$

thus

$$\text{(3.9)} \qquad E_n^{(m)}(G_a; C[0,1]) \geqslant a^{2r+m+1}(n^{-2m} a^{-m} - 1) .$$

Now the Bernstein polynomials ($k \in \mathbb{N}$)

$$(B_k f)(x) := \sum_{j=o}^{k} f(\tfrac{j}{k}) \binom{k}{j} x^j (1-x)^{k-j}$$

are shape preserving in the sense that (cf. [13, p. 12])

$$\text{(3.10)} \qquad f \in K_m \Rightarrow B_k f \in K_m \qquad (k \geqslant m),$$

$$\text{(3.11)} \qquad f \in C^{(r)}[0,1] \Rightarrow \| (B_k f)^{(r)} \|_C \leqslant \| f^{(r)} \|_C \qquad (k \geqslant r) .$$

Since $\| B_j f - f \|_C = o_f(1)$, $j \to \infty$, for each $f \in C[0,1]$, there exists $k = k_a \in \mathbb{N}$ such that

$$\|B_k G_a - G_a\|_C \leqslant a^{2r+m+1} .$$

Setting $p_a := B_k G_a \in K_m$ (cf. (3.10)), one has by (3.7 - 9,11) that for $n \geqslant 2r+m+1$

$$\|p_a^{(j)}\|_C \leqslant (2r+m+1)!/(2r+m+1-j)! \qquad (0 \leqslant j \leqslant r+m), \tag{3.12}$$

$$E_n(p_a;C[0,1]) \leqslant \|p_a - G_a\|_C + E_n(G_a;C[0,1]) \leqslant 2\,a^{2r+m+1} , \tag{3.13}$$

$$\begin{aligned} E_n^{(m)}(p_a;C[0,1]) &\geqslant E_n^{(m)}(G_a;C[0,1]) - \|p_a - G_a\|_C \\ &\geqslant a^{2r+m+1}(n^{-2m}a^{-m}-2) . \end{aligned} \tag{3.14}$$

With $a_n := n^{-2}(M_n+2)^{-1/m} < 1$, thus $n^{-2m}a_n{}^{-m}-2 = M_n$, set

$$X = C[0,1],\; h_n(x) = p_{a_n}(x) + x^n \in K_m ,$$

$$S_t f = \omega_r(t,f;C[0,1]),\; T_n f = S_{1/n} f ,$$

$$R_n f = E_n^{(m)}(f;C[0,1])/a_n^{2r+m+1} M_n ,$$

$$V_n f = E_n(f;C[0,1])/a_n^{2r+m+1} .$$

Then (2.3,4) follow by (3.12) with $\sigma(t) = t^r$, $\varphi_n = n^{-r}$ since

$$S_t h_n \leqslant t^r \|h_n^{(r)}\|_C \leqslant t^r((2r+m+1)!/(r+m+1)! + n^r) .$$

Moreover, (2.5) holds true in view of

$$T_n p_{a_n} \leqslant n^{-r}\|p_{a_n}^{(r)}\|_C = o(1) \qquad (n \to \infty) ,$$

$$T_n(x^n) \geqslant \Big|\sum_{j=0}^{r}(-1)^j\binom{r}{j}\Big(1+\frac{j-r}{n}\Big)^n\Big| \to |1-e^{-1}|^r .$$

Concerning (2.6-8) note that $R_n(x^n) = 0 = V_m(x^n)$ for $m \geq n$. Therefore (2.6,7) follow by (3.13,14) whereas (2.8) is obvious since h_n is a polynomial, thus $V_m h_n = 0$ for m large enough. Hence an application of Theorem 2.1 in connection with Remark 2.2 establishes the assertion. □

3.4 <u>On the comparison of moduli of continuity.</u> For $1 \leq p < \infty$ let $L^p_{2\pi}$ and $L^p(\mathbb{R})$ be the space of 2π-periodic and measurable functions f, *pth* power integrable over $[0,2\pi]$ and $\mathbb{R}$ with norm

$$\{\int_0^{2\pi} |f(u)|^p \, du\}^{1/p}, \quad \{\int_{-\infty}^{\infty} |f(u)|^p \, du\}^{1/p},$$

respectively. Let X be one of the spaces $C_{2\pi}$, $L^p_{2\pi}$, or $L^p(\mathbb{R})$, $1 \leq p < \infty$. Then the *rth* modulus of continuity $\omega_r(t,f;X)$ as well as the Lipschitz classes $\mathrm{Lip}(\beta,r;X)$ are defined as in (1.1,5).

For $0 < \beta < r \in \mathbb{N}$ the following question arises as an inverse theorem in approximation theory: Given a positive sequence $\{\delta_n\}$, tending monotonically to zero, does $\|\Delta^r_{\delta_n} f\|_X = O(\delta_n^\beta)$ for $n \to \infty$ already imply $\|\Delta^r_\delta f\|_X = O(\delta^\beta)$ for $\delta \to 0$, provided

$$\delta_n = O(\delta_{n+1}) \qquad (n \to \infty)\ ? \tag{3.15}$$

It seems that this problem was posed by DeVore [3, Chapter 8] who treated the case $X = C_{2\pi}$, $r = 2$. For $X = L^p_{2\pi}$ Freud [12] discussed $p = 2$, $r \in \mathbb{N}$ and Ditzian [9] $1 \leq p < \infty$, $r = 2$. In a general setting Boman [1] gave an affirmative answer for all of the above spaces and $r \in \mathbb{N}$.

Our contribution to the problem consists in a unified approach towards the necessity of condition (3.15) which was shown by H.S. Shapiro (cf. [3, p. 239]) for $X = C_{2\pi}$, $r = 2$ and by Freud [12] for $X = L^p_{2\pi}$, $r \in \mathbb{N}$ (the latter treatment may be compared with Section 2).

COROLLARY 3.4. <u>Let</u> $0 < \beta < r \in \mathbb{N}$, $0 < \delta_{n+1} \leq \delta_n = o(1)$. <u>On</u> X <u>the assertion</u> $(n \to \infty)$

(3.16) $$\|\Delta^r_{\delta_n} f\|_X = \mathcal{O}_f(\delta_n^\beta) \Longleftrightarrow f \in \mathrm{Lip}(\beta,r;X)$$

is equivalent to (3.15).

Proof. For the sufficiency see [1;3;9;12]. To show the necessity of (3.15), consider the function

$$g_t(x) := \begin{cases} G(x/t)\,, & X = C_{2\pi}\,, L^p_{2\pi} \\ t^{-1/p}\,G(x/t)\,, & X = L^p(\mathbb{R})\,, \end{cases}$$

$$G(x) := \begin{cases} e^{ix}\,, & X = C_{2\pi}\,, L^p_{2\pi} \\ e^{-x^2}\,, & X = L^p(\mathbb{R})\,, \end{cases}$$

which satisfies $(t > 0)$

(3.17) $$\|g_t\|_X = \|G\|_X = A\,,$$

(3.18) $$\|g_t^{(r)}\|_X = t^{-r}\|G^{(r)}\|_X = B_r\, t^{-r}\,,$$

(3.19) $$\|\Delta^r_t g_t\|_X = \|\Delta^r_1 G\|_X = C_r > 0\,.$$

Now assume that (3.15) does not hold, i.e., there exists a subsequence $\{k_n\} \subset \mathbb{N}$ such that $\alpha_n := \delta_{k_n}/\delta_{k_n+1}$ tends to infinity. Choose s such that $\beta/r < s < 1$, and set

$$h_n = g_{\varepsilon_n}\,, \quad \varepsilon_n := \delta^s_{k_n}\,\delta^{1-s}_{k_n+1}\,,$$

$$S_n f = \|\Delta^r_{\delta_n} f\|_X\,, \quad T_n f = \omega_r(\varepsilon_n, f; X)\,,$$

upon replacing the continuous parameter $t \to 0+$ by a discrete one. Obviously,(2.3,5) are fulfilled by (3.17,19). Moreover, for

$k_n < m \in \mathbb{N}$ one has by (3.18) (cf. (3.3))

$$S_m h_n \leqslant \delta_m^r \| h_n^{(r)} \|_X \leqslant B_r \delta_m^r \varepsilon_n^{-r}$$

$$= B_r (\delta_m/\delta_{k_n})^{sr} (\delta_m/\delta_{k_n+1})^{(1-s)r} \leqslant B_r (\delta_m/\delta_{k_n})^{sr} \leqslant B_r ,$$

and for $k_n \geqslant m$ by (3.17)

$$S_m h_n \leqslant 2^r \| h_n \|_X \leqslant A 2^r \leqslant A 2^r (\delta_m/\delta_{k_n})^{sr} ,$$

thus (2.4) with $\sigma_m = \delta_m^{sr}$, $\varphi_n = \delta_{k_n}^{sr}$. Then the relevant assertions (2.9,10) already deliver the existence of some $f_o \in X$ such that with $\omega(t) = t^{\beta/sr}$ (cf. (2.1,2))

$$\| \Delta_{\delta_n}^r f_o \|_X = O(\omega(\delta_n^{sr})) = O(\delta_n^{\beta}) ,$$

$$\omega_r(\varepsilon_n, f_o; X) \neq o(\delta_{k_n}^{\beta}) .$$

Since $\delta_{k_n}^{\beta} / \varepsilon_n^{\beta} = \alpha_n^{\beta(1-s)}$ tends to infinity, this is a contradiction to (3.16), proving the necessity. Note that there is no need for functionals R_n, V_n so that conditions (2.6-8) as well as assertion (2.11) cancel. □

Analogously, one may discuss the necessity of (3.15) for the whole class of multiplier operators, considered by Boman [1] (in connection with the sufficiency of (3.15)).

REFERENCES

1. J. Boman, On a problem concerning moduli of smoothness. In: Fourier Analysis and Approximation Theory, Vol. I (Proc. Conf. Budapest, 1976, Eds. G. Alexits and P. Turán). North-Holland Publ. Comp., Amsterdam 1978, 175 - 179.

2. J. Boman, A problem of Steckin on trigonometric approximation. In: Constructive Funktion Theory '77 (Proc. Conf. Blagoevgrad, 1977, Eds. Bl. Sendov and D. Vacov). Publ. House Bulg. Acad. Sci., Sofia 1980, 269 - 273.

3. R.A. DeVore, The Approximation of Continuous Functions by Positive Linear Operators. Lecture Notes in Mathematics 293. Springer, Berlin 1972.

4. R.A. DeVore, S.D. Riemenschneider, and R.C. Sharpley, Weak interpolation in Banach spaces. J. Functional Analysis 33 (1979), 58 - 94.

5. W. Dickmeis and R.J. Nessel, Condensation principles with rates. Studia Math. 75 (1982), 55 - 68.

6. W. Dickmeis, R.J. Nessel, and E. van Wickeren, On the sharpness of estimates in terms of averages. Math. Nachr. (in print).

7. W. Dickmeis, R.J. Nessel, and E. van Wickeren, A general approach to quantitative negative results in approximation theory (in print).

8. W. Dickmeis, R.J. Nessel, and E. van Wickeren, A general approach to counterexamples in numerical analysis (in print).

9. Z. Ditzian, Inverse theorems for functions in L_p and other spaces. Proc. Amer. Math. Soc. 54(1976), 80-82.

10. J. Favard, Sur la comparaison des procédés de summation. In: On Approximation Theory (Proc. Conf. Oberwolfach, 1963, Eds. P.L. Butzer and J. Korevaar, ISNM 5), Birkhäuser, Basel 1964, 4 - 11.

11. J. Favard, On the comparison of the processes of summation. SIAM J. Numer. Anal. 1(1964), 38 - 52.

12. G. Freud, On the problem of R. DeVore. Canad. Math. Bull. 17(1974), 39 - 44.

13. G.G. Lorentz, Bernstein Polynomials. Univ. of Toronto Press, Toronto 1953.

14. G.G. Lorentz and K. Zeller, Degree of approximation by monotone polynomials I;II. J. Approx. Theory 1(1968), 501 - 504; 2(1969), 265 - 269.

15. R.J. Nessel and E. van Wickeren, Negative results in connection with Favard's problem on the comparison of approximation processes. (Proc. Conf. Oberwolfach on Functional Analysis and Approximation, 1983, Eds. P.L. Butzer, R.L. Stens, and B. Sz.-Nagy) Birkhäuser, Basel 1984 (in print).

16. A.F. Timan, Theory of Approximation of Functions of a Real Variable. Pergamon Press, New York 1963.

R.J. Nessel and E. van Wickeren, Lehrstuhl A für Mathematik, RWTH Aachen, Templergraben 55, D - 5100 Aachen, Germany

Functional Inequalities

Deciduous Plantation

International Series of
Numerical Mathematics, Vol. 71

ON SCHUR-CONCAVE t-NORMS AND TRIANGLE FUNCTIONS

Claudi Alsina

Abstract. In this paper various functional inequalities which arise in the study of Schur-concavity of t-norms and triangle functions are considered and solved.

1. INTRODUCTION

The aim of this paper is to solve various functional inequalities which arise in the study of the Schur-concavity of t-norms and triangle functions. Schur-concave functions have been studied in detail in [5]. t-norms, as topological semigroups of the closed unit interval, and triangle functions, as topological semigroups of the space of probability distribution functions, have been studied in recent years in relation with the theory of probabilistic metric spaces (for a complete reference see [8]).

The major purpose of the present work is to characterize, in Section 2, Schur-concave t-norms: any t-norm which is indeed a copula is always Schur-concave. In Section 3 we analyse when the triangle functions π_C , τ_C and σ_C , induced by a copula and t-norm C, are Schur-concave. In this context it is quite surprising that while π_C are <u>always</u> Schur-concave, operations σ_C and τ_C are <u>never</u> at the same time Schur-concave and associative.

Throughout the paper we assume that the reader is familiar with the basic definitions and concepts of the theory of t-norms (see,e.g.[1], [4] and [8]), copulas (see,e.g.[8]) and triangle functions (see,e.g.[6], [7], [8], [3], and [2]).From now on 'I' will denote the closed unit interval [0,1], $\bar{\alpha}$ will indicate the number $1-\alpha$ whenever α is in I, and M, M* and W will denote, respectively, the operations on I defined by M(a,b)= Min(a,b), M*(a,b)= Max(a,b) and W(a,b)= M*(a+b-1,0).

2. ON SCHUR-CONCAVE t-NORMS

We need to recall first the following definitions (see, e.g. [5]):

DEFINITION 2.1. Given (a,b) and (c,d) in IxI we say that (a,b) is _majorized_ by (c,d) and we write $(a,b) \preccurlyeq (c,d)$, if $\mathrm{Max}(a,b) \leqslant \mathrm{Max}(c,d)$ and $a+b = c+d$, i.e., if there exists α in I such that $a = \alpha c + \overline{\alpha} d$ and $b = \overline{\alpha} c + \alpha d$.

DEFINITION 2.2. A two-place function T from IxI into I is said to be _Schur-concave_ if $T(a,b) \geqslant T(c,d)$ whenever $(a,b) \preccurlyeq (c,d)$ in IxI.

In the next lemmas we will characterize Schur-concave t-norms.

LEMMA 2.1. _M and W are Schur-concave t-norms. If T is a Schur concave t-norm then_ $W \leqslant T \leqslant M$. _Thus, M is the strongest Schur-concave t-norm and W is the weakest._

Proof. It is easy to see that M and W are Schur-concave. Let T be a Schur-concave t-norm. If x and y are in I and $x+y \geqslant 1$ then $(x,y) \preccurlyeq (x+y-1,1)$ and we obtain $T(x,y) \geqslant T(x+y-1,1) = x+y-1 = W(x,y)$. Of course if $x+y \leqslant 1$ then $W(x,y) = 0 \leqslant T(x,y)$. Whence $W \leqslant T$. The inequality $T \leqslant M$ is satisfied by any t-norm.

LEMMA 2.2. _Let T be a strict t-norm with additive generator t. Then T is Schur-concave if and only if t is convex_, i.e., _if and only if T is a copula_.

Proof. For any a,b in I it is $(\frac{a+b}{2}, \frac{a+b}{2}) \preccurlyeq (a,b)$. Thus, if T is a Schur-concave strict t-norm with additive generator t we will get

$$t^{(-1)}(t(a)+t(b)) = T(a,b) \leqslant T(\frac{a+b}{2}, \frac{a+b}{2}) = t^{(-1)}(2t(\frac{a+b}{2})),$$

and since $t^{(-1)}$ is a continuous strictly decreasing function, we have

$$t(a) + t(b) \geqslant 2t(\frac{a+b}{2}),$$

i.e., t is convex. To prove sufficiency, assume that T has a convex additive generator t and let $(a,b) \preccurlyeq (c,d)$ in IxI. Then $a = \alpha c + \overline{\alpha} d$ and $b = \overline{\alpha} c + \alpha d$, for some α in I, and the convexity of t yields

$$t(a) = t(\alpha c + \bar{\alpha} d) \leqslant \alpha t(c) + \bar{\alpha} t(d),$$

$$t(b) = t(\bar{\alpha} c + \alpha d) \leqslant \bar{\alpha} t(c) + \alpha t(d),$$

consequently adding both inequalities we have

$$t(a) + t(b) \leqslant t(c) + t(d),$$

whence

$$T(a,b) = t^{-1}(t(a) + t(b)) \geqslant t^{-1}(t(c) + t(d)) = T(c,d),$$

and T is Schur-concave.

LEMMA 2.3. *Let T be a non-strict Archimedean t-norm with additive generator t. Then T is Schur-concave if and only if t satisfies the following inequality:*

$$t(\alpha x + \bar{\alpha} y) + t(\bar{\alpha} x + \alpha y) \leqslant t(x) + t(y) \tag{2.1}$$

for all x,y in I such that $t(x) + t(y) \leqslant 1$, and for every α in I.

The proof. of Lemma 2.3 is an immediate consequence of Definitions 2.1 and 2.2 and the representation for t-norms. We remark that condition (2.1) is a convexity relation for t in a restricted domain. Obviously if the function t is convex on I then (2.1) is satisfied and we have:

LEMMA 2.4. *If T is a non-strict Archimedean t-norm which is a copula then T is Schur-concave.*

Unfortunally the converse of Lemma 2.4 is not true. To see this consider the following example suggested by A. Sklar :

EXAMPLE 2.1. Let t be the continuous strictly decreasing function from I into I defined by:

$$t(x) = \frac{1 + \cos \pi x}{2} .$$

Then t is an additive generator for the non-strict Archimedean t-norm:

$$T(x,y) = \begin{cases} \frac{1}{\pi} \text{ arcos } (2\cos \pi x + 2\cos \pi y + 1), & \text{if } x + y \geqslant 1, \\ 0 & , \text{if } x + y \leqslant 1. \end{cases}$$

Since t is not convex T cannot be a copula. But t satisfies (2.1) so that T is Schur-concave. This last property is easily seen by using the properties of the cosine function and the well-known trigonometric identity

$$\cos \pi A + \cos \pi B = 2\cos \frac{\pi}{2}(A+B) \cos \frac{\pi}{2}(A-B).$$

Finally we can consider ordinal sums of Archimedean t-norms and a tedious study yields:

LEMMA 2.5. *Let T be an ordinal sum of Archimedean t-norms* $\{T_i \mid i \in I\}$. *Then T is Schur-concave if and only if each t-norm* T_i, $i \in I$, *is Schur-concave.*

As consequence of the above lemmas we have proved the following

THEOREM 2.1. *If T is a t-norm and a copula then T is Schur-concave. Any Schur-concave strict t-norm is a copula.*

The above theorem gives an interesting geometrical property of a copulas: if $z=C(x,y)$ is the surface associated to a copula C which is also a t-norm the intersections of the surface with all the vertical planes of the form $x+y = a$ $(0 \leqslant a \leqslant 2)$ are curves which are increasing from $(0,a)$ (resp. $(a-1,1)$) to $(\frac{a}{2}, \frac{a}{2})$, decreasing from $(\frac{a}{2}, \frac{a}{2})$ to $(a,0)$ (resp. $(1,a-1)$). Hence the maximum is obtained at $(\frac{a}{2}, \frac{a}{2})$ which is a point located on the diagonal of IxI.

3. ON SCHUR-CONCAVE TRIANGLE FUNCTIONS

In this section we will apply the above study on copulas and Schur-concavity to analyse the Schur-concavity of some families of triangle functions.

DEFINITION 3.1. Given two pairs of distribution functions (F,G) and (H,K) in $\Delta^+ \times \Delta^+$ we say that (F,G) is *majorized* by (H,K) if there exists an α in I such that $F = \alpha H + \overline{\alpha}K$ and $G = \overline{\alpha}H + \alpha K$. In this case we write $(F,G) \leqslant (H,K)$. *A triangle function* τ *is Schur-concave* if $\tau(F,G) \geqslant \tau(H,K)$ whenever $(F,G) \leqslant (H,K)$ in $\Delta^+ \times \Delta^+$. Thus the Schur-concavity of a triangle function is equivalent to the inequality:

$$\tau(H,K) \leqslant \tau(\alpha H + \overline{\alpha}K, \overline{\alpha}H + \alpha K),$$

for all H,K in Δ^+ and for all α in I.

THEOREM 3.1. <u>For any t-norm and copula</u> C <u>the triangle function</u> π_C <u>is Schur-concave. In particular,</u> π_M <u>is the strongest Schur-concave triangle function.</u>

<u>Proof</u>. By Theorem 2.1 the associative copula C is necessarily Schur-concave on IxI. Assume $(F,G) \leq (H,K)$, i.e., $F = \alpha H + \bar{\alpha} K$ and $G = \bar{\alpha} H + \alpha K$, for some α in I. Applying the Schur-concavity of C we obtain

$$\pi_C(F,G)(x) = C(F(x),G(x)) = C(\alpha H(x) + \bar{\alpha} K(x), \bar{\alpha} H(x) + \alpha K(x)) \geq C(H(x),K(x)) = \pi_C(H,K)(x).$$

Moreover, any triangle function τ satisfies $\tau \leq \pi_M$, i.e., π_M is the strongest Schur-concave triangle function.

Now we will analyse the Schur-concavity of τ_C and σ_C operations.

THEOREM 3.2. <u>For any t-norm</u> C <u>the triangle function</u> τ_C <u>is never Schur-concave</u>.

<u>Proof</u>. If there would exists an operation τ_C Schur-concave then we would obtain for any α in [0,1]:

$$1 = \varepsilon_1(2) = \tau_C(\varepsilon_1,\varepsilon_0)(2) \leq \tau_C(\alpha\varepsilon_1 + \bar{\alpha}\varepsilon_0, \bar{\alpha}\varepsilon_1 + \alpha\varepsilon_0)(2) = M^*(\alpha,1-\alpha) < 1,$$

which is a contradiction.

THEROEM 3.3. <u>There exists no t-norm copula</u> C <u>such that the operation</u> σ_C <u>is a Schur-concave triangle function</u>.

<u>Proof</u>. Suppose that there would exists an associative copula C such that the operation σ_C would be Schur-concave.Then for any a,b in I it is

$$(G_a,G_b) \geq (G_{(a+b)/2}, G_{(a+b)/2})$$

where for any x in I the function G_x is defined by

$$G_x(t) = \begin{cases} 0, & t \leq 0, \\ x, & 0 < t \leq 1, \\ 1, & 1 < t. \end{cases}$$

According with the Schur-concavity of σ_C and the fact that

$$\sigma_C(G_x, G_y)(t) = \begin{cases} 0, & t \leq 0, \\ C(x,y), & 0 < t \leq 1, \\ x+y-C(x,y) & 1 < t \leq 2, \\ 1, & 2 < t. \end{cases}$$

we would have

$$\sigma_C(G_a, G_b)(2) \leq \sigma_C(G_{(a+b)/2}, G_{(a+b)/2})(2),$$

i.e.,

$$a + b - C(a,b) \leq \frac{a+b}{2} + \frac{a+b}{2} - C\left(\frac{a+b}{2}, \frac{a+b}{2}\right),$$

so that

$$C\left(\frac{a+b}{2}, \frac{a+b}{2}\right) \leq C(a,b) \tag{3.6}$$

But by Theorem 2.1 any copula is Schur-concave whence

$$C(a,b) \leq C\left(\frac{a+b}{2}, \frac{a+b}{2}\right). \tag{3.7}$$

It follows from (3.6) and (3.7) that C must necessarily be a solution of the functional equation

$$C\left(\frac{a+b}{2}, \frac{a+b}{2}\right) = C(a,b) \tag{3.8}$$

If x,y in I are such that $x + y > 1$ we have by (3.8):

$$W(x,y) = x + y - 1 = C(x + y - 1, 1) = C\left(\frac{x+y-1+1}{2}, \frac{x+y-1+1}{2}\right) = C(x,y),$$

whence C=W on $\{(x,y) \in I^2 \mid x + y \geq 1\}$ and consequently C=W on IxI. But we know that σ_W is not a triangle function because σ_W does not satisfy the associativity equation (see [3]). In conclusion no operation σ_C can be a Schur-concave triangle function and the theorem is proved.

Finally we will prove a strong result for a large collection of triangle functions.

THEOREM 3.4. <u>Let τ be a triangle function such that there exist a t-norm T for which the following property is satisfied:</u>

$\tau(\lambda\varepsilon_a,\mu\varepsilon_b) = T(\lambda,\mu).\varepsilon_{Max(a,b)}$ _for all_ $\lambda,\mu\in I$ _and_ $a,b\geqslant 0$.(3.8) _Then, if_ τ _is Schur-concave necessarily_ $\tau\geqslant\pi_W$, i.e., π_W _is the weakest Schur-concave triangle function satisfying_ (3.8).

Proof. First we show that if τ is Schur-concave and satisfies (3.8) then necessarily T is Schur-concave. In effect, if $a = \alpha c+\overline{\alpha}d$ and $b = \overline{\alpha}c+\alpha d$ are numbers in I, then:

$$T(a,b) = T(a,b).\varepsilon_0(1) = T(a,b)\varepsilon_{Max(0,0)}(1) = \tau(a\varepsilon_0,b\varepsilon_0)(1)$$

$$= \tau(\alpha(c\varepsilon_0) + \overline{\alpha}(d\varepsilon_0),\ \overline{\alpha}(c\varepsilon_0) + \alpha(d\varepsilon_0))(1)$$

$$\geqslant \tau(c\varepsilon_0,\ d\varepsilon_0)(1) = T(c,d),$$

and it follows from Lemma 3.1 that $T\geqslant W$. Next consider any F,G in Δ^+ and $x,y>0$. For all δ in $[0,M(x,y)]$ it is easily seen that the following inequalities hold:

$$F\geqslant F(x-\delta)\varepsilon_{x-s} \quad\text{and}\quad G\geqslant G(x-\delta)\varepsilon_{x-s}\ .$$

Then we have:

$$\tau(F,G)(x) \geqslant \tau(F(x-\delta)\varepsilon_{x-\delta},\ G(x-\delta)\varepsilon_{x-\delta})(x)$$

$$= T(F(x-\delta),G(x-\delta))\varepsilon_{x-\delta}(x)$$

$$= T(F(x-\delta),\ G(x-\delta) \geqslant W(F(x-\delta),G(x-\delta))$$

which in turn implies (using the left-continuity of F and G and the arbitrariety of δ):

$$\tau(F,G)(x) \geqslant W(F(x),G(x)),$$

whence $\tau \geqslant \pi_W$.

REFERENCES

1. J.Aczél, Functional equations and their applications. Academic Press, New York, 1966.

2. C.Alsina, Some functional equations in the space of uniform distribution functions. Aequationes Math. 22(1981),153-164.

3. M.J.Frank, Associativity in a class of operations on spaces of distribution functions. Aequationes Math. 12(1975),121-144.

4. C.H.Ling, Representation of associative functions. Publ.Math. Debrecen 12 (1965), 189-212.

5. A.W.Marshall and I.Olkin, Inequalities: Theory of majorization and its applications. Academic Press, New York, 1979.

6. R.Moynihan, B.Schweizer and A.Sklar, Inequalities among operations on probability distribution functions. General Inequalities I, ed. by E.F.Beckenbach, Birkhäuser Verlag Basel, 1978, pp.133-149.

7. B.Schweizer, Multiplications on the space of probability distribution functions. Aequationes Math. 12(1975), 151-183.

8. B.Schweizer and A.Sklar, Probabilistic metric spaces. Elsevier North-Holland, New York, 1982.

Claudi Alsina, Departament de Matemàtiques i Estadística (ETSAB), Universitat Politècnica de Barcelona, Diagonal 649, Barcelona 28, Spain.

International Series of
Numerical Mathematics, Vol. 71

STABILITY OF SOME ITERATIVE FUNCTIONAL EQUATIONS

Bogdan Choczewski

Abstract. The purpose of this paper is to present a brief survey of known notions of and results on the stability of linear functional equations of the iterative type.

1. INTRODUCTION

We start with the inhomogeneous functional equation

$$\phi(f(x)) = g(x)\phi(x) + h(x)\,, \quad x \in I := |0,A|\,, \quad 0 < A \leq \infty\,, \tag{1}$$

where $|0,A|$ stands for any of the four intervals with the ends 0 and A. We shall assume that

(H) $f\colon I \to I$ is a continuous and strictly increasing function on I, $0 < f(x) < x$ in $I \setminus \{0\}$; $g\colon I \to R$ is continuous in I, $g(x) \neq 0$ in $I \setminus \{0\}$; $h\colon I \to R$ is also continuous in I.

Let us denote for short

$$S\phi := (g\circ f^{-1})\cdot(\phi\circ f^{-1}) + h\circ f^{-1}\,,$$

so that equation (1) becomes

$$\phi(f(x)) = (S\phi)(f(x))\,, \quad x \in I\,. \tag{2}$$

2. STABILITY AND ITERATIVE STABILITY

The following notion of stability is due to D. Brydak [1].

DEFINITION 1. Equation (2) is said to be *stable* in the interval I if there is a positive K such that for every positive ε and every continuous function $\psi\colon I \to R$ satisfying the inequality

$$|\psi(f(x))-(S\psi)(f(x))| \leq \varepsilon\,, \quad x \in I\,,$$

there exists a continuous solution $\phi\colon I \to R$ of (2) such that

(3) $$|\phi(x) - \psi(x)| \le K\varepsilon , \quad x \in I .$$

This is an adaptation of the definition proposed by D.H. Hyers in [5] for the Cauchy's additive functional equation.

It turns out that equation (2) is, in general, unstable.

EXAMPLE 1. Consider the equation

$$\phi(x/(x+1)) = \phi(x) , \quad x \in I := (0,\infty) .$$

Continuous solutions in I of the equation are all bounded, while the corresponding inequality

$$|\psi(x/(x+1))-\psi(x)| \le \varepsilon , \quad x \in I ,$$

is satisfied also by the unbounded function $\psi(x) = \varepsilon/x$, $x \in I$.

This fact motivates the following modification of Definition 1, which has also been introduced by D. Brydak [1]. Before reproducing it, let us observe that (2) is equivalent to the sequence of equations

$$\phi(f^n(x)) = (S^n\phi)(f^n x)) , \quad x \in I , \quad n \in N ,$$

where the upper indices denote iterates of the function f and the operator S, respectively.

DEFINITION 2. Equation (2) is called *iteratively stable* if there exists a constant $K > 0$ with the following property: For every $\varepsilon > 0$ and every function ψ continuous in I and satisfying the system of inequalities

$$|\psi(f^n(x))-(S^n\psi)(f^n(x))| \le \varepsilon , \quad x \in I , \quad n \in N ,$$

there exists a continuous solution ϕ of (2) in I satisfying (3).

3. HOMOGENEOUS EQUATION

The stability problem for equation (2) can be reduced to that for the linear homogeneous equation

(4) $$\phi(f(x)) = g(x)\phi(x) , \quad x \in I .$$

This is our

THEOREM 1 ([3]). _Assume_ (H) _and that there exists a continuous solution of_ (2) _in_ I. _If equation_ (4) _is stable_ (_resp. iteratively stable_) _in_ I, _then the same is true of equation_ (2).

The proof of this fact is straightforward. It makes use of the form of solutions of any linear inhomogeneous equation.

Let (H_o) stand for (H) in the case where $h = 0$. To study stability of equation (4) we need some information on its continuous solutions in I.

If $0 \notin I$ then equation (4) possesses a family of continuous solutions in I, depending on an arbitrary function.

For the case where $0 \in I$ the following facts have been proved in [2]; cf. also [6; Chap. II]. The set of those solutions of (4) which remain continuous at the origin depends on the behaviour of the sequence

$$G_n(x) := \prod_{i=0}^{n-1} g(f^i(x)) \quad (x \in I\,,\ n \in N)\,, \quad G_o := 1\,.$$

This sequence appears after iterating equation (4) n times, i.e. (with $h = 0$ in the definition of S),

$$(S^n\phi)(f^n(x)) = G_n(x)\,, \quad x \in I\,, \quad n \in N\,.$$

There are three cases to be considered (solution means continuous solution in I).

(A) If G_n converges pointwise in I to a continuous function $G \not\equiv 0$, then there exists a one-parameter family of solutions of (4), given by $\phi(x) = c/G(x)$, where $x \in I$ and $c \in R$.

(B) If G_n approaches the limit zero almost uniformly in a subinterval of I, then the solution depends on an arbitrary function.

(C) If neither (A) nor (B) applies, (4) has only the trivial solution.

4. SOME RESULTS ON STABILITY

In the first case (A) we have the following result.

THEOREM 2 ([1]). _Assume that_ (H_o) _and_ (A) _hold. If_ G _is_ *bounded below by a positive constant, then equation (4) is iteratively stable in* I.

In case (B) there exists always an open set, say $U \subset I$, which is maximal for almost uniform convergence of G_n to zero, cf. [13].

THEOREM 3 ([3]). *Assume that* (H_o) *and* (B) *hold. If there exist positive numbers* C_1, C_2, C_3, *an interval* $I_o := [f(x_o), x_o] \subset I \setminus \{0\}$ *and a point* $x_1 \geq x_o$ *such that for every positive integer* n *we have* $C_1 \leq |G_n(x)|$ *in* $I \setminus (U \cup \{0\})$, $|G_n(x)| \leq C_2$ *in* $I \cap U$ *and* $|G_n(f^{-n}(x))| \geq C_3$ *for* $x \in [x_1, A|$, *then equation* (4) *is iteratively stable in* I.

Another result valid in case (B) is formulated in

THEOREM 4 ([1]). *If* (H_o) *holds and if, for some positive* $x_o \in I$, *the sequence* G_n *converges to zero, uniformly in* I_o *(*I_o *is defined in Theorem* 3), *then equation* (4) *is stable in* $[0, x_o)$.

If $|g(0)| > 1$, then obviously case (C) applies. We have

THEOREM 5 ([9]). *Assume* (H_o) *and* $|g(0)| > 1$. *Then equation* (4) *is iteratively stable in* I.

The subsequent example shows that in this case equation (4) need not be stable (cf. Definition 1).

EXAMPLE 2 ([9]). Let $I = [0,1)$, $f(x) = x/2$ and $g(x) = 2(1-x)$. The sequence G_n tends to infinity on I and by (C) $\phi = 0$ is the unique continuous solution of equation (4) in I. But the associated inequality has the unbounded solution $\psi(x) = \varepsilon/(1-x)$ in I.

Other results on stability and/or iterative stability of linear equations can be found in [1], [9], [10]. Stability of non-linear equations has been dealt with by E. Turdza in [8], [11].

5. SET STABILITY

We conclude the paper with remarks on the stability of sets with respect to equation (1). The notion of set stability for a difference equation has been introduced by G.A. Shanholt [7]. An analogous notion for an iterative functional equation makes only sense, when the equation has a set of solutions which is large enough. This actually is the case if $0 \notin I$.

We restrict consideration to equation (4) and to the case where

$$(H_1) \qquad 0 \notin I, \quad g(x) > 0 \quad \text{in } I.$$

LEMMA 1 ([6; p.46]). *Assume that* (H_0) *and* (H_1) *hold. The solutions of equation* (4) *depend on an arbitrary function, i.e., every function* $\phi_o\colon I_o \to R$ $(I_o := [f(x_o),x_o]$, x_o *arbitrarily fixed in* $I)$, *which is continuous in* I_o *and such that*

$$(5) \qquad \phi_o(x_o) = y_o, \qquad \phi_o(f(x_o)) = g(x_o)y_o,$$

can be uniquely extended onto the whole of I *as a continuous solution*, $\phi(\cdot,x_o,y_o,\phi_o)$, *of equation* (4).

Now, let $|a,b|$ be any real interval, $-\infty < a \le b < \infty$.

DEFINITION 3. The interval $|a,b|$ is said to be stable (with respect to equation (4) if for every $x_o \in I$ and $\varepsilon > 0$ there exists a $\delta > 0$ such that for every function $\phi_o\colon I_o \to R$, continuous in I_o, satisfying (5), the graph of which lies in $I_o \times (a-\delta,b+\delta)$, the corresponding solution-extension $\phi_o(\cdot,x_o,y_o,\phi_o)$ of (4) (cf. Lemma 1) has its graph in $(0,x_o] \times (a-\varepsilon,b+\varepsilon)$. If δ does not depend on x_o, then the interval $|a,b|$ is called uniformly stable.

We do not supply natural modifications of Definition 3 for the case of improper intervals. Some necessary and sufficient conditions for stability of intervals have been found by M. Czerni [4].

THEOREM 6. *Assume that* (H_o) *and* (H_1) *hold. Then:*

(i) *Either* $a > 0$ *or* $b < 0$. *A non-degenerate proper interval* $|a,b|$ *is stable iff (in the whole of* I) *either* $g(x) = 1$ *or* $g(x) > \max\{a/b, b/a\}$ *or* $g(x) < \min\{a/b,b/a\}$.

(ii) $0 \in (a,b)$. *The intervals* $(-\infty,b|$ *and* $|a,+\infty)$ *are stable iff* $g(x) \geq 1$ *in* I.

(iii) $0 \in |a,b| \neq R$. *A non-degenerate proper or improper interval* $|a,b|$ *is stable iff* $g(x) \leq 1$ *in* I.

(iv) *Each of the intervals* $\{0\}$, $|0,+\infty)$, $(-\infty,0|$ *is stable iff* $\sup_{n\in N} \max_{[f(x),x]} G_n(t) < \infty$ *for* $x \in I$.

In the case where $g(x) < 0$ in I there are unstable intervals with respect to equation (4), and no other assumptions than (H_o) and $0 \notin I$ are imposed, cf. [4].

REMARK. Stability of compact, connected subsets of a Banach space with respect to a non-linear iterative functional equation has been considered by E. Turdza [12]. The result obtained is similar to one by G.A. Shanholt [7].

REFERENCES

1. D. Brydak, On the stability of the functional equation $\phi(f(x)) = g(x)\phi(x) + F(x)$. Proc. Amer. Math. Soc. 26 (1970), 455-460.

2. B. Choczewski and M. Kuczma, On the "indeterminate case" in the theory of a linear functional equation. Fund. Math. 58 (1966), 163-175.

3. B. Choczewski, E. Turdza and R. Węgrzyk, On the stability of a linear functional equation. Rocznik Nauk.-Dydak. WSP w Krakowie, Prace Mat IX, 69 (1979), 15-21.

4. M. Czerni, Interval stability for linear homgeneous functional equations. Uniw. Śląski w Katowicach Prace Nauk.-Prace Mat. (to appear).

5. D.H. Hyers, On the stability of the linear functional equation. Proc. Nat. Acad. Sci. U.S.A. 27 (1941), 222-224.

6. M. Kuczma, Functional equations in a single variable. Polish Scientific Publishers, Warszawa, 1968.

7. G.A. Shanholt, Set stability for difference equations. Int. J. Control 19 (1974), 309-314.

8. E. Turdza, On the stability of the functional equation of the first order. Ann. Polon. Math. 24 (1970), 35-38.

9. E. Turdza, Some remarks on stability of the functional equation $\phi(f(x)) = g(x)\phi(x) + F(x)$ (in Polish). Rocznik Nauk.-Dydak. WSP w Krakowie Prace Mat. VI, 41 (1970), 155-164.

10. E. Turdza, On the stability of the functional equation $\phi(f(x)) = g(x)\phi(x) + F(x)$. Proc. Amer. Math. Soc. 30 (1971), 484-486.

11. E. Turdza, Some remarks on the stability of the non-linear functional equation of the first order. Demonstratio Math. 6 (1973), 883-889.

12. E. Turdza, Set stability for a functional equation of iterative type. ibidem 15 (1982), 443-448.

13. R. Węgrzyk, Continuous solutions of a linear homogeneous functional equation. Ann. Polon. Math. 35 (1977), 15-20.

B. Choczewski, Institute of Mathematics, University of Mining and Metallurgy, 30-059 Kraków, Poland.

International Series of
Numerical Mathematics, Vol. 71

CHARAKTERISIERUNGEN UND UNGLEICHUNGEN FÜR DIE q-FACTORIAL-FUNKTIONEN

Hans-Heinrich Kairies

Abstract. In this paper we present characterizations of the q-factorial functions (which have been recently discussed by Askey, Moak, Muldoon, Kairies) as special solutions of their functional equation

$F(x+1) = (q^x-1)(q-1)^{-1} F(x)$.

Furthermore we prove some inequalities for these functions which give detailed information about their behaviour near $x=1$.

1. EINLEITUNG

Wir betrachten die "q-factorial" - oder "q-Gamma" - Funktionen Γ_q für positive reelle Argumente. Sie sind gegeben durch

$$(1)\qquad \Gamma_q(x) = (1-q)^{1-x} \prod_{n=0}^{\infty} \frac{1-q^{n+1}}{1-q^{n+x}} \qquad \text{für } q\in(0,1),$$

$$(2)\qquad \Gamma_q(x) = (q-1)^{1-x}\, q^{x(x-1)/2} \prod_{n=0}^{\infty} \frac{1-q^{-(n+1)}}{1-q^{-(n+x)}} \qquad \text{für } q\in(1,\infty).$$

Die Funktionen Γ_q erfüllen sowohl für $q\in(0,1)$ als auch für $q\in(1,\infty)$ die Funktionalgleichung

$$(3)\qquad \forall x\in\mathbb{R}_+ : F(x+1) = (q^x-1)(q-1)^{-1} F(x).$$

Die Eulersche Γ-Funktion ist eingebettet in die Schar der Γ_q : $\lim_{q\to 1} \Gamma_q = \Gamma$.

In diesem Zusammenhang sei hingewiesen auf $\lim_{q\to 1} (q^x-1)(q-1)^{-1} = x$.

R. Askey [1], D.S. Moak [8], [9], M.E. Muldoon und H.-H. Kairies [4] haben die Γ_q in neueren Arbeiten untersucht und unter anderm verschiedene Charakterisierungen der Γ_q mit Hilfe von (3) oder von anderen Funktionalgleichungen angegeben. So ist Γ_q für $q \in (0,1)$ als normierte, logarithmisch konvexe Lösung von (3) charakterisiert, verhält sich also genau wie die Γ-Funktion. Dagegen ist Γ_q für $q > 1$ zwar normierte, logarithmisch konvexe Lösung von (3), aber nicht die einzige Lösung mit dieser Eigenschaft (Moak [8]). Eine Erklärung für dieses unterschiedliche Verhalten und auch für das bisher nicht gedeutete Auftreten des zusätzlichen Faktors $q^{x(x-1)/2}$ in (2) wird von uns in 2. gegeben. Dort werden die Funktionen $\gamma_q := \log \Gamma_q$ als Krullsche Normallösungen der Differenzengleichung

$$(4) \qquad \forall\, x \in \mathbb{R}_+ : \quad f(x+1) - f(x) = \log \frac{q^x - 1}{q - 1}$$

charakterisiert. In 3. werden einige Ungleichungen für die γ_q bewiesen; damit werden auch Ergebnisse aus [3] über die Γ-Funktion auf die γ_q übertragen.

2. DIE γ_q ALS KRULL-NORMALLÖSUNGEN

W. Krull hat in [5] und [6] sein Konzept einer Normallösung von

$$(5) \qquad \forall\, x \in \mathbb{R}_+ : \quad f(x+1) - f(x) = \varphi(x)$$

entwickelt. Wir sagen, daß die Störfunktion $\varphi : \mathbb{R}_+ \to \mathbb{R}$ den Krull-Bedingungen der Ordnung $m \in \mathbb{N} \cup \{0\}$ genügt, falls

$$(K_m) \qquad \varphi^{(m)} \text{ schließlich monoton und } \lim_{x \to \infty} \varphi^{(m)}(x) = 0$$

gilt. Der Krullsche Hauptsatz aus [6] besagt dann, daß es unter der Voraussetzung (K_m) genau eine Normallösung $f_K : \mathbb{R}_+ \to \mathbb{R}$ von (5) gibt, für die $f_K(1) = 0$ und

$$\forall\, h \in (0,\infty) \quad : \lim_{x \to \infty} \left[f_K^{(m)}(x+h) - f_K^{(m)}(x) \right] = 0$$

gilt.

Die Krullsche Normallösung f_K der Ordnung m ist gegeben durch die kompakt konvergente Reihe

$$(6) \qquad f_K(x) = \lim_{j\to\infty} \Big[\sum_{n=1}^{m} \frac{1}{n!} (B_n(x-1) - B_n)\, \varphi^{(n-1)}(j+2) + \sum_{n=0}^{j} (\varphi(n+1)-\varphi(n+x)) \Big].$$

Die dabei auftretenden Bernoulli-Polynome sind erklärt durch

$$se^{sx}(e^s-1)^{-1} = \sum_{n=0}^{\infty} B_n(x) \frac{1}{n!} s^n, \; B_n = B_n(0).$$

Ist nun $\varphi_q(x) = \log \dfrac{q^x-1}{q-1}$, so liegt folgende Situation vor : Im Fall $q \in (0,1)$ erfüllt φ_q die Krull-Bedingung (K_1), aber nicht (K_o). Im Fall $q \in (1,\infty)$ erfüllt φ_q die Krull-Bedingung (K_2), aber weder (K_o) noch (K_1). Dieser Sachverhalt ist wesentlich für den folgenden

SATZ 1. _Im Fall_ $q\in(0,1)$ _ist_ γ_q _Krullsche Normallösung der Ordnung_ $m=1$ _von_ (4). _Im Fall_ $q\in(1,\infty)$ _ist_ γ_q _Krullsche Normallösung der Ordnung_ $m=2$ _von_ (4).

Beweis. Im Fall $q\in(0,1)$ liefert die Darstellung (6)

$$\begin{aligned} f_K(x) &= \lim_{j\to\infty} [(x-1)\,\varphi_q(j+2) + \sum_{n=0}^{j} (\varphi_q(n+1) - \varphi_q(n+x))] \\ &= (x-1)\varphi_q(1) + \sum_{n=0}^{\infty} \{(x-1)[\varphi_q(n+2)-\varphi_q(n+1)]+[\varphi_q(n+1)-\varphi_q(n+x)]\} \\ &= \sum_{n=0}^{\infty} [(x-1)\log \frac{q^{n+2}-1}{q^{n+1}-1} + \log \frac{q^{n+1}-1}{q^{n+x}-1}] \\ &= (1-x)\log(1-q) + \log \prod_{n=0}^{\infty} \frac{1-q^{n+1}}{1-q^{n+x}} \\ &= \gamma_q(x). \end{aligned}$$

Für $q\in(1,\infty)$ ergibt sich aus (6)

$$f_K(x) = \lim_{j\to\infty} [(x-1)\varphi_q(j+2) + \tfrac{1}{2}(x^2-3x+2)\varphi_q'(j+2) + \sum_{n=0}^{j} (\varphi_q(n+1)-\varphi_q(n+x))]$$

$$= \lim_{j\to\infty} [(x-1)\log\frac{q^{j+2}-1}{q-1} + \frac{1}{2}(x^2-3x+2)\frac{q^{j+2}}{q^{j+2}-1}\log q + \sum_{n=0}^{j}\log\frac{q^{n+1}-1}{q^{n+x}-1}]$$

$$= \frac{1}{2}(x^2-3x+2)\log q + \lim_{j\to\infty}[(x-1)\log\frac{q^{j+2}-1}{q-1} + \sum_{n=0}^{j}(\log\frac{1-q^{-(n+1)}}{1-q^{-(n+x)}} + \log q^{1-x})]$$

$$= \frac{1}{2}(x^2-3x+2)\log q + (x-1)\log\frac{q}{q-1} + \log\prod_{n=0}^{\infty}\frac{1-q^{-(n+1)}}{1-q^{-(n+x)}}$$

$$= \frac{1}{2}x(x-1)\log q + (1-x)\log(q-1) + \log\prod_{n=0}^{\infty}\frac{1-q^{-(n+1)}}{1-q^{-(n+x)}}$$

$$= \gamma_q(x). \quad -$$

Die aus (1) bzw. (2) resultierenden Darstellungen für $\gamma_q = \log \Gamma_q$ ergeben sich also unmittelbar aus der Darstellung (6) der Krullschen Normallösung.

Wir machen nun Gebrauch von einer Ergänzung der Krullschen Theorie. Die Anwendung eines Resultats von Kuczma (Theorem 5.13 aus [7]) liefert den

SATZ 2. <u>Im Fall</u> $q\in(0,1)$ <u>ist</u> γ_q <u>die eindeutig bestimmte konvexe und durch</u> $f(1) = 0$ <u>normierte Lösung von</u> (4). <u>Im Fall</u> $q\in(1,\infty)$ <u>ist</u> γ_q <u>die eindeutig bestimmte, zu</u> $M^2(\mathbb{R}_+)$ <u>gehörende und durch</u> $f(1) = 0$ <u>normierte Lösung von</u> (4). -

($f\in M_+^2(\mathbb{R}_+)$[$M_-^2(\mathbb{R}_+)$] gilt genau dann, wenn die dividierte Differenz $[x_1,x_2,x_3,x_4;f]$ für $x_\nu\in\mathbb{R}_+$ nichtnegativ [nichtpositiv] ist.
$M^2(\mathbb{R}_+) = M_+^2(\mathbb{R}_+)\cup M_-^2(\mathbb{R}_+)$; vergl. [7], p. 22.) Das Ergebnis liefert für $q\in(0,1)$ die schon in der Einleitung erwähnte, in [1] bewiesene Erweiterung des Satzes von Bohr/Mollerup. Im Fall $q\in(1,\infty)$ ist zu beachten, daß

$\varphi_q\in M_-^1(\mathbb{R}_+)$, aber $\lim_{x\to\infty} \Delta_1^1 \varphi_q(x) = \log q \neq 0$ und

$\varphi_q\in M_+^2(\mathbb{R}_+)$ sowie $\lim_{x\to\infty} \Delta_1^2 \varphi_q(x) = 0$ gilt.

Wir erwähnen noch eine weitere Charakterisierungsmöglichkeit für die Funktionen γ_q als spezielle Lösungen von (4): In [2] wurde von L. Büsing gezeigt, daß unter gewissen Voraussetzungen an die Störfunktion φ, die in unserem Fall $\varphi(x) = \log\frac{q^x-1}{q-1}$ erfüllt sind, die Nörlundsche und die Schrothsche Hauptlösung von (5) bis auf einen konstanten Summanden mit der Krullschen

Normallösung übereinstimmt. Die γ_q lassen sich daher auch als Hauptlösungen im Sinn von Nörlund bzw. Schroth der Differenzengleichung (5) charakterisieren. Damit ist für die Behandlung der γ_q die Nörlundsche bzw. Schrothsche Theorie der Hauptlösungen einsetzbar.

3. EINIGE UNGLEICHUNGEN FÜR DIE γ_q

In [3] haben wir die Ungleichung $\Gamma(x) \cdot \Gamma(\frac{1}{x}) \geqq 1$, $x \geqq 1$, in ganz elementarer Weise unter Benutzung des Krullschen Satzes bewiesen. Wir zeigen nun, daß auch $\Gamma_q(x) \cdot \Gamma_q(\frac{1}{x}) \geqq 1$ gilt (und etwas mehr) und zwar in Satz 3 für $q\in(0,1)$ und in Satz 4 für $q\in(1,\infty)$. Durch die Formulierung der Sätze wird ihre Verankerung in der Krullschen Theorie deutlich gemacht.

SATZ 3. Es sei $f : \mathbb{R}_+ \to \mathbb{R}$ konvex, $f(1) = 0$ und f erfülle (4) für $q\in(0,1)$. Ferner sei $F(x) := f(x) + f(\frac{1}{x})$.
Dann folgt aus $1 \leqq x < y$: $F(x) \leqq F(y)$.

Beweis. Nach Satz 2 ist $f = \gamma_q$, und es genügt, den Beweis für $x\in[1,2)$ zu führen.
Wir haben

$$F(x) = (1-x)\log(1-q) + \log \prod_{n=0}^{\infty} \frac{1-q^{n+1}}{1-q^{n+x}} + (1-\frac{1}{x})\log(1-q)+\log \prod_{n=0}^{\infty} \frac{1-q^{n+1}}{1-q^{n+1/x}}$$

$$= \lim_{N\to\infty} F_N(x), \text{ wobei}$$

$$F_N(x) = \log(1-q)^{2-(x+1/x)} \prod_{n=0}^{N} \frac{(1-q^{n+1})^2}{(1-q^{n+x})(1-q^{n+1/x})} .$$

Wir zeigen nun, daß für $x\in(1,2)$ und für hinreichend großes $N \geqq N(x,q)$

$$(7) \qquad F_N'(x) > 0$$

erfüllt ist.

$$F_N'(x) = (x^{-2}-1)\log(1-q)+ \log q \sum_{n=0}^{N} \frac{q^{n+x}}{1-q^{n+x}} - x^{-2} \log q \sum_{n=0}^{N} \frac{q^{n+1/x}}{1-q^{n+1/x}} .$$

Es ist $F_N'(1) = 0$, und $F_N'(x) > 0$ für $x \in (1,2)$ ist genau dann erfüllt, wenn

$$(8) \qquad h_N\left(\frac{1}{x}\right) < h_N(x)$$

gilt, wobei

$$h_N(x) = x\left[\log q \sum_{n=0}^{N} \frac{q^{n+x}}{1-q^{n+x}} - \log(1-q)\right].$$

Durch diese "Symmetrisierung" läßt sich das bisher störende gleichzeitige Auftreten von x und x^{-1} eliminieren. Denn (8) ist gezeigt, falls $h_N'(x) > 0$, also falls

$$(9) \qquad -\log q \sum_{n=0}^{N} \frac{q^{n+x}}{1-q^{n+x}} < -\log(1-q) + x(\log q)^2 \sum_{n=0}^{N} \frac{q^{n+x}}{(1-q^{n+x})^2}.$$

Durch Flächenvergleich folgt nun

$$G(x,N) := \frac{q^x}{1-q^x} + \int_x^{x+N} \frac{q^t}{1-q^t}\,dt - \sum_{n=0}^{N} \frac{q^{x+n}}{1-q^{x+n}} > 0 \qquad \text{und}$$

$$H(x,N) := \sum_{n=0}^{N} \frac{q^{n+x}}{(1-q^{n+x})^2} - \int_x^{x+N+1} \frac{q^t}{(1-q^t)^2}\,dt > 0.$$

Da beide Funktionen $x \mapsto q^x(1-q^x)^{-1}$ und $x \mapsto q^x(1-q^x)^{-2}$ konvex sind und streng monoton fallen, gilt sogar für $x \in [1,2]$ und unabhängig von N :

$$G(x,N) \geqq G := \int_2^3 \frac{q^t}{1-q^t}\,dt - \frac{q^3}{1-q^3} > 0 \qquad \text{und}$$

$$H(x,N) \geqq H := \frac{q^2}{(1-q^2)^2} - \int_2^3 \frac{q^t}{(1-q^t)^2}\,dt > 0.$$

Die bei der Definition von $G(x,N)$ und $H(x,N)$ auftretenden Integrale lassen sich elementar auswerten, und damit läßt sich nun (9) in der folgenden äquivalenten Form darstellen :

(10)
$$- \log q \cdot \frac{q^x}{1-q^x} + \log \frac{1-q^{x+N}}{1-q^x} + \log q \cdot G(x,N)$$
$$< - \log(1-q) + x \log q \cdot \left(\frac{q^{x+N+1}}{1-q^{x+N+1}} - \frac{q^x}{1-q^x} \right) + x(\log q)^2 H(x,N).$$

Als nächstes beweisen wir

(11) $$\lim_{N\to\infty} \left[\log \frac{1-q^{x+N}}{1-q^x} - \log q \cdot \frac{q^x}{1-q^x}\right] < \lim_{N\to\infty} \left[- \log(1-q) + x \log q \left(\frac{q^{x+N+1}}{1-q^{x+N+1}} - \frac{q^x}{1-q^x}\right)\right],$$

also

$$\log \frac{1-q}{1-q^x} < \log q \cdot (1-x) \frac{q^x}{1-q^x} .$$

Diese Ungleichung kann man so bestätigen:

$$\log \frac{1-q}{1-q^x} \leqq \frac{q^x-q}{1-q^x} \qquad \text{und} \qquad q^x - q < (1-x)q^x \log q$$

läßt sich zurückführen auf $q^{x-1}[1 - (1-x) \log q] < 1$, was wegen $q\epsilon(0,1)$ und $x\epsilon(1,2)$ klar ist. Aus (11) folgt nun, daß für hinreichend großes $N(x,q)$ und $N \geqq N(x,q)$ gilt :

(12)
$$- \log q \cdot \frac{q^x}{1-q^x} + \log \frac{1-q^{x+N}}{1-q^x} + G \log q$$
$$< - \log(1-q) + x \log q\left(\frac{q^{x+N+1}}{1-q^{x+N+1}} - \frac{q^x}{1-q^x}\right) + (\log q)^2 H .$$

Daher ist erst recht (10) und damit (9), also auch (7) erfüllt. Es läßt sich nun nachrechnen, daß wegen

$$\gamma_q'(x) = - \log(1-q) + \log q \cdot \sum_{n=0}^{\infty} \frac{q^{n+x}}{1-q^{n+x}}$$

$F'(x) = \lim_{N\to\infty} F_N'(x)$ gilt. Daher ist $F'(x) \geqq 0$ für $x \geqq 1$, was die Behauptung des Satzes beweist. -

Die entsprechende Behauptung für $q > 1$ kann nun leicht nachgewiesen werden, wenn man die Beziehung

$$(13) \qquad \forall\, x \in \mathbb{R}_+ : \gamma_q(x) = \gamma_{1/q}(x) + \tfrac{1}{2}(x-1)(x-2)\log q \quad , \ q > 1$$

berücksichtigt. (13) folgt durch einfache Rechnung aus (1) und (2).

SATZ 4. <u>Es sei</u> $f \in M^2(\mathbb{R}_+)$, $f(1) = 0$ <u>und</u> f <u>erfülle</u> (4) <u>für</u> $q \in (1,\infty)$. <u>Ferner sei</u> $F(x) := f(x) + f(\frac{1}{x})$. <u>Dann folgt aus</u> $1 \leq x < y : \quad F(x) < F(y)$.

<u>Beweis.</u> Nach Satz 2 ist $f = \gamma_q$, und es genügt, den Beweis für $x \in [1,2)$ zu führen. Wir haben mit $\alpha := 1/q \in (0,1)$ wegen (13)

$$F(x) = \gamma_\alpha(x) + \gamma_\alpha(\tfrac{1}{x}) + \tfrac{1}{2}\log q\,[x^2 - 3x + x^{-2} - 3x^{-1} + 4] \quad , \text{ also}$$

$$F'(x) = \gamma'_\alpha(x) - x^{-2}\gamma'_\alpha(\tfrac{1}{x}) + \tfrac{1}{2}\log q\,[2x - 3 - 2x^{-3} + 3x^{-2}] \;.$$

Nach Satz 3 ist $\gamma'_\alpha(x) - x^{-2}\gamma'_\alpha(\frac{1}{x}) \geq 0$ für $x \in (1,2)$. Es bleibt also nur noch zu zeigen, daß $2x - 3 - 2x^{-3} + 3x^{-2} > 0$ für $x \in (1,2)$ ist. Das ist gleichwertig mit $2(x^2-1)(x^2+1) > 3x(x^2-1)$, also klar für $x \in (1,2)$. –

In beiden Fällen $q \in (0,1)$ und $q \in (1,\infty)$ ist $F(1) = 0$ und damit auch $F(x) \geq 0$ für $x \geq 1$, was mit $\Gamma_q(x)\cdot\Gamma_q(\frac{1}{x}) \geq 1$ äquivalent ist. Wir werden in den beiden folgenden Sätzen das Verhalten von γ_q in der Umgebung von $x = 1$ noch durch eine andere Ungleichung beschreiben. Dazu sei für $t > 0$

$$F_t(x) := f(x) + f(1 - t(x-1)) \quad , \quad x \in [1, 1 + t^{-1}).$$

Ist nun $f = \gamma_q$, so folgt : $\lim_{x\to 0+} f(x) = +\infty$, $f(1) = f(2) = 0$ und f ist konvex. Daher ist $F_t(1) = 0$ für beliebiges $t \in \mathbb{R}_+$ und $x \mapsto F_t(x)$ ist streng monoton wachsend für $t \geq 1$, insbesondere ist $F_t(x) > 0$ für $x > 1$ und $t \geq 1$. Andererseits kann für kein $t \in (0,1)$ und alle $x \in [1,1+t^{-1})$ die Ungleichung $F_t(x) \geq 0$ erfüllt sein, denn $F'_t(1) = (1-t)f'(1)$ ist dann negativ. Die folgenden Sätze besagen nun, daß $F_t(x) > 0$ für $x > 1$ auch mit $t \in (0,1)$ möglich ist, allerdings ist dann t in Abhängigkeit von x einzuschränken.

SATZ 5. Es sei f konvex, $f(1) = 0$ und f erfülle (4) mit $q \in (0,1)$. Ferner sei $\tau(x,q) := \log(2-q^{x-1})/\log q^{1-x}$. Dann ist $F_t(x) > 0$, falls $x \in (1, 1+\frac{1}{t})$ und $t \geqq \tau(x,q)$.

Beweis. Es ist $F_t(x) = \gamma_q(x) + \gamma_q(1-t(x-1))$. Für beliebiges $t > 0$ und $x \geqq 2$ ist $F_t(x) > 0$ trivialerweise erfüllt. Nach der Vorbemerkung ist daher die Behauptung nur für $t \in (0,1)$ und $x \in (1,2)$ zu beweisen. Wir haben

$$F_t(x) = [1 - x + t(x-1)]\log(1-q) + \log \prod_{n=0}^{\infty} \frac{(1-q^{n+1})^2}{(1-q^{n+x})(1-q^{n+1+t-tx})}$$

$$= (x-1)(t-1)\log(1-q) + \log \prod_{n=0}^{\infty} c_n(x,t,q) .$$

Der erste Summand ist positiv, da $x > 1$, $t \in (0,1)$, $q \in (0,1)$. Wir bestimmen nun für festes $q \in (0,1)$ die Paare (x,t), für die $c_n(x,t,q) \geqq 1$ gilt. $c_n(x,t,q) \geqq 1$ ist gleichwertig mit

$$(1-q^{n+1})^2 \geqq (1-q^{n+x})(1-q^{n+1+t-tx})$$

und das wiederum (wie durch kurze Rechnung bestätigt werden kann) mit

$$2q \leqq q^x + q^{1+t-tx} + q^{n+2}(1-q^{(1-t)(x-1)}). \tag{14}$$

Der letzte Summand ist positiv und strebt gegen Null für $n \to \infty$. $c_n(x,t,q) \geqq 1$ ist also erfüllt, falls

$$w(x,t) := q^x + q^{1+t-tx} \geqq 2q \quad . \tag{15}$$

Wir fixieren nun $x \in (1,2)$. Dann gilt $w(x,0) = q^x + q < 2q$ und wegen der Konvexität von $x \mapsto q^x$: $w(x,1) = q^x + q^{2-x} > 2q$. Weiter ist $t \mapsto w(x,t)$ streng monoton wachsend. Daher existiert ein eindeutig bestimmtes $\tau(x,q)$ mit $w(x,\tau(x,q)) = 2q$. Aus $q^x + q^{1+\tau-\tau x} = 2q$ folgt nun

$$\tau(x,q) = \log(2-q^{x-1}) / \log q^{1-x}. \tag{16}$$

Dach Definition von $\tau(x,q)$ ist klar, daß $t \geqq \tau(x,q)$ zur Folge hat: $w(x,t) \geqq 2q$ und damit auch $c_n(x,t,q) \geqq 1$. Damit ist der Satz bewiesen. Daß wir eine erhebliche Verbesserung der Aussage aus der Vorbemerkung erhalten haben, möge belegt werden durch Angabe einiger Eigenschaften von $\tau(x,q)$.

Durch (16) sei $\tau(x,q)$ auch für den Randwert $x=2$ erklärt. Dann ist $\lim_{x\to 1+} \tau(x,q) = 1$ und $\tau(2,q) < 1$. Für festes q ist $x \mapsto \tau(x,q)$ streng monoton fallend, und es gilt $\lim_{x\to 1+} \tau'(x,q) = \log q < 0$. Für festes x ist $q \mapsto \tau(x,q)$ streng monoton steigend und als Beispiele seien die folgenden Randwerte erwähnt: $\tau(2,\frac{1}{30}) = 0{,}198\ldots$, $\tau(2,\frac{1}{4}) = 0{,}404\ldots$, $\tau(2,\frac{1}{2}) = 0{,}585\ldots$. -

Wir benutzen wieder die Identität (13), um den Fall $q \in (1,\infty)$ auf den Fall $q \in (0,1)$ zurückzuführen. Dabei stellt sich jedoch eine Komplikation ein, was auch bei der Formulierung des Satzes zum Ausdruck kommt.

SATZ 6. Es sei $f \in M^2(\mathbb{R}_+)$, $f(1) = 0$ und f erfülle (4) mit $q \in (1,\infty)$. Ferner sei $\tau(x) := \frac{1}{2(x-1)}\left(\sqrt{1-4(x-1)(x-2)} - 1\right)$. Dann ist $F_t(x) > 0$, falls $x \in (1,1+\frac{1}{t})$ und $t \geq \max\left\{\tau(x,\frac{1}{q}),\ \tau(x)\right\}$.

Beweis. Es ist $F_t(x) = \gamma_q(x) + \gamma_q(1-t(x-1))$. Wie bei Satz 5 genügt es, die Behauptung für $t \in (0,1)$ und $x \in (1,2)$ zu beweisen. Aus (13) folgt mit $\alpha = \frac{1}{q}$

$$F_t(x) = \gamma_\alpha(x) + \gamma_\alpha(1-t(x-1)) + \frac{1}{2}\log q \cdot [x^2 - 3x + 2 + t^2(x-1)^2 + t(x-1)] .$$

Nach Satz 5 ist $\gamma_\alpha(x) + \gamma_\alpha(1-t(x-1)) > 0$ für $x \in (1,1+\frac{1}{t})$ und $t \geq \tau(x,\alpha)$. Es bleibt noch

$$W(x,t) := t^2(x-1)^2 + t(x-1) + (x-1)(x-2)$$

zu untersuchen. Es ist $W(x,0) = (x-1)(x-2) < 0$ für $x \in (1,2)$ und $W(x,1) = 2(x-1)^2 > 0$. Ferner ist $t \mapsto W(x,t)$ streng monoton wachsend. Daher existiert genau ein $\tau(x) \in (0,1)$ mit $W(x,\tau(x)) = 0$. Man rechnet leicht nach, daß

$$\tau(x) = \frac{1}{2(x-1)}\left(\sqrt{1-4(x-1)(x-2)} - 1\right). \tag{17}$$

Für $t \geq \tau(x)$ ist $W(x,t) \geq 0$ und damit $F_t(x) > 0$ für $x \in (1,1+\frac{1}{t})$ und $t \geq \max\{\tau(x),\ \tau(x,\alpha)\}$. Wir geben noch einige Eigenschaften von $\tau(x)$ an und verweisen auf die analogen Aussagen über $\tau(x,q)$. Durch (17) sei $\tau(x)$ auch für den Randwert $x=2$ erklärt.
Dann ist $\lim_{x\to 1+} \tau(x) = 1$, $\tau(\frac{7}{5}) = \frac{1}{2}$, $\tau(\frac{1}{2}) = \sqrt{2} - 1$ und $\tau(2) = 0$. τ ist streng monoton fallend und $\lim_{x\to 1+} \tau'(x) = -2$. -

BEMERKUNG. a) Für $q \in (1, \exp 2)$ ist $\tau(x) \leqq \tau(x,1/q)$ und für diese q ist daher $F_t(x) > 0$, falls $x \in (1,1+1/t)$ und $t \geqq \tau(x,1/q)$, ganz entsprechend wie in Satz 5.

b) Man kann zeigen, daß aus Satz 5 nicht Satz 3 und daß aus Satz 3 nicht Satz 5 folgt. Es läßt sich jedoch beweisen, daß für $q \in [1-\varepsilon,1)$ und für hinreichend kleines $\varepsilon \in \mathbb{R}_+$ Satz 5 eine Folge von Satz 3 ist. Ebenso läßt sich für $q \in (0,\varepsilon]$ und für hinreichend kleines $\varepsilon \in \mathbb{R}_+$ nachweisen, daß aus Satz 5 folgt: $f(x) + f(\frac{1}{x}) > 0$ für $x > 1$, also eine Verschärfung einer Teilaussage von Satz 3.

LITERATUR

1. R. Askey, The q-Gamma and q-Beta functions. Applicable Analysis 8(1978), 125-141.

2. L. Büsing, Vergleich von Hauptlösungsbegriffen für Nörlundsche Differenzengleichungen. Eingereicht bei Manuscripta Mathematica.

3. H.-H. Kairies, An inequality for Krull solutions of a certain difference equation. In: ISNM 64, Proceedings of the Third International Conference on General Inequalities, 277-280, Birkhäuser Verlag, 1983.

4. H.-H. Kairies and M.E. Muldoon, Some characterizations of q-factorial functions. Erscheint in Aequationes Mathematicae.

5. W. Krull, Bemerkungen zur Differenzengleichung $g(x+1) - g(x) = \varphi(x)$. Math. Nachr. 1(1948), 365-376.

6. W. Krull, Bemerkungen zur Differenzengleichung $g(x+1) - g(x) = \varphi(x)$, II. Math. Nachr. 2(1949), 251-262.

7. M. Kuczma, Functional equations in a single variable. Polish Scientific Publishers, Warszawa, 1968.

8. D.S. Moak, The q-Gamma function for $q > 1$. Aequationes Math. 20(1980), 278-285.

9. D.S. Moak, The q-Gamma function for $x < 0$. Aequationes Math. 21(1980), 179-191.

Hans-Heinrich Kairies
Institut für Mathematik der Technischen Universität Clausthal
3392 Clausthal-Zellerfeld, Bundesrepublik Deutschland

International Series of
Numerical Mathematics, Vol. 71

A NOTE ON t-CONVEX FUNCTIONS

Norbert Kuhn

Abstract. In this note we discuss convexity properties of functions $f:I\to[-\infty,\infty[$ defined on an interval $I\subset\mathbb{R}$. One of our main results is the following: If the function $f:I\to[-\infty,\infty[$ satisfies the convexity inequality for some $t\in]0,1[$, then the same holds true for all rational $t\in[0,1]$.

1. INTRODUCTION

Let $f:I\to[-\infty,\infty[$ be a function defined on an interval $I\subset\mathbb{R}$.

DEFINITION 1. For $t\in[0,1]$, the function f is called _t-convex_ iff one has $f(tu+(1-t)v)\leq tf(u)+(1-t)f(v)$ for all $u,v\in I$, with $0\cdot(-\infty):=0$. And f is called _convex_ iff f is t-convex for all $t\in[0,1]$. Furthermore we define $K(f):=\{t\in[0,1]\text{: f is t-convex}\}$.

DEFINITION 2. For $t\in[0,1]$, the function f is called _t-affine_ iff $f(tu+(1-t)v)=tf(u)+(1-t)f(v)$ for all $u,v\in I$. And f is called _affine_ iff f is t-affine for all $t\in[0,1]$. Furthermore we define $A(f):=\{t\in[0,1]\text{: f is t-affine}\}$.

Evidently $0,1\in A(f)\subset K(f)$. Also it is well known that $\frac{1}{2}\in K(f)$ $\Rightarrow$ $\mathbb{Q}\cap[0,1]\subset K(f)$; but $\mathbb{Q}\cap[0,1]\subset K(f)$ does not imply $[0,1]=K(f)$. In this paper we shall prove a result of an apparently new type. For a subset $M\subset\mathbb{R}$ define $[M]$ to be the subfield generated by M. Then our main result reads as follows.

THEOREM 1K. _Let_ $K(f)\neq\{0,1\}$. _Then_ $[K(f)]\cap[0,1]\subset K(f)$, _hence_ $[K(f)]\cap[0,1]=K(f)$. _In particular we have_ $\mathbb{Q}\cap[0,1]\subset K(f)$.

One could perhaps expect that the above conclusion is true under the assumption $\mathbb{Q}\cap]0,1[\cap K(f)\neq\emptyset$, but in full generality it seems to be a surprise. Our methode of proof is quite simple. First we prove the analogous result for $A(f)$ in a completely elementary way (see Section 2).

THEOREM 1A. Let $A(f)\neq\{0,1\}$. Then $[A(f)]\cap[0,1]\subset A(f)$, hence $[A(f)]\cap[0,1]=A(f)$.

Then for the implication 1A ⇒ 1K we make use of an abstract Hahn-Banach-theorem due to Rodé [2]. We formulate the special case of Rodé's theorem which is relevant for us.

THEOREM 2. For $f:I\to[-\infty,\infty[$ and $a\in I$ there exists $\varphi:I\to[-\infty,\infty[$ with $\varphi\leq f$, $\varphi(a)=f(a)$ and $K(f)\subset A(\varphi)$.

Proof of Theorem 2. The mappings $\tau_t:I\times I\to I,(u,v)\to tu+(1-t)v$ $(t\in[0,1])$ are commuting in the sense of [2]. We define a function $g:I\to[-\infty,\infty[$ to be $g(a):=f(a)$ and $g(u):=-\infty$ if $u\in I$, $u\neq a$. Then

$$f(\tau_t(u,v))\leq tf(u)+(1-t)f(v)\quad \forall u,v\in I \text{ and } t\in K(f),$$

$$g(\tau_t(u,v))\geq tg(u)+(1-t)g(v)\quad \forall u,v\in I \text{ and } t\in[0,1].$$

Now the abstract Hahn-Banach-theorem of Rodé gives a function $\varphi:I\to[-\infty,\infty[$ with $g\leq\varphi\leq f$ and $K(f)\subset A(\varphi)$. Thus the assertion follows.

Proof of 1A ⇒ 1K via Theorem 2. Let $f:I\to[-\infty,\infty[$ and $t\in[K(f)]\cap[0,1]$. In order to prove $t\in K(f)$ we fix $u,v\in I$ and put $a:=tu+(1-t)v\in I$. Let $\varphi:I\to[-\infty,\infty[$ as in Theorem 2. Then Theorem 1A gives $t\in[A(\varphi)]\cap[0,1]=A(\varphi)$. It follows

$$f(a)=\varphi(a)=t\varphi(u)+(1-t)\varphi(v)\leq tf(u)+(1-t)f(v).$$

This proves the assertion.

By the usual induction proof we obtain from Theorem 1K a natural version of Jensen's inequality. The analogous result for $A(f)$ follows from Theorem 1A.

COROLLARY 3K. We have

$$f\left(\sum_{l=1}^{n} t_l u_l\right)\leq\sum_{l=1}^{n} t_l f(u_l)$$

for all $u_1,\dots,u_n \in I$ and all $t_1,\dots,t_n \in K(f)$ with $\sum_{l=1}^{n} t_l = 1$.

COROLLARY 3A. We have

$$f\left(\sum_{l=1}^{n} t_l u_l\right) = \sum_{l=1}^{n} t_l f(u_l)$$

for all $u_1,\dots,u_n \in I$ and all $t_1,\dots,t_n \in A(f)$ with $\sum_{l=1}^{n} t_l = 1$.

2. PROOF OF THEOREM 1A

We start with two well known remarks.

REMARK 4K. i) $s \in K(f) \Rightarrow 1-s \in K(f)$; ii) $\alpha,\beta,s \in K(f) \Rightarrow s\alpha+(1-s)\beta \in K(f)$; iii) $s,t \in K(f) \Rightarrow st \in K(f)$ (put $\beta=0$ in ii); iv) $K(f) \neq \{0,1\} \Rightarrow \overline{K(f)} = [0,1]$.

REMARK 4A. i) $s \in A(f) \Rightarrow 1-s \in A(f)$; ii) $\alpha,\beta,s \in A(f) \Rightarrow s\alpha+(1-s)\beta \in A(f)$; iii) $s,t \in A(f) \Rightarrow st \in A(f)$; iv) $A(f) \neq \{0,1\} \Rightarrow \overline{A(f)} = [0,1]$.

REMARK 5. Assume that there exists an $a \in I$ with $f(a) = -\infty$.

K) Let $K(f) \neq \{0,1\}$. Then $K(f) = [0,1]$; and $f|\text{int } I = -\infty$, where int I denotes the interior of I.

A) Let $A(f) \neq \{0,1\}$. Then $A(f) = [0,1]$; and $f(x) = -\infty$ in at most one point $x \in I$ which then must be a boundary point of I.

Proof of K). Let $u \in \text{int } I$. Then $u + \frac{t}{1-t}(u-a) \in I$ for all sufficiently small $t>0$. Because of 4K.iv) there exists some positive $t \in K(f)$ so that $b := u + \frac{t}{1-t}(u-a) \in I$. It follows $u = ta+(1-t)b$ and consequently $f(u) \leq tf(a)+(1-t)f(b) = -\infty$. In view of $f|\text{int } I = -\infty$ the function f is convexe.

Proof of A). The assumption $A(f) \neq \{0,1\}$ implies that $f(x) \neq -\infty$ in at most one point $x \in I$. Because of K) this is a boundary point of I. Now $A(f) = [0,1]$ is clear.

LEMMA 6. i) $s,t \in A(f)$ with $s<t \Rightarrow \frac{s}{t} \in A(f)$

ii) $s,t \in A(f)$ with $s \leq t \Rightarrow t-s \in A(f)$

iii) $s,t \in A(f)$ with $s+t \leq 1 \Rightarrow s+t \in A(f)$

<u>Proof.</u> i) We may assume that $A(f)\neq\{0,1\}$; and then because of 5.A) that $f(x)\neq-\infty$ $\forall x\in I$. In order to prove $\alpha:=\frac{s}{t}\in A(f)$ let $u,v\in I$. Then $t(\alpha u+(1-\alpha)v)+(1-t)v=su+(1-s)v$ and hence

$$tf(\alpha u+(1-\alpha)v)+(1-t)f(v)=sf(u)+(1-s)f(v).$$

It follows that $f(\alpha u+(1-\alpha)v)=\alpha f(u)+(1-\alpha)f(v)$.

ii) We may assume $s<t$. Then $t-s=t(1-\frac{s}{t})\in A(f)$ by i) and 4.A).

iii) From $s+t\leqq 1$ we have $s\leqq 1-t$ and therefore $(1-t)-s=1-(s+t)$ $\in A(f)$ by ii). Thus $s+t\in A(f)$.

<u>LEMMA 7.</u> <u>If</u> $A(f)\neq\{0,1\}$, <u>then</u> $E:=\{\pm\frac{s}{t}:\ s,t\in A(f),t>0\}$ <u>is</u> <u>a</u> <u>field</u>.

<u>Proof.</u> Evidently i) $0,1\in E$; ii) $u\in E\Rightarrow -u\in E$; iii) $u\in E,u\neq 0\Rightarrow\frac{1}{u}\in E$; iv) $u,v\in E\Rightarrow uv\in E$. It remains to show v) $u,v\in E\Rightarrow u+v\in E$. Let α,β,s,t $\in A(f)$ with $\beta,t>0$. Then we have to show $\frac{\alpha}{\beta}-\frac{s}{t}\in E$ and $\frac{\alpha}{\beta}+\frac{s}{t}\in E$. Clearly $\frac{\alpha}{\beta}-\frac{s}{t}=\frac{\alpha t-\beta s}{\beta t}\in E$ by 4.A) and Lemma 6. To prove $\frac{\alpha}{\beta}+\frac{s}{t}\in E$ we fix some positive $\delta\in A(f)$ with $(\alpha t+\beta s)\delta\leqq 1$. Then $\frac{\alpha}{\beta}+\frac{s}{t}=\frac{(\alpha t+\beta s)\delta}{\beta t\delta}\in E$ by 4.A) and Lemma 6. This proves the lemma.

<u>Proof of Theorem 1A.</u> Evidently $E\subset[A(f)]$, and $[A(f)]\subset E$ by Lemma 7. Hence $E=[A(f)]$. But $[A(f)]\cap[0,1]=E\cap[0,1]\subset A(f)$ by Lemma 6.i). Thus the assertion is proved.

3. FURTHER RESULTS ON t-CONVEX FUNCTIONS

The following result is a second application of Theorem 1K.

THEOREM 8. <u>Let</u> $J\subset I$ <u>be</u> <u>an</u> <u>interval</u> <u>with</u> $\operatorname{int} J\neq\emptyset$; <u>let</u> $f:I\to[-\infty,\infty[$ <u>and</u> $g:J\to[-\infty,\infty[$ <u>be</u> <u>functions</u> <u>with</u> $f\leqq g$ <u>on</u> J. <u>Then</u>
A) $A(f)\neq\{0,1\}\Rightarrow K(g)\subset A(f)$;
K) $K(f)\neq\{0,1\}\Rightarrow K(g)\subset K(f)$.

<u>Proof.</u> A) 1) Because of 5.A) we may assume that $f(x)\neq-\infty$ for all $x\in I$. We fix some $t\in K(g)$ and a 0-1-sequence $(\alpha_l)_l$ with

$t=\sum_{l=1}^{\infty}\alpha_l\cdot 2^{-l}$. Let $s_n:=\sum_{l=1}^{\infty}\alpha_{n+l}\cdot 2^{-l}$ $\forall n\geqq 0$. Then $s_0=t\in K(g)$ and $s_{n+1}=$ $2s_n-\alpha_{n+1}$ $\forall n\geqq 0$. Theorem 1K gives $s_n\in K(g)$ $\forall n\geqq 0$.

2) We prove that f is t-convexe. Let $u,v\in I$. We fix $w\in \text{int}\, J$. Then there exists a positive $\alpha\in A(f)$ with $x:=\alpha u+(1-\alpha)w$, $y:=\alpha v+(1-\alpha)w\in J$. For all $n\in\mathbb{N}$ then

$$\alpha f(tu+(1-t)v)+(1-\alpha)f(w)=f(tx+(1-t)y)=f\left(\sum_{l=1}^{\infty}2^{-l}(\alpha_l x+(1-\alpha_l)y)\right)$$
$$=f\left(\sum_{l=1}^{n}2^{-l}(\alpha_l x+(1-\alpha_l)y)+2^{-n}(s_n x+(1-s_n)y)\right);$$

in view of Corollary 3A and $Q\cap[0,1]\subset A(f)$ this is

$$=\sum_{l=1}^{n}2^{-l}(\alpha_l f(x)+(1-\alpha_l)f(y)) + 2^{-n}f(s_n x+(1-s_n)y).$$

Furthermore we have

$$f(s_n x+(1-s_n)y)\leqq g(s_n x+(1-s_n)y)\leqq s_n g(x)+(1-s_n)g(y).$$

By combining the above it follows that

$$\alpha f(tu+(1-t)v)+(1-\alpha)f(w)$$
$$\leqq \sum_{l=1}^{n}2^{-l}(\alpha_l f(x)+(1-\alpha_l)f(y)) + 2^{-n}(s_n g(x)+(1-s_n)g(y)).$$

For $n\to\infty$ we obtain

$$\alpha f(tu+(1-t)v)+(1-\alpha)f(w)\leqq tf(x)+(1-t)f(y)=\alpha f(tu+(1-t)v)+(1-\alpha)f(w).$$

Therefore $f(tu+(1-t)v)\leqq tf(u)+(1-t)f(v)$. This proves $t\in K(f)$.

3) It remains to show that f is t-concave, that is

$$f(tu+(1-t)v)\geqq tf(u)+(1-t)f(v)\quad \forall u,v\in I.$$

Let $u,v\in I$ and $w:=\frac{1}{2}u+\frac{1}{2}v$. Then Theorem 2 gives a t-affine function

$\varphi:I\to[-\infty,\infty[$ with $\varphi\leqq f$ and $\varphi(w)=f(w)$. Since φ and f are $\frac{1}{2}$-affine, it follows $f(u)+f(v)=2f(w)=2\varphi(w)=\varphi(u)+\varphi(v)$, and therefore $\varphi(u)=f(u)$ and $\varphi(v)=f(v)$. It follows that $f(tu+(1-t)v)\geqq\varphi(tu+(1-t)v)=t\varphi(u)+(1-t)\varphi(v)=tf(u)+(1-t)f(v)$. This proves 3), hence A).
K) follows directly from A) and Theorem 2.

COROLLARY 9. For a function $f:I\to[-\infty,\infty[$ we have

$$A(f)\neq\{0,1\} \Rightarrow A(f)=K(f).$$

COROLLARY 10. Let $J\subset I$ be an interval with $\operatorname{int} J\neq\emptyset$. Then for every function $f:I\to[-\infty,\infty[$ we have

A) $A(f)\neq\{0,1\} \Rightarrow A(f)=A(f|J)$;

K) $K(f)\neq\{0,1\} \Rightarrow K(f)=K(f|J)$.

The following theorem shows that $K(f)\neq\{0,1\}$ is a local property. The conjecture that this theorem is valid goes back to a conversation with Professor Heinz König and influenced the final version of Theorem 8. For this and for his interest in this paper I want to thank him warmly.

THEOREM 11. Let $f:I\to[-\infty,\infty[$ be a function with the property:

(*) For every $x\in I$ there exists an interval $I_x\subset\mathbb{R}$ with $x\in\operatorname{int} I_x$ and $K(f|I\cap I_x)\neq\{0,1\}$.

Then $K(f|I\cap I_x)=K(f)$ $\forall x\in I$.

Proof. By Corollary 10.K) we have to show that $K(f)\neq\{0,1\}$. For this we may assume that I is compact. Thus there exist $x(1),\dots,x(n)\in I$ with $I\subset\bigcup_{l=1}^{n}\operatorname{int} I_{x(l)}$. An obvious induction argument shows that we may assume n=2. Then we have the following situation:

(**) There exist $a,b\in I$ with $a<b$ so that $K(f|I\cap]-\infty,b[)\neq\{0,1\}$ and $K(f|I\cap]a,\infty[)\neq\{0,1\}$.

Let $\alpha,\beta\in\mathbb{R}$ with $I=[\alpha,\beta]$. Without loss of generality we may assume that $b<\beta$. Let $\varepsilon\in]0,\frac{(b-a)^2}{\beta-b}[$ and $a_n:=a-(n-1)\varepsilon$ $\forall n\in\mathbb{N}$. Then $a_n\downarrow-\infty$.

Let $T_n:=I\cap]a_n,\infty[$ $\forall n\in\mathbb{N}$ and $S:=I\cap]-\infty,b[$. Now we proof by induction that $K(f|T_n)\neq\{0,1\}$ $\forall n\in\mathbb{N}$. For n=1 this is clear. In the induction step $n\Rightarrow n+1$ we note that $\frac{\beta-b}{\beta-a_n}<\frac{b-a_n}{b-a_{n+1}}$ and fix a rational t with $\frac{\beta-b}{\beta-a_n}<t<\frac{b-a_n}{b-a_{n+1}}$. To prove $t\in K(f|T_{n+1})$ we fix $u,v\in T_{n+1}$ with $u<v$. In view of $t\in K(f|T_n)\cap K(f|S)$ we may assume that $a_{n+1}<u\leqq a_n$ and $b\leqq v\leqq\beta$. Let $w:=tu+(1-t)v$. Then $a_n<w<b$. We fix a rational $r\in]0,t[$ with $c:=w+\frac{r}{1-r}(w-u)\in]w,b[$ and put $s:=\frac{t-r}{t(1-r)}\in\mathbb{Q}\cap]0,1[$. Then we have $w=ru+(1-r)c$ and $c=sw+(1-s)v$. It follows:

$$f(w)\leqq rf(u)+(1-r)f(c) \quad \text{since } [u,c]\subset S \text{ and } r\in K(f|S)$$

$$f(c)\leqq sf(w)+(1-s)f(v) \quad \text{since } [c,v]\subset T_n \text{ and } s\in K(f|T_n)$$

By combining this we obtain $f(w)\leqq rf(u)+(1-r)(sf(w)+(1-s)f(v))$ and then $f(w)\leqq tf(u)+(1-t)f(v)$. Thus the assertion follows.

4. FUNCTIONS DEFINED ON CONVEX SETS

In this section we transfer some of the results obtained to functions defined on a convex subset X of a real vector space. For a function $f:X\to[-\infty,\infty[$ we define the sets A(f) and K(f) as before.

THEOREM 12. *For every function* $f:X\to[-\infty,\infty[$ *we have*
A) $A(f)\neq\{0,1\} \Rightarrow [A(f)]\cap[0,1]=A(f)$;
K) $K(f)\neq\{0,1\} \Rightarrow [K(f)]\cap[0,1]=K(f)$.

Proof. We prove only K). Let $u,v\in X$ and $t\in[K(f)]\cap[0,1]$. Then the function $\varphi:[0,1]\to[-\infty,\infty[, s\to f(su+(1-s)v)$ has the property $K(f)\subset K(\varphi)$. It follows $t\in[K(\varphi)]\cap[0,1]=K(\varphi)$ and therefore
$f(tu+(1-t)v)=\varphi(t)=\varphi(t\cdot 1+(1-t)\cdot 0)\leqq t\varphi(1)+(1-t)\varphi(0)=tf(u)+(1-t)f(v)$.
This proves the assertion.

In [1] proved Ger that Theorem 12 is sharp. With our notations his result reads as follows.

THEOREM 13. *If* $K\subset\mathbb{R}$ *is a field and if* X *has more than one point then there is a function* $f:X\to K$ *with* $A(f)=K(f)=K\cap[0,1]$.

The following theorem can be proved without difficulties. It can serve to generalize known Hahn-Banach-theorems.

THEOREM 14. *Let* $f,g:X\to[-\infty,\infty[$ *be functions with* $g\leqq f$. *Then*

A) $A(g)\neq\{0,1\} \Rightarrow K(f)\subset A(g)$;

K) $K(g)\neq\{0,1\} \Rightarrow K(f)\subset K(g)$.

REFERENCES

1. R.Ger, Homogeneity sets for Jensen-convex functions. General Inequalities 2 (Oberwolfach 1978), 193-201, Internat.Ser.Numer.Math. 47, Birkhäuser, Basel-Stuttgart, 1980.

2. G.Rodé, Eine abstrakte Version des Satzes von Hahn-Banach. Arch.Math. 31(1978), 474-481.

Norbert Kuhn, Fachbereich Mathematik, Universität des Saarlandes, D-6600 Saarbrücken, Fed.Rep.Germany

International Series of
Numerical Mathematics, Vol. 71

THE STABILITY OF AN ITERATIVE LINEAR EQUATION

Erwin Turdza

Abstract. This paper cxtends the results of B. Choczewski, E. Turdza, R. Wegrzyk [2] and E. Turdza [5] on the stability of the linear functional equation of iterative type to the case when the values of the unknown function are in a Banach space.

1. INTRODUCTION

The problem of stability of a functional equation descends from D.H. Hyers [3] who has formulated and solved this problem for the Cauchy functional equation. The first paper on the stability of iterative equations is due to D. Brydak [1]. The problem was next examined by E. Turdza [5], [6] and B. Choczewski, E. Turdza and E. Wegrzyk [2]. All results obtained by these authors are related to the case when given and unknown functions are real-valued. In this paper we shall consider functions with values in a Banach space.

We shall deal with the linear functional equations

(1) $$\Phi(x) = G(x)\Phi(f(x)) + H(x) ,$$

(2) $$\phi(f(x)) = g(x)\phi(x) + h(x) ,$$

where f, g, G, h, H are given functions and Φ and ϕ are unknown functions.

2. EQUATION (1)

In dealing with equation (1) we shall assume the following hypothesis

(H_1) X is a topological space, K is the field of real or complex numbers, Y is a Banach space over K, $f: X \to X$; $G: X \to K$; $H, \Phi: X \to Y$.

DEFINITION 1. The equation (1) is called stable [iteratively stable] in a class C of functions defined on X if there exists a positive constant s such that for each number $\varepsilon > 0$ and each solution $\Psi \in C$ of the inequality [system of inequalities]

(3) $$|\Psi(x) - G(x)\Psi(f(x)) - H(x)| < \varepsilon \quad \text{for } x \in X$$

[(4) $$|\Psi(x) - G_n(x)\Psi(f^n(x)) - \sum_{i=0}^{n-1} G_i(x)H(f^i(x))| < \varepsilon \quad \text{for } x \in X] ,$$

where f^n denotes the n-th iterate of the function f, $|\cdot|$ is the norm in the Banach space Y and

$$G_n(x) := \prod_{i=0}^{n-1} G(f^i(x)) , \quad G_0(x) := 1 \quad \text{for } x \in X ,$$

there exists a solution $\Phi \in C$ of the equation (1) satisfying the inequality

(5) $$|\Psi(x) - \Phi(x)| \le s \cdot \varepsilon \quad \text{for } x \in X .$$

The follwoing result, which we quote here as Lemma 1, is well known in the theory of iterative functional equations; it can be found in [4].

LEMMA 1. Let us assume that hypothesis (H_1) is fulfilled and that there exists a point $\xi \in X$ such that the functions f, G, H are continuous at ξ, for each $x \in X$ we have $\lim_{n\to\infty} f^n(x) = \xi$, and for each neighbourhood U_ξ of the point ξ there exists a compact neighbourhood C of the point ξ such that $f(C) \subset C \subset U_\xi$. If $|G(\xi)| < 1$, then there exists the unique solution $\Phi: X \to Y$ of equation (1). The solution Φ is bounded in a neighbourhood of the point ξ and continuous at ξ. It is given by the formula

(6) $$\Phi(x) = \sum_{n=0}^{\infty} G_n(x)H(f^n(x)) .$$

LEMMA 2. Assume that hypothesis (H_1) holds and the function $\Psi: X \to Y$ satisfies inequality (3). Then for each integer n the function Ψ satisfies the inequality

(7) $$|\Psi(x) - G_n(x)\Psi(f^n(x)) - \sum_{i=0}^{n-1} G_i(x)H(f^i(x))| < \varepsilon \sum_{i=0}^{n-1} |G_i(x)| \quad \text{for } x \in X.$$

Proof by induction. For $n = 1$ inequality (7) reduces to inequality (4), which is true by assumption. Let us assume that inequality (7) holds for a fixed $n \geq 1$ and let us put $f(x)$ in place of x in (7). Then we have

$$|\Psi(f(x)) - G_n(f(x))\Psi(f^{n+1}(x)) - \sum_{i=0}^{n-1} G_i(f(x))H(f^{i+1}(x))| < \varepsilon \sum_{i=0}^{n-1} |G_i(f(x))| \quad \text{for } x \in X ,$$

whence

$$|\Psi(f(x)) - \frac{1}{G(x)} G_{n+1}(x)\Psi(f^{n+1}(x)) - \sum_{i=0}^{n-1} \frac{1}{G(x)} G_{i+1}(x)H(f^{i+1}(x))| < \varepsilon \sum_{i=0}^{n-1} \left|\frac{1}{G(x)} G_{i+1}(x)\right| ,$$

and in the sequel

$$|\Psi(x) - G_{n+1}(x)\Psi(f^{n+1}(x)) - \sum_{i=0}^{n} G_i(x)H(f^i(x)) + H(x) - G(x)\Psi(f(x)) - \Psi(x)| < \varepsilon\left(\sum_{i=0}^{n} |G_i(x)| - |G_o(x)|\right) .$$

The last inequality implies, by virtue of properties of the norm, that

$$|\Psi(x) - G_{n+1}(x)\Psi(f^{n+1}(x)) - \sum_{i=0}^{n} G_i(x)H(f^i(x))| < \varepsilon \sum_{i=0}^{n} |G_i(x)| ,$$

i.e. inequality (7) is satisfied for the integer $n+1$, and the proof of the lemma is complete.

THEOREM 1. Let the assumptions of Lemma 1 be fulfilled. If there exists a constant $M > 0$ such that for all positive integers n

$$\sum_{i=0}^{n-1} |G_i(x)| \leq M \quad \text{for } x \in X , \tag{8}$$

then equation (1) is stable in the class of all functions bounded at the point ξ.

Proof. Assume that inequality (3) holds. Then, by virtue of Lemma 2, we have (7). Because $f^n(x)$ tends to ξ as $n \to \infty$ for $x \in X$ and $|G(\xi)| < 1$, then $G_n(x)$ tends to 0 as $n \to \infty$. It implies that

$G_n(x)$ tends to 0 as $n \to \infty$. Now the boundedness of the function Ψ at ξ implies that

$$\lim_{n\to\infty} G_n(x)\Psi(f^n(x)) = 0 \quad \text{for } x \in X .$$

It follows from Lemma 1 that $\lim_{n\to\infty} \sum_{i=0}^{n-1} G_i(x)H(f^i(x)) = \Phi(x)$, and the function Φ is a solution of equation (1) bounded at the point ξ. Whence and from inequalities (7) and (8) we obtain

$$|\Psi(x) - \Phi(x)| \le \varepsilon M \qquad \text{for } x \in X ,$$

which ends the proof.

3. EQUATION (2)

We make the following assumptions.

(H_2) I is the interval $[0,b)$, where $0 < b \le \infty$, K is the field of real or complex numbers, Y is a Banach space over K, the functions $f: I \to I$, $g: I \to K$ and $h: I \to Y$ are continuous, and ϕ: $I \to Y$. Furthermore, $0 < f(x) < x$ for $x \in (0,b)$, $f(0) = 0$, f is strictly increasing in I, $g(x) \neq 0$ for $x \in I$.

DEFINITION 2. Equation (2) is called stable [iteratively stable] in $I_o \subset I$ in a class $\mathcal{C}$ of functions, if there exists a positive constant s such that for each number $\varepsilon > 0$ and each solution ψ of the inequality

$$(9) \qquad |\psi(f(x)) - g(x)\psi(x) - h(x)| < \varepsilon \qquad \text{for } x \in I$$

[of the system of inequalities

$$(10) \qquad \left|\psi(f^n(x)) - \bar{G}_n(x)\psi(x) - \bar{G}_n(x) \sum_{i=0}^{n-1} \frac{h(f^i(x))}{\bar{G}_{i+1}(x)}\right| < \varepsilon$$

$$\text{for } x \in I , \quad n = 1,2,\dots ,$$

where

$$\bar{G}_n(x) := \prod_{i=0}^{n-1} g(f^i(x)) , \quad G_o \equiv 1 \;]$$

there exists a solution $\phi \in \mathcal{C}$ of equation (2) defined on I_o and satisfying inequality (5) (with Ψ and Φ replaced by ψ and ϕ, respectively) for $x \in I_o$.

As it follows from Lemma 3, it is enough to examine the stability [iterative stability] of the simpler, homogeneous equation

$$\phi(f(x)) = g(x)\phi(x) . \tag{11}$$

LEMMA 3. *If equation* (11) *is stable [iteratively stable] in* $I_o \subset I$ *in the class* $C[I_o]$ *of functions continuous on* I_o *and if there exists a solution* $\bar{\phi} \in C[I_o]$ *of equation* (2), *then equation* (2) *is also stable [iteratively stable] in the class* $C[I_o]$. (*We obtain the definition of stability of equation* (11) *putting* $h \equiv 0$ *in* (9) *and* (10) *in Definition* 2.)

The proof of Lemma 3 is quite similar to that of Theorem 1 in [2] and is therefore omitted.

In the theory of equation (11) the basic role is played by the sequence $\bar{G}_n$. There are three possible cases:

(A) There is a continuous function $D: I \to K$, $D(x) \neq 0$ for $x \in I$ such that

$$\lim_{n\to\infty} \bar{G}_n(x) = D(x) \quad \text{for } x \in I .$$

(B) There is an interval $J \subset I$ such that

$$\lim_{n\to\infty} \bar{G}_n(x) = 0$$

uniformly in J.

(C) Neither case (A) nor case (B) is fulfilled.

We quote here as Lemma 4 a well known result in the theory of equation (11); it can be found in [4].

LEMMA 4. *In case* (A) *the continuous solutions of equation* (11) *form one parameter family of functions*

$$\phi(x) = \frac{\eta}{D(x)}, \quad \eta \in Y .$$

In case (B) (*see also* [7]) *for each continuous function* ϕ_o: $[f(x_o),x_o] \to Y$, $x_o \in (0,b)$ *such that* $\phi_o(f(x_o)) = g(x_o)\phi_o(x_o)$, $\phi_o(x) = 0$ *for* $x \in I_o \setminus U$ *and* $\lim_{\substack{x\to a\\ x\in I_o\setminus U}} m(x)\phi_o(x) = 0$ *for* $a \in I_o \cap (\bar{U} \setminus U)$, *where* $m(x) = \sup_n |\bar{G}_n(x)|$, $I_o := [f(x_o), x_o]$ *and* U *is the maximal open set on which the sequence* $\bar{G}_n$ *tends to zero a.u.* (*such a set existsts; see* [7]), *there exists the unique solution* ϕ *of equation* (11) *such that* $\phi|_{I_o} = \phi_o$. *Each continuous solution of equation* (11) *can be obtained in the described manner.*

In case (C) *the unique continuous solution of equation* (11) *is the function* $\phi(x) = 0$ *for* $x \in I$.

THEOREM 2. *Let us assume that hypothesis* (H_2) *is fulfilled and that case* (A) *occurs. If there exists a continuous solution of equation* (2) *on* I *and the function* $\frac{1}{D(x)}$ *is bounded, then equation* (2) *is iteratively stable in* I *in the class of functions continuous on* I. *If moreover the series* $\sum_{i=1}^{\infty} \left|\frac{\bar{G}_n(x)}{\bar{G}_i(x)}\right|$ *is bounded, then equation* (2) *is stable in the class of functions continuous on* I.

Proof. It follows from Lemma 3 that it is enough to prove Theorem 2 for equation (11). Let a function ψ be a continuous solution of inequality (10). Then passing with n to infinity in (10) we have

$$|\psi(0) - D(x)\psi(x)| \leq \varepsilon ,$$

whence

$$\left|\frac{\psi(0)}{D(x)} - \psi(x)\right| \leq \frac{\varepsilon}{|D(x)|} ,$$

and putting $\phi(x) = \frac{\psi(0)}{D(x)}$ we have

$$|\phi(x) - \psi(x)| \leq M_1\varepsilon \quad \text{for } x \in I ,$$

where $\left|\frac{1}{D(x)}\right| \leq M_1$ for $x \in I$, i.e. (5) (with ϕ and ψ in place of Φ and Ψ) is satisfied with $s = M_1$. Now, if inequality (9) holds, then (an inductive proof) for each positive integer n we have

$$|\psi(f^n(x)) - \bar{G}_n(x)\psi(x)| < \varepsilon \sum_{i=1}^{n} \left|\frac{\bar{G}_n(x)}{\bar{G}_i(x)}\right| ,$$

and (passing with n to infinity) we have

$$|\psi(0) - D(x)\psi(x)| \leq \varepsilon M ,$$

where $\sum_{i=1}^{n} \left|\frac{\bar{G}_n(x)}{\bar{G}_i(x)}\right| \le M$ for $x \in I$ and $n = 1,2,\dots$, and in the sequel (5) (with ϕ and ψ in place of Φ and Ψ) holds with $\phi(x) = \frac{\psi(0)}{D(x)}$ and $s = \frac{M}{M_1}$, which ends the proof.

THEOREM 3. *Let us assume that hypothesis* (H_2) *holds and that case* (B) *occurs. If there exists a continuous solution of equation* (2) *and constants* L *and* L_1 *such that for all positive integers* n *we have*

$$\left|\frac{1}{\bar{G}_n(x)}\right| \le L \quad \text{for } x \in I \quad \text{and} \quad |\bar{G}_n(x)| \le L_1 \quad \text{for } x \in I \ ,$$

then equation (2) *is iteratively stable in each interval* $[0,x_o]$, $x_o \in I$, *in the class of functions continuous on* $[0,x_o]$.

Proof. It follows from Lemma 3 that it is enough to prove that Theorem 3 is true for equation (11). Let $x \in I$ such that $\lim_{n\to\infty} \bar{G}_n(x) = 0$. Then, passing with n to infinity in (10), we have

$$|\psi(0)| \le \varepsilon \ .$$

Let us denote

$$\underline{s}(x) = \liminf_{n \to \infty} |\bar{G}_n(x)| \ .$$

We have $\frac{1}{\underline{s}(x)} \le L$ for $x \in I$. It follows from inequality (10) that

$$||\psi(f^n(x))| - |\bar{G}_n(x)||\psi(x)|| < \varepsilon \quad \text{for } x \in I \ ,$$

which implies

$$|\psi(0) - \underline{s}(x)|\psi(x)|| \le \varepsilon \quad \text{for } x \in I \ ,$$

whence

$$|\underline{s}(x)||\psi(x)| \le 2\varepsilon \quad \text{for } x \in I$$

and in the sequel

$$|\psi(x)| \le \frac{2\varepsilon}{\underline{s}(x)} \le \frac{2}{L}\varepsilon \quad \text{for } x \in I \, .$$

Now let us take an arbitrary $x_o \in (0,b)$ and a continuous function ϕ_o defined on the interval $I_o := [f(x_o),x_o]$ such that $\phi_o(x_o) = \psi(x_o)$, $\phi_o(f(x_o)) = g(x_o)\phi_o(x_o)$, $\phi_o(x) = 0$ for $x \in I_o \setminus U$,

$$\lim_{\substack{x\to a\\ x\in I_o\setminus U}} m(x)\phi_o(x) = 0 \text{ for } a\in I_o\cap(\bar{U}\setminus U) \text{ and}$$

(12) $$|\psi(x)-\phi_o(x)|\le r\varepsilon \quad \text{for } x\in I_o ,$$

where $r = \max\left(1,\frac{2}{L}\right)$ (such a function ϕ_o exists, because Y is an arcly connected space, the set $\bigcup_{x\in I_o} k(\psi(x),r)$ is a connected set in Y and $|\psi(f(x_o)) - g(x_o)\psi(x_o)| < \varepsilon$). Let the function ϕ be a continuous solution of equation (11) in $[0,x_o]$ such that $\phi|_{I_o} = \phi_o$ (by virtue of Lemma 4 such a solution ϕ exists). Now, for $x\in [0,x_o]\cap U$ there exists a $t\in I_o$ and the integer n such that $x = f^n(t)$, whence

$$|\psi(x)-\phi(x)| = |\psi(f^n(t)) - \phi(f^n(t))|$$

$$= |\psi(f^n(t))-G_n(t)\psi(t)+G_n(t)\psi(t)-G_n(t)\phi(t)+G_n(t)\phi(t)-\phi(f^n(t))|$$

$$\le \varepsilon + L_1\varepsilon = (1+L_1)\varepsilon .$$

For $x\in[0,x_o]\setminus U$ we have $\phi(x) = 0$, whence

$$|\psi(x)-\phi(x)| \le r\varepsilon .$$

The last inequality implies that

(13) $$|\psi(x)-\phi(x)| \le s\varepsilon \quad \text{for } x\in[0,x_o] ,$$

where $s := \max(1+L_1,r)$, which ends the proof.

THEOREM 4. <u>Let us assume hypothesis</u> (H_2). <u>If</u> $\lim_{n\to\infty}|\bar{G}_n(x)| = \infty$ <u>for</u> $x\in I$ <u>and if there exists a continuous solution of equation</u> (11), <u>then equation</u> (2) <u>is iteratively stable in</u> I <u>in the class of continuous functions</u>.

<u>Proof.</u> By virtue of Lemma 3 it is enough to prove Theorem 4 for equation (11). It follows from inequality (10) and the continuity of ψ at zero that $|\bar{G}_n(x)\psi(x)|$ is bounded for all positive integers n sufficiently large. This is possible only when $|\psi(x)| = 0$ for $x\in I$. Hence inequality (13) follows with $\phi(x) = 0$ for $x\in I$ and $s = 1$, which ends the proof.

REFERENCES

1. B. Brydak, On the stability of the functional equation $\phi(f(x)) = g(x)\phi(x) + F(x)$. Proc. Amer. Math. Soc. 26 (1970), 455-460.

2. B. Choczewski, E. Turdza and R. Wegrzyk, On the stability of a linear functional equation. Rocznik Nauk.-Dydak. WSP w Krakowie, Prace Mat. IX (1979), 15-21.

3. D.H. Hyers, On the stability of the linear functional equation. Proc. Nat. Acad. Sci. USA 27 (1941), 222-222.

4. M. Kuczma, Iterative functional equations (to appear).

5. E. Turdza, On the stability of the functional equation $\phi(f(x)) = g(x)\phi(x) + F(x)$. Proc. Amer. Math. Soc. 30 (1971), 484-486.

6. E. Turdza, Pewne uwagi o stabilności liniowego równania funkcyjnego $\phi(f(x)) = g(x)\phi(x) + F(x)$. Rocznik Nauk.-Dydak. WSP w Krakowie, Prace Mat. VI, 41 (1970), 155-164.

7. R. Wegrzyk, Continuous solution of a linear homogeneous equation. Ann. Polon. Math. 35 (1977), 15-20.

Erwin Turdza, ul. Koniewa 59/70, 30-150 Kraków, Poland.

Inequalities for Differential Operators

Walke Homestead

International Series of
Numerical Mathematics, Vol. 71

ISOPERIMETRIC INEQUALITY FOR THE EFFECTIVENESS IN SEMILINEAR PARABOLIC PROBLEMS

Catherine Bandle and Ivar Stakgold

To the memory of Professor E. Beckenbach

Abstract. The effectiveness of a reaction is the ratio of the actual average reaction rate in a region to the rate corresponding to reference values of reactant concentration and temperature. An isoperimetric inequality for the effectiveness, previously derived for the steady-state problem, is extended to the parabolic problem with suitable initial conditions. Of all regions with equal volume, the ball is shown to have the least effectiveness.

1. INTRODUCTION

A simple model for a reaction-diffusion process leads to the following non-linear problem [2,4,5]

$$(P)\quad \begin{cases} u_t(x,t) - \Delta u(x,t) = \lambda g(u) & \text{in } D \times \mathbb{R}^+ , \quad \lambda > 0 \\ u(x,t) = 0 & \text{in } \partial D \times \mathbb{R}^+ \\ u(x,0) = 0 & . \end{cases}$$

Here, $D \subset \mathbb{R}^N$ is a bounded domain where the reaction takes place and $1 - u =: c$ represents the concentration of the reactant. The function g is subject to the following conditions

(A-1) $g: \mathbb{R}^+ \to [0,1]$ is continuous and $g \in C^2(0,1)$

(A-2) $g(0) = 1$ and $g(\sigma) = 0$ for $\sigma \geq 1$

(A-3) $g'(\sigma) \leq 0$ in $(0,1)$

(A-4) $g''(\sigma) \leq 0$ in $(0,1)$.

Typical examples for such problems are:

<u>p-th order reactions</u> $g(\sigma) = \{\max\,(1-\sigma,0)\}^p =: (1-\sigma)_+^p$, $0 < p \leq 1$.

The existence of a unique weak solution has been established in

[5] by means of generalized version of the method of upper and lower solutions [6]. The solution turns out to be continuous and to take its values in [0,1]. The special feature of this problem is, that depending on the behavior of $g(\sigma)$ and in particular $g'(\sigma)$ in a neighborhood of $\sigma = 1$, $u(x,t)$ reaches its maximum value in finite time.

The set

$$\Omega(t) := \{x \in D: \ u(x,t) = 1\}$$

is called the <u>dead core</u> at time t. It describes the region where no reaction takes place, and has been studied in a series of papers [4,5,7,8]. Another quantity of interest in applications is the so called <u>effectiveness</u>, defined by

$$\eta(t) := \frac{1}{\text{vol } D} \int_D g[u(x,t)] \, dx \ .$$

The purpose of this note is to prove

THEOREM 1. <u>Among all domains</u> D <u>of given volume, the sphere has the lowest effectiveness</u>.

This minimum property of the sphere was conjectured by Aris for the stationary problem with linear g, and later on proved by Amundson and Luss [1] by means of variational techniques. It was then extended to linear parabolic problems in [3] and to the non-linear stationary case [4] without assuming g to be concave in $(0,1)$.

As for the stationary problem, the main tools are rearrangement techniques. We shall follow closely the arguments developed in [3] where similar results have been derived for parabolic equations with an increasing non-linearity.

2. PRELIMINARY RESULTS

The following concavity result will be needed in the proof of Theorem 1.

LEMMA 2.1. *Let* $y\colon [a_0,M] \to \mathbb{R}$ *be a non-increasing function and* $y^*\colon [a_0,M] \to \mathbb{R}$ *be such that*

$$\int_{a_0}^{a} y(\alpha)\, d\alpha \le \int_{a_0}^{a} y^*(\alpha)\, d\alpha \qquad \forall a \in [a_0,M] .$$

Then for any non-increasing, concave function g

$$\int_{a_0}^{a} g[y(\alpha)]\, d\alpha \ge \int_{a_0}^{a} g[y^*(\alpha)]\, d\alpha \qquad \forall a \in [a_0,M] .$$

Proof. For simplicity, we assume $g \in C^1(a_0,M)$. The general case then follows by approximating g by smooth functions. Since g is concave, we have

$$\int_{a_0}^{a} [g(y)-g(y^*)]\, d\alpha \ge \int_{a_0}^{a} g'(y)[y-y^*]\, d\alpha \tag{2.1}$$

If we set $w(a) = g'[y(a)]$, the right-hand side of (2.1) can be written as

$$w(a_0)\int_{a_0}^{\xi} [y-y^*]d\alpha + w(a)\int_{\xi}^{a} [y-y^*]d\alpha \quad \text{for some } \xi \in [a_0,a] \tag{2.2}$$

or equivalently

$$w(a)\int_{a_0}^{a} [y-y^*]\, d\alpha + \{w(a_0)-w(a)\}\int_{a_0}^{\xi} [y-y^*]\, d\alpha \quad .$$

Since by our assumptions, $w(a) \le 0$ and $w(a_0) \le w(a)$, the assertion is established.

Remark. A counterpart of this lemma for convex functions g is found in [3; p. 174].

Let us now introduce the following notation:

M measure of D

$D(\mu,t) := \{x \in D\colon\ u(x,t) \ge \mu\}$, (u solution of (P))

$m(\mu,t) := \text{meas}\,\{D(\mu,t)\}$.

1) We are indebted to the referee who pointed out that this lemma follows from theorems 249 and 250 of Hardy-Littlewood-Pólya's book "Inequalities".

Definition. The function $\tilde{u}: [0,M] \to [0,1]$ defined by

$$\tilde{u}(a,t) = \sup \{\mu : m(\mu,t) \geq a\}$$

is the decreasing rearrangement of u.

The following properties of the decreasing rearrangement are well-known [3,9] .

(A) For any positive continuous function $\psi(\sigma)$

$$\int_0^a \psi[\tilde{u}(\alpha,t)]\, d\alpha = \int_{D(a,t)} \psi[u(x,t)]\, dx , \quad \text{where}$$

$$D(a,t) := D(\tilde{u}(a,t),t) .$$

(B) For any subdomain $B \subset D$ of volume a, we have

$$\int_B u(x,t)\, dx \leq \int_{D(a,t)} u(x,t)\, dx = \int_0^a \tilde{u}(\alpha,t)\, d\alpha .$$

(C) $$q(a)\tilde{u}_a(a,t) + \int_{D(a,t)} [\lambda g(u)-u_t]\, dx \geq 0 \quad \text{in} \quad (0,M) ,$$

where $q^{1/2}(a)$ is the surface area of the sphere of volume a.

Let us now consider problem (P) in the sphere D^* centered at the origin and having the same volume M as D. The quantities related to this problem will be designated by an $*$. Since $u^*(x,\cdot)$ is radially symmetric and decreasing in $|x|$

(C^*) $$q(a)\tilde{u}^*_a(a,t) + \int_{D^*(a,t)} [\lambda g(u^*)-u^*_t]\, dx = 0 \quad \text{in} \quad (0,M) .$$

It was observed in [3] that, for classical solutions, the function

$$H(a,t) := \int_{D(a,t)} u(x,t)\, dx = \int_0^a \tilde{u}(\alpha,t)\, d\alpha$$

satisfies

(D) $$H_t(a,t) = \int_{D(a,t)} u_t(x,t)\,dx \quad \text{a.e.}\ .$$

This identity is proved in [5] for weak solutions. By inserting (A) and (D) into (C), we obtain

LEMMA 2.2. *If* $u(x,t)$ *is a solution of problem* (P), *then*

$$q(a)H_{aa}(a,t) + \lambda \int_0^a g(\tilde{u})\,d\alpha - H_t(a,t) \geq 0 \quad \text{in} \quad (0,M) \times \mathbb{R}^+$$

$$H(0,t) = 0\ , \qquad H_a(M,t) = 0 \quad \text{and} \quad H(a,0) = 0\ .$$

Equality holds for the solution of problem (P*).

3. PROOF OF THEOREM 1

We now are in a position to prove Theorem 1. Let

$$E(a,t) = \int_{D(a,t)} g[u(x,t)]\,dx = \int_0^a g[\tilde{u}(\alpha,t)]\,d\alpha \quad ,$$

and

$$\delta E(a,t) := E(a,t) - E^*(a,t)\ .$$

Note that $\eta(t) = E(M,t)/M$. The theorem will be proved by contradiction. Suppose that $\delta E(M,t_0) < 0$ for some $t_0 > 0$. Then there exist a number $\tau > 0$ and a function $0 \leq a(t) < M$ in $(t_0-\tau,t_0)$ such that

(3.1) $$\delta E(a,t) < 0 \quad \text{in} \quad U := \{(a,t):\ a(t) < a < M,\ t_0 - \tau < t < t_0\}$$

and

(3.2) $$\delta E(a(t),t) = 0\ .$$

By Lemma 2.2

(3.3) $$q(a)\ \delta H_{aa} - \delta H_t > 0 \quad \text{in} \quad U$$

and, in view of a slightly more general version of the maximum principle [5], δH achieves its maximum on ∂U. Since

$\delta H_t \geq 0$ at this point, it must lie in $(a(t),t)$ or (M,t), $t_0 - \tau \leq t \leq t_0$.

Let us first consider the case where ∂D is analytic. Then Lemma 2.2 and therefore also (3.3) hold for $a = M$. This prevents δH from taking its maximum in $a = M$. Suppose now that the supremum of δH in U is achieved at $(a(t_1),t_1)$. Consequently, $\delta H(a(t_1),t_1) \geq \delta H(a,t_1)$, which is equivalent to

$$\int_{a(t_1)}^{a} \tilde{u}(\alpha,t_1)\, d\alpha \leq \int_{a(t_1)}^{a} \tilde{u}^*(\alpha,t_1)\, d\alpha \qquad \forall\, a \in (a(t_1),M) .$$

Lemma 2.1 with $y = \tilde{u}(\cdot,t)$, $y^* = \tilde{u}^*(\cdot,t_1)$ and $a_0 = a(t_1)$ implies that

$$\int_{a(t_1)}^{a} g[\tilde{u}]\, d\alpha \geq \int_{a(t_1)}^{a} g[\tilde{u}^*]\, d\alpha$$

and thus

$$E(a,t_1) - E(a(t_1),t_1) \geq E^*(a,t_1) - E^*(a(t_1),t_1)$$

which by (3.2) implies that

$$\delta E(a,t_1) \geq 0 \quad \text{in} \quad (a(t_1),M)$$

contradicting our assumption. If ∂D is not analytic, we approximate D by domains with analytic boundaries and use the continuous dependence of u on D.

Remarks. (1) The same techniques have been used in [5] to establish the isoperimetric inequalities

$$\max_{x\in D} u(x,t) \leq \max_{x\in D} u^*(x,t) = u^*(0,t)$$

and

$$\text{meas } \Omega(t) \leq \text{meas } \Omega^*(t) .$$

(2) The estimate remains valid if Δ is replaced by a second order operator of divergence form with the ellipticity constant 1.

REFERENCES

1. N. R. Amundson and D. Luss, On a conjecture of Aris: proof and remarks. A. I. Ch. E. J. 13 (1967).

2. R. Aris, The Mathematical Theory of Diffusion and Reaction in Permeable Catalysts. Clarendon Press, Oxford (1975).

3. C. Bandle, Isoperimetric Inequalities and Applications. Pitman (1980).

4. C. Bandle, R. Sperb and I. Stakgold, Diffusion-Reaction with Monotone Kinetics. To appear in Nonlinear Analysis TMA.

5. C. Bandle and I. Stakgold, The onset and formation of a dead core. Trans. Amer. Math. Soc. (submitted).

6. J. Deuel and P. Hess. Nonlinear parabolic boundary value problems with upper and lower solutions. Israel J. Math. 29 (1978), 92-104.

7. J. Diaz and J. Hernandez, Some results on the existence of free boundaries for parabolic reaction-diffusion systems (to appear).

8. A. Friedman and D. Phillips, The free boundary of a semilinear elliptic equation (to appear).

9. G. Talenti, Elliptic equations and rearrangements. Ann. Scuola Norm. Sup. Pisa 3 (1976), 697-718.

Catherine Bandle
Universität Basel
Rheinsprung 21
Ch-4051 Basel
Switzerland

Ivar Stakgold
Dept. of Mathematical Sciences
University of Delaware
Newark, Delaware 19711
U.S.A.

Acknowledgement. The research of the second author was supported by an NSF Grant.

International Series of
Numerical Mathematics, Vol. 71

APPLICATIONS OF GENERALIZED CONVEX FUNCTIONS TO SECOND ORDER DIFFERENTIAL INEQUALITIES

Dobiesław Brydak

Abstract. In this paper we first prove some theorems on generalized convex functions, introduced by E.F. Beckenbach [1]. Next we show that a function ψ is a solution of the differential inequality $\psi''(x) \geq f(x,\psi(x),\psi'(x))$ iff it is convex with respect to the family of all solutions to the first boundary value problem for the associated equation $y'' = f(x,y,y')$. Our proof is based on the notion of the first integral, and we do not assume the continuity of ψ''.

M.M. Peixoto proved in his paper [7] that a function $\psi \in C^2(a,b)$ is a generalized convex function with respect to the two-parameters family of solutions of the equation $y'' = f(x,y,y')$ if and only if it satisfies the inequality $\psi''(x) \geq f(x,\psi(x),\psi'(x))$. Since one of the lemmas in [7] appeared to be incorrect, L. Jackson proved the theorem of Peixoto in [5] by a quite different method. A similar theorem was proved by E. Moldovan in [6] for n-th order differential inequalities. In each of those papers continuity of the second derivative of ψ was assumed. F.F. Bonsall was the only author who proved the above theorem without that assumption (for the second order homogeneous inequalities only). In this paper we shall prove the same theorem for second order nonlinear differential inequalities without the assumption of continuity of the second derivative of ψ. The proof will be based on the notion of first integral of n-th order differential equation. This method has first been used in [4] for differential equations, but the idea of using the first integral for the investigation of properties of a function with respect to a one-parameter family of functions was first applied in [3]. In the present paper, before dealing with differential inequalities, we are going to use this method to study the properties of functions with respect to a two-parameter family of functions, i.e., to prove some theorems on generalized convex functions defined by E.F. Beckenbach [1].

2. GENERALIZED CONVEX FUNCTIONS

In this section we shall give the definition of generalized convex functions due to E.F. Beckenbach [1] and an equivalent definition for functions with continuous derivatives, and we shall also prove some of their properties. The proofs of these properties, very simple as they are, were never, as to my knowledge, published.

In the sequel, I will denote an open interval of reals. Following E.F. Beckenbach, we shall assume the following hypothesis

H_1. Let F be a two-parameter family of functions defined in an interval I and satisfying the following conditions:

(i) Every function $\phi \in F$ is continuous in I;

(ii) The graph of each $\phi \in F$ lies in a given region $D \subset \mathbb{R}^2$;

(iii) For every two points $(x_1, y_1), (x_2, y_2) \in D$, $x_1 \neq x_2$, there is a unique member ϕ of the family F such that $\phi(x_1) = y_1$ and $\phi(x_2) = y_2$.

DEFINITION 1. The function ψ will be called a subfunction (convex function) with respect to the family F provided hypothesis H_1 is fulfilled and $\psi(x) \le \phi(x)$ for all $x_1, x_2, x \in I$ with $x_1 < x < x_2$, where

$$(1) \qquad \psi(x_1) = \phi(x_1) \quad \text{and} \quad \psi(x_2) = \phi(x_2), \quad \phi \in F.$$

Similarly we define a superfunction (concave function) with respect to the family F by reversing the inequality sign. A strictly convex or strictly concave function with respect to the family F is defined in exactly the same way, except that we take the strong inequality sign.

To deal with differentiable subfunctions, we have to assume another hypothesis

H_2. Let F be a two-parameter family of functions defined in an interval I and satisfying the following conditions:

(i) Every function $\phi \in F$ belongs to $C^1(I)$;

(ii) The graph of each $\phi \in F$ lies in a given region D;

(iii) For every $(x_o, y_o) \in D$ and every real number y_o' there is a unique member ϕ of F such that $\phi(x_o) = y_o$ and $\phi'(x_o) = y_o'$.

Now we are able to give a necessary and sufficient condition for a differentiable function to be a subfunction with respect to F.

THEOREM 1. _Let hypotheses_ H_1 _and_ H_2 _be satisfied. A function_ $\psi \in C^1(I)$ _is a subfunction with respect to the family_ F, _if and only if for every_ $x_o \in I$

(2) $$\psi(x) \geq \phi_o(x) \quad \text{for } x \in I ,$$

where $\phi_o \in F$ _is such that_

(3) $$\phi_o(x_o) = \psi(x_o) , \quad \phi_o'(x_o) = \psi'(x_o) .$$

Proof. Let $\psi \in C^1(I)$ be a subfunction with respect to the family F and let $x_o \in I$. By virtue of hypothesis H_2 there is a unique member ϕ_o of F satisfying (3). Let us assume that inequality (2) does not hold, i.e., that there exists a point $x_1 \in I$ such that $\psi(x_1) < \phi_o(x_1)$. Let ϕ be the unique member of F satisfying $\phi(x_o) = \psi(x_o)$ and $\phi(x_1) = \psi(x_1)$. Without loss of generality we may assume that $x_o < x_1$. It follows from hypothesis H_1 that $\phi(x) \leq \phi_o(x)$ for $x \in [x_o, x_1]$. Hence

$$\psi(x) \leq \phi(x) \leq \phi_o(x) \quad \text{for } x \in [x_o, x_1] ,$$

in view of definition 1. Therefore hypothesis H_2 implies that

$$\psi'(x_o) \leq \phi'(x_o) < \phi_o'(x_o) = \psi'(x_o) .$$

This contradiction proves the first part of the theorem.

Now let us assume that $\psi \in C^1(I)$ and inequality (2) is satisfied whenever $\phi_o \in F$ satisfies (3). Let x_o, x_1 and x_2 be arbitrary points of I such that $x_1 < x_o < x_2$ and $\psi(x_1) = \phi(x_1)$, $\psi(x_2) = \phi(x_2)$ and $\psi(x_o) > \phi(x_o)$, where $\phi \in F$. By virtue of hypothesis H_2 there exists a function $\phi_o \in F$ satisfying (3). Due to inequality (2), $\phi_o(x) \leq \psi(x)$ in I. Since ϕ satisfies (1), there are two points x_1', x_2' with $x_1 \leq x_1' < x_o < x_2' \leq x_2$ and $\phi(x_i') = \phi_o(x_i')$, $i = 1,2$. This contradiction to hypothesis H_1 ends the proof of the theorem.

Theorem 1 is an analogue of a well known theorem on convex functions. The following theorem is another analogue.

THEOREM 2. *Let hypotheses* H_1 *and* H_2 *be satisfied and let* $\psi \in C^1(I)$ *be a subfunction with respect to the family* F. *If*

$$\psi(x_1) = \phi(x_1), \quad \psi'(x_1) = \phi'(x_1) \quad \text{and} \quad \psi(x_2) = \phi(x_2),$$

where $x_1, x_2 \in I$, $\phi \in F$, *then*

$$\psi(x) = \phi(x) \qquad \text{for } x \in [x_1, x_2] \qquad (\text{or } x \in [x_2, x_1]) .$$

Proof. Let us assume that $x_1 < x_2$. If $x_2 < x_1$ the proof is similar. If the theorem fails, then there exists a point x_o such that $x_1 < x_o < x_2$ and $\psi(x_o) \neq \phi(x_o)$. Since ψ is a subfunction with respect to F, we have $\psi(x) \geq \phi(x)$ for $x \in I$, by virtue of theorem 1, thus $\psi(x_o) > \phi(x_o)$. But ϕ is a unique member of F satisfying (1), therefore $\psi(x) \leq \phi(x)$ for $x \in [x_1, x_2]$, in view of definition 1. This contradiction ends the proof of the theorem.

2. FIRST INTEGRAL OF THE FAMILY F

Per analogiam to the first integral for a differential equation we may formulate the following

DEFINITION 2. A function $R: D \times \mathbb{R} \to \mathbb{R}$ will be called a first integral of the family F if and only if $R[x, \phi(x), \phi'(x)] = \text{const}$ for $x \in I$, $\phi \in F$.

The notion of first integral provides a very useful characterization of subfunctions:

THEOREM 3. *Let $\psi \in C^1(I)$ and let hypotheses H_1 and H_2 be satisfied. Denote*

(4) $$\eta(x) := R[x,\psi(x),\psi'(x)] \quad \textit{for } x \in I ,$$

where R is a first integral of the family F strictly increasing with respect to the third variable in $D \times \mathbb{R}$.

The function ψ is (strictly) convex with respect to the family F in I, if and only if the function η is (strictly) increasing in I.

The function ψ is (strictly) concave with respect to the family F in I, if and only if the function η is (strictly) decreasing in I.

Proof. We are going to prove the theorem for convex functions. The proof for strictly convex and (strictly) concave functions is similar.

Let ψ be a convex function with respect to the family F. Assume that the theorem fails, i.e., there exist two points $x_1, x_2 \in I$ such that $x_1 < x_2$ and $\eta(x_1) > \eta(x_2)$. It follows from hypothesis H_1 that there exists a unique function $\phi \in F$ such that $\phi(x_1) = \psi(x_1)$, $\phi(x_2) = \psi(x_2)$. Hence, by virtue of definition 1 and hypothesis H_2, we have $\psi'(x_1) \le \phi'(x_1)$, $\psi'(x_2) \ge \phi'(x_2)$, whence $\eta(x_1) = R[x_1,\psi(x_1),\psi'(x_1)] \le R[x_1,\phi(x_1),\phi'(x_1)] = R[x_2,\phi(x_2),\phi'(x_2)] \le R[x_2,\psi(x_2),\psi'(x_2)] = \eta(x_2)$, by virtue of (4), definition 2 and monotonicity of R with respect to the last variable. Therefore η is an increasing function.

Now let us assume that η increases in I, but ψ is not a convex function with respect to F, i.e., there exist two points $x_1, x_2 \in I$ and a function $\phi \in F$ such that $\phi(x_1) = \psi(x_1)$, $\phi'(x_1) = \psi'(x_1)$ and $\phi(x_2) > \psi(x_2)$. Without loss of generality we may assume that $x_1 < x_2$. First we are going to prove that there exists a point x_o belonging to the interval $[x_1,x_2]$ such that $\psi(x_o) \le \phi(x_o)$ and $\psi'(x_o) < \phi'(x_o)$. Indeed, either $\psi(x) \le \phi(x)$ in $[x_1,x_2]$, thus there must be a point x_o between x_1 and x_2 such that $\psi'(x_o) < \phi'(x_o)$ or, if this is not the case, i.e., there is a point $t \in (x_1,x_2)$ such that $\psi(t) > \phi(t)$, then there must exist a point $x_o \in (t,x_2)$ such that $\psi(x_o) = \phi(x_o)$ and $\psi'(x_o) < \phi'(x_o)$, in view of hypothesis H_2. Therefore, by virtue of (4), definition 2 and the monotonicity of R with respect to the last variable, we have

$$\eta(x_o) = R[x_o,\psi(x_o),\psi'(x_o)] < R[x_o,\phi(x_o),\phi'(x_o)]$$
$$= R[x_1,\phi(x_1),\phi'(x_1)] = R[x_1,\psi(x_1),\psi'(x_1)] = \eta(x_1)\ ,$$

which is a contradiction, because η has been assumed to be an increasing function. This contradiction ends the proof of the theorem.

3. ORDINARY DIFFERENTIAL INEQUALITIES

Let us consider the differential equation

$$y^{(n)} = f(x,y,y',\dots,y^{(n-1)}) \quad \text{for } x \in I\ , \tag{5}$$

where I is an interval. We assume that the function f satisfies the following hypothesis

H_3. (i) f is defined and continuous in a region $S \subset I \times \mathbb{R}^n$;

(ii) Any initial value problem with initial values in S is uniquely solvable (in S), and the solution exists in I;

(iii) There exists a first integral $R(x,y,y',\dots,y^{(n-1)})$ of equation (5) defined, continuous and differentiable in S and such that $\partial R/\partial y^{(n-1)} > 0$ for $x \in I$.

As an example of the close relations between a first integral and differential inequalities we are going to prove the following

THEOREM 4. *Let hypothesis H_3 be satisfied and let ψ be a function defined and n times differentiable in I such that its graph lies in S. The function ψ is a solution of the differential inequality*

$$\psi^{(n)}(x) \geq f[x,\psi(x),\psi'(x),\dots,\psi^{(n-1)}(x)] \quad \text{for } x \in I\ , \tag{6}$$

if and only if the function

$$\eta(x) := R[x,\psi(x),\psi'(x),\dots,\psi^{(n-1)}(x)] \quad \text{for } x \in I \tag{7}$$

is an increasing function in I.

The function ψ is a solution of the inequality opposite to (6), if and only if the function η defined by formula (7) is a decreasing function in I.

Proof. We are going to prove the theorem for inequality (6). The proof for the opposite inequality is similar.

Let ψ be n times differentiable in I and let $x \in I$. In view of hypothesis H_3, there exists a unique solution ϕ of equation (5) satisfying the conditions

$$\phi(x) = \psi(x), \quad \phi^{(i)}(x) = \psi^{(i)}(x), \qquad i = 1,2,\dots,n-1 . \tag{8}$$

It follows from H_3 and (7) that

$$R_x + R_y\psi'(x) + \dots + R_{y^{(n-1)}}\psi^{(n)}(x) = \eta'(x)$$

and

$$R_x + R_y\phi'(x) + \dots + R_{y^{(n-1)}}\phi^{(n)}(x) = 0 ,$$

because R is a first integral of equation (5). Thus

$$\eta'(x) = R_{y^{(n-1)}}[\psi^{(n)}(x) - \phi^{(n)}(x)] .$$

Hence, if ψ is a solution of (6), then

$$\eta'(x) \geq R_{y^{(n-1)}}[f[x,\psi(x),\psi'(x),\dots,\psi^{(n-1)}(x)] - f[x,\phi'(x),\dots,\phi^{(n-1)}]]$$
$$\geq 0$$

because of (6), (5) and (8), thus η is an increasing function in I.

Conversely, if η is an increasing function in I, then

$$R_{y^{(n-1)}}[\psi^{(n)}(x) - \phi^{(n)}(x)] = \eta'(x) \geq 0 ,$$

whence

$$\psi^{(n)}(x) \geq \phi^{(n)}(x) = f[x,\phi(x),\phi'(x),\dots,\phi^{(n-1)}(x)]$$
$$= f[x,\psi(x),\psi'(x),\dots,\psi^{(n-1)}(x)]$$

because ϕ satisfies (5) and conditions (8). Therefore the theorem has been proved.

A similar theorem can be proved in exactly the same way for the strong version of inequality (6) and the opposite inequality provided we assume that η is a strictly monotonic function in I.

Now let us consider the second order differential inequality

(9) $\psi''(x) \geq f[x,\psi(x),\psi'(x)]$ for $x \in I$.

We assume the following hypothesis

H_4. (i) f is defined and continuous in a region $S \subset I \times \mathbb{R}^2$;

(ii) Any initial value problem as well as any first boundary value problem for the differential equation

(10) $y'' = f(x,y,y')$ for $x \in I$

is uniquely solvable in S on I;

(iii) There exists a first integral R of equation (10) defined, continuous and differentialbe in S such that $\partial R/\partial y' > 0$ in S.

As an application of all foregoing results we can prove the following

THEOREM 5. *Let hypothesis* H_4 *be satisfied and let* ψ *be a function defined and twice differentiable in* I *such that its graph lies in* S. *The function* ψ *is a solution of inequality* (9), *if and only if it is convex with respect to the family* F *of all solutions of equation* (10).

Proof. It follows from theorem 4, applied to the case where $n = 2$, that ψ satisfies inequality (9), if and only if the function $\eta(x) := R[x,\psi(x),\psi'(x)]$ is an increasing function in I. Next η is an increasing function, if and only if ψ is a convex function with respect to the family F, by virtue of theorem 3, because hypothesis H_4 implies that hypotheses $H_1 - H_3$ are satisfied. Thus the theorem has been proved.

In a similar way we may prove theorem 5 in case of the inequality opposite to inequality (9) as well as in case of strong inequalities.

In this place I want to thank Mr. J. Krzyszkowski for some improvements of theorem 3.

REFERENCES

1. E.F. Beckenbach, Generalized convex functions. Bull. Amer. Math. Soc. 43 (1937), 363-371.

2. F.F. Bonsall, The characterization of generalized convex functions. Quart. J. Math. 1 (1950), 100-111.

3. D. Brydak, On functional inequalities in a single variable. Dissertationes Math. CLX (1979), 1-48.

4. D. Brydak, A generalization of Pólya's theorem. In: E.F. Beckenbach and W. Walter, General Inequalities 3, pp. 427-430. Birkhäuser Verlag, Basel - Boston - Stuttgart, 1983.

5. L. Jackson, On second order differential inequalities. Proc. Amer. Math. Soc. 17 (1966), 1023-1027.

6. E. Moldovan, Applications des fonctions convexes généralisées. Mathematica (Cluj), Vol. 1 (24) 2 (1959), 281-286.

7. M.M. Peixoto, Generalized convex functions and second order differential inequalities. Bull. Amer. Math. Soc. 55 (1949), 563-572.

Dobiesław Brydak, Institute of Mathematics, Pedagogical University, 30-011 Kraków, Poland

International Series of
Numerical Mathematics, Vol. 71

INEQUALITIES RELATED TO PÓLYA MATRICES

Achim Clausing

Abstract. A Pólya matrix P is a 0-1-matrix representing a certain differential system. By means of a minimum principle, this system induces the class of P-concave functions for which we derive a variety of integral inequalities. The proofs are based on the total positivity properties of the extended Green's kernel of the underlying differential system.

1. INTRODUCTION

In the realm of inequalities, the equivalent to integrating a differential equation is, roughly, to take a differential *in*equality and extract an integral inequality out of it. If the differential equation is linear, its solution may be obtained via a Green's kernel, and then it is natural that the corresponding approach to inequalities should be based on positivity properties of the Green's kernel. This idea has been used, explicitly or implicitly, by many authors, cf. [1], Ch. 4.

The present paper is another variation of this theme. We shall consider functions f satisfying $f^{(p)}(x) \geq 0$ or $f^{(p)}(x) \leq 0$ in some interval (a,b) plus - and this is the key point - an array of boundary inequalities of the form $\pm f^{(j)}(a) \geq 0$, $\pm f^{(j)}(b) \geq 0$, the signs chosen so as to make f positive. For these functions, we derive integral inequalities similar to those considered in [7] and [3].

The boundary conditions that we employ are governed by a two-row Pólya matrix (cf. [9]). In [4] and [5], we have investigated the Green's kernel of the differential system associated with a Pólya matrix. Our present results are based on this analysis and in particular on the fact that a certain kernel embracing the Green's kernel is totally positive. These matters are reviewed in Section 2 of this paper as far as we need them here.

In Section 3, we are concerned with quotients $Q(f) = \int f\,d\mu / \int f\,d\nu$, μ and ν being finite Borel measures on (a,b). We shall derive sharp upper and lower bounds on $Q(f)$ for f satisfying a differential inequality related to a Pólya matrix. The inequalities of Section 4 are multivariate versions of the main result, Theorem 3.2, of Section 3.

For ease of exposition, we have confined ourselves to the basic differential inequality $f^{(p)}(x) \geq 0$ (or ≤ 0) in this paper. It should be noted, however, that all our results can be extended to inequalities of the type $Lf \geq 0$ where L is a disconjugate differential operator since the basic total positivity result (Theorem 2.4) is valid in this more general situation.

Finally, we mention that terms referring to inequalities like positive or increasing are used in the non-strict sense. We write $\int f$ instead of $\int_a^b f(x)\,dx$ where there is no danger of confusion.

2. PÓLYA MATRICES AND THEIR ASSOCIATED FUNCTION CONES

We shall consider differential systems

$$(1)\qquad \begin{aligned} y^{(p)} &= f \qquad (f \in C[a,b],\ -\infty < a < b < \infty) \\ \delta_i(y) &= a_i \qquad (a_i \in \mathbb{R},\ i = 1,\ldots,p)\ , \end{aligned}$$

where $p \geq 2$ and $\delta_1,\ldots,\delta_p$ are boundary conditions of the form $y_a^{(j)}$ or $y_b^{(j)}$ $(0 \leq j \leq p-1)$. It is convenient to represent (1) by a $(2 \times p)$-matrix $P = (e_{ij})_{i=0,1;\,j=0,\ldots,p-1}$ in which $e_{0j} = 1$ if $y_a^{(j)}$ is contained in (1), $e_{1j} = 1$ if $y_b^{(j)}$ is contained in (1), and $e_{ij} = 0$ otherwise.

Pólya [8] has shown that the system (1) has a unique solution $y \in C^p[a,b]$ for every right-hand side $f, a_1,\ldots,a_p$ if and only if P has at least j entries 1 in the first j columns $(j = 1,\ldots,p)$. A $(2 \times p)$-matrix P containing p ones and p zeros and satisfying this condition is called a *Pólya matrix* of order p. It is used as a shorthand notation for solvable systems (1). For example, $\binom{1\,1\,0}{1\,0\,0}$ is a Pólya matrix, representing $\{y_a, y_a', y_b\}$, while $\binom{0\,0\,1}{1\,0\,1}$, representing $\{y_a'', y_b, y_b''\}$, violates the condition.

If P is a Pólya matrix, the unique solution of (1) is

(2) $$y(x) = \sum_{i=1}^{p} a_i G_i(x) + \int_a^b G(x,t) f(t)\,dt,$$

where $G(x,t)$ is the Green's kernel of the system (1) and $G_1,\ldots,G_p$ are the polynomials of degree $\leq p-1$ specified by

(3) $$\delta_i(G_j) = \delta_{ij} \quad (i,j = 1,\ldots,p).$$

In [4], [5] we have investigated, in a more general framework, the system (1). We review now a few of its basic properties.

A key fact is that the "basic polynomials" G_i and the Green's kernel G do not change sign in (a,b), resp. in $(a,b) \times (a,b)$. The corresponding signs can be read from P. To that purpose we index the boundary conditions in (1) in such a way that the leftmost entry 1 in the top row of P corresponds to δ_1 and the other entries 1 in P, counted clockwise, correspond to $\delta_2,\ldots,\delta_p$. For example, $P = \binom{101110}{010010}$ yields $\delta_1 = y_a$, $\delta_2 = y''_a$, $\delta_3 = y'''_a$, $\delta_4 = y_a^{(4)}$, $\delta_5 = y_b^{(4)}$, $\delta_6 = y'_b$. The number of conditions at $x = a$ will be denoted by q (hence q is the number of ones in the top row of P). If j_i is the differentiation order of $\delta_i = y_{x_i}^{(j_i)}$ $(i = 1,\ldots,p)$, then the signs σ_i of G_i in (a,b) and the sign σ of G are

(4) $$\begin{aligned} \sigma &= (-1)^{p-q} \\ \sigma_i &= (-1)^{i+j_i+1} \quad (i = 1,\ldots,q) \\ \sigma_i &= (-1)^{p+1} \quad (i = q+1,\ldots,p)\,. \end{aligned}$$

E.g., for $P = \binom{101110}{010010}$ we find $\sigma = \sigma_1 = \sigma_6 = 1$, $\sigma_2 = \sigma_3 = \sigma_4 = \sigma_5 = -1$.

Using (2) and (4) it is clear that the following *minimum principle* holds:

(5) $$\begin{gathered} \sigma y^{(p)} \geq 0\,,\ \sigma_i \delta_i(y) \geq 0 \quad (i = 1,\ldots,p) \\ \text{implies } y \geq 0 \quad (y \in C^p[a,b])\,. \end{gathered}$$

In the case $P = \binom{10}{10}$, this says that $y'' \leq 0$, $y(a) \geq 0$, $y(b) \geq 0$ implies $y \geq 0$, thus (5) amounts to the familiar fact that a concave function (in $C^2[a,b]$) attains its minimum at a boundary point.

The minimum principle is the main reason why one wants to consider the following functions:

2.1 DEFINITION. *Let* P *be a Pólya matrix of order* p . *A function* $f : (a,b) \to \mathbb{R}$ *is called* *P-concave* *if it is the restriction of some* $f \in C^p[a,b]$ *satisfying*

$$\sigma f^{(p)} \geq 0, \quad \sigma_i \delta_i(f) \geq 0 \qquad (i = 1,\ldots,p)$$

or is the pointwise limit of such functions. The convex cone of all P-*concave functions is denoted by* C_P . *The pointwise closure in* C_P *of all* $f \in C_P \cap C^p[a,b]$ *satisfying*

$$\delta_i(f) = 0 \qquad (i = 1,\ldots,p)$$

is denoted by C_P^o .

Clearly, the polynomials $\sigma_i G_i$ are contained in C_P . It is a little less obvious that the functions σG_t $(t \in (a,b))$, where $G_t(x) = G(x,t)$, also lie in C_P ([5], Lemma 2.6). Indeed, the functions $\sigma_i G_i$ and σG_t generate C_P , and since they will be rather ubiquitous in the sequel we free them of the signs σ_i, σ by altering the system (1) into

$$(1^+) \qquad \begin{aligned} \sigma y^{(p)} &= f \qquad (f \in C[a,b]) \\ \sigma_i \delta_i(y) &= a_i \qquad (a_i \in \mathbb{R},\ i = 1,\ldots,p) \ . \end{aligned}$$

The Green's kernel G belonging to (1^+) and the polynomials G_i of degree $\leq p-1$ satisfying $\delta_i(G_j) = \delta_{ij}$ $(i,j = 1,\ldots,p)$ then obviously are positive in (a,b) and contained in C_P .

The structure of C_P is described in the following result quoted from [5]. We need to exclude the cases $P = \begin{pmatrix} 0\,0 \cdots 0 \\ 1\,1 \cdots 1 \end{pmatrix}$ and $P = \begin{pmatrix} 1\,1 \cdots 1 \\ 0\,0 \cdots 0 \end{pmatrix}$; the remaining Pólya matrices will be called *two-sided*.

2.2 THEOREM. *Let* P *be a two-sided Pólya matrix. Then every* $f \in C_P$, $f \neq 0$, *is strictly positive,* $p-2$ *times continuously differentiable, and* $\delta_i(f)$ *exists for* $i = 1,\ldots,p$. *If* $\nu \neq 0$ *is a finite, positive Borel measure on* (a,b) , *then*

$$\tilde{C}_P = \{f \in C_P \mid \int f \, d\nu = 1\}$$

is a convex base for C_P . *On* C_P , *the topologies of pointwise and locally uniform convergence in* (a,b) *coincide. In this topology,* $\tilde{C}_P$ *is a metrizable, compact set. Its extremal points are the functions*

$$\tilde{G}_t = G_t / \int G_t \, d\nu \quad (t \in \{1,\dots,p\} \cup (a,b)).$$

The set of these functions is closed:

$$\tilde{G}_q = \lim_{t \downarrow a} \tilde{G}_t \ , \quad \tilde{G}_{q+1} = \lim_{t \uparrow b} \tilde{G}_t \ .$$

(q is the number of ones in the top row of P .)

An easy consequence of this theorem is

2.3 COROLLARY. *The compact convex set*

$$\tilde{C}_P^o = \tilde{C}_P \cap C_P^o$$

is a base for C_P^o . *Its extremal points are the functions*

$$\tilde{G}_t \quad (t \in \{q, q+1\} \cup (a,b)).$$

The crucial step in the proof of Theorem 2.2 is a total positivity argument. The inequalities in this paper will likewise depend on it, so we cite it explicitly. For information about totally positive (TP) kernels, cf. [6].

Again, P will be a two-sided Pólya matrix of order p with q ones in its first row. By J_q , we denote the set $\{1,\dots,p\} \cup \cup (a,b)$, provided with the ordering

(6) $\quad 1 < \dots < q < s < t < q+1 < \dots < p \quad (s,t \in (a,b),\ s < t).$

2.4 THEOREM ([4], Theorem 5.2). *The extended Green's kernel*

$$\Gamma_P : (a,b) \times J_q \to \mathbb{R},$$

given by $\Gamma_P(x,t) = G(x,t)$ *if* $t \in (a,b)$, $\Gamma_P(x,i) = G_i(x)$ *if* $i \in \{1,\dots,p\}$, *is* TP *with respect to the order* (6).

For example, Γ_P is TP_2 , i.e. $s,t \in J_q$, $s < t$, implies that G_t/G_s is increasing. This property will be used extensively in this paper. It also justifies the seemingly arbitrary way in which we index the boundary conditions associated with a Pólya matrix.

2.5 REMARK.

For a Pólya matrix P, the explicit values of the constants in the inequalities treated in this paper can be calculated on a microcomputer.

The crucial step, the determination of the basic polynomials, is done as follows: Given P, use the sign formula (4) to form the matrix $D = (\delta_j(x^{i-1}))_{i,j=1,\dots,p}$. Then D is nonsingular, and the entries in the line j of D^{-1} are the coefficients of $G_j(x)$, that is, $G_j(x) = D^{-1}\cdot(1,\dots,x^{p-1})^t$. The numbers in tables 1 - 3 below have been obtained in this way.

3. MONOTONICITY PROPERTIES

The subsequent inequalities owe their existence to the fact that convolution with a TP kernel preserves certain monotonicity properties.

In the following lemma, J denotes a set which is (isomorphic to) a Borel subset of $\mathbb{R}$. $H : J \times (a,b) \to (0,\infty)$ is a continuous kernel, and $Kh : J \to \mathbb{R}$ is defined by

$$Kh(j) = \int_a^b H(j,x)h(x)\,dx \tag{7}$$

for all h for which the integral is finite. We recall that a function F on J is *quasiconcave* if it is monotone or there is some $t \in J$ such that $F(x)$ is increasing for $x \leqq t$ and decreasing for $x \geqq t$. If $-F$ is quasiconcave, F is called *quasiconvex*.

3.1 LEMMA. Let $f,g \in C(a,b)$, $g > 0$, be such that f/g is increasing [resp. decreasing, quasiconcave, quasiconvex] on (a,b). Then the same monotonicity property holds for Kf/Kg on J, provided that the kernel H is TP_3.

Proof. Let us prove the claim concerning quasiconcavity. Since f/g is quasiconcave, $h = f - cg$ has at most two sign changes in (a,b) for any $c \in \mathbb{R}$, and if it has two, then $h \geq 0$ between them. Now H is TP_3, so by [6], Ch. 5, Theorem 3.1, $Kh = Kf - cKg$ has at most two sign changes in J, and if there are two, then $Kf - cKg \geq 0$ and thus $Kf/Kg - c \geq 0$ between them. Since Kf/Kg is continuous, this yields that it is quasiconcave. For the claim concerning the preservation of monotonicity, only the TP_2 property is needed. □

The continuity assumption on f and g in the lemma is by far too stringent. If μ is a σ-finite Borel measure on (a,b) such that $\int_a^b H(j,x)\,d\mu(x) = K\mu(j) < \infty$, and ν is a measure of the same type, then we agree to say that μ/ν is increasing [resp. decreasing, quasiconcave, quasiconvex] if there are sequences $f_n, g_n \in C(a,b)$, $g_n > 0$, such that for all n, f_n/g_n has the respective property, and $\mu = \lim f_n$, $\nu = \lim g_n$ holds in the distribution sense. The monotonicity properties of μ/ν in this sense are then obviously inherited to $K\mu/K\nu$.

We now return to the situation of section 2, that is, we consider a two-sided Pólya matrix P and its associated cones of P-concave functions.

3.2 THEOREM. Let μ and ν be two finite Borel measures on (a,b), $\nu > 0$. For $f \in C_P \setminus \{0\}$, we put

$$Q(f) = \frac{\int f\,d\mu}{\int f\,d\nu}. \tag{8}$$

Also, let $c_i = Q(G_i)$ $(i = 1,\dots,p)$. Then the following inequalities hold for all $f \in C_P$, $f \neq 0$:

a) $c_1 \leq Q(f) \leq c_p$ if μ/ν is increasing.

b) $c_p \leq Q(f) \leq c_1$ " " " decreasing.

c) $\min(c_1,c_p) \leq Q(f)$ " " " quasiconcave.

d) $Q(f) \leq \max(c_1,c_p)$ " " " quasiconvex.

The following inequalities hold for $f \in C_P^o \setminus \{0\}$:

a°) $c_q \leq Q(f) \leq c_{q+1}$ if μ/ν is increasing.

b°) $c_{q+1} \leq Q(f) \leq c_q$ _if_ μ/ν _is_ _decreasing_.
c°) $\min(c_q, c_{q+1}) \leq Q(f)$ " " " _quasiconcave_.
d°) $Q(f) \leq \max(c_q, c_{q+1})$ " " " _quasiconvex_.

Proof. Consider the base $\tilde{C}_P$ of C_P as defined in Theorem 2.2. Since it is compact and $Q|_{\tilde{C}_P}$ is linear, Q attains its extrema over $\tilde{C}_P$ (and hence over C_P) on the closed set $\{\tilde{G}_t \mid t \in J_q\}$ of extremals of $\tilde{C}_P$. Thus we have to find the extrema of the function $\tilde{Q}(t) \equiv Q(\tilde{G}_t)$ $(t \in J_q)$. Now

$$\tilde{Q}(t) = \frac{\int G_t \, d\mu}{\int G_t \, d\nu} = \frac{\int_a^b \Gamma_P(x,t) \, d\mu(x)}{\int_a^b \Gamma_P(x,t) \, d\nu(x)} ,$$

thus if we choose $H(t,x) = \Gamma_P(x,t)$ in Lemma 3.1 (by Theorem 2.4, Γ_P is TP_3), we find that $\tilde{Q}$ has the same monotonicity properties on J_q as μ/ν on (a,b). The particular order (6) of J_q now yields inequalities a) - d). The other inequalities follow in the same way from Corollary 2.3. □

Various choices of μ and ν in this theorem lead to numerous inequalities for P-concave functions. If, for example, ν equals the Lebesgue measure λ on (a,b), then μ can be chosen to be $\mu = \delta_t$ (the Dirac measure at t), or $\mu = 1_{(s,t)}(x) \cdot \lambda$ $(a < s < t < b)$, or, if $(a,b) = (0,1)$, $\mu = \sin \pi x \cdot \lambda$. Thus we obtain sharp estimates for

$f(t)$, $\int_s^t f$, or for the Fourier coefficient $\int_0^1 f(x) \sin \pi x \, dx$.

The following result is one example of how to use Theorem 3.2. Others are given in the next section.

3.3 PROPOSITION. _Let_ P _be a two-sided Pólya matrix of order_ $p \geq 3$ _such that the first column is_ $\binom{1}{0}$ _and_ $q \geq 2$, _and let_ $s \geq 1$. _Then there are constants_ $c,d > 0$ _such that_

$$c \int f' \leq \|f\|_s \leq d \int f'$$

holds for all $f \in C_P$ _satisfying_ $f(a) = 0$. _The best values of_

<u>these constants are</u>

$$c = \frac{\|G_p\|_s}{\int G_p'} , \quad d = \frac{\|G_2\|_s}{\int G_2'} .$$

<u>Proof.</u> The assumptions on P guarantee that the columns 2 through p of P form a two-sided Pólya matrix M of order $p-1$. To this matrix, we apply Theorem 3.2.b with $\nu = \lambda$ and $\mu = 1_{(a,t)}(x)\cdot\lambda$ and obtain

$$Q(G_{p-1}^M) \leq Q(f) = \frac{\int_a^t g}{\int_a^b g} \leq Q(G_1^M) ,$$

where $g \in C_M$, $g \not\equiv 0$. (The superscript M indicates that the G_i^M refer to M instead of P.) Now $\int_a^b g \equiv f(t) \in C_P$ and $f(a) = 0$. Conversely, for each $f \in C_P$, $f(a) = 0$, f' lies in C_M as is easily seen. Using (4), it is also readily checked that $G_{i+1} = G_i^M$ holds for $i = 1,\ldots,p-1$. Thus we have for all $t \in (a,b)$, $f \in C_P \setminus \{0\}$, $f(a) = 0$:

$$\frac{G_p(t)}{\int G_p'} \leq \frac{f(t)}{\int f'} \leq \frac{G_2(t)}{\int G_2'} .$$

By integration, the proof follows. □

In Table 1, we have compiled the values of c and d for $s = 2$ and a few examples of Pólya matrices.

P	c	d
$\begin{pmatrix} 1&1&0 \\ 0&1&0 \end{pmatrix}$	0.4472	0.7303
$\begin{pmatrix} 1&1&0 \\ 0&0&1 \end{pmatrix}$	0.4472	0.5774
$\begin{pmatrix} 1&1&1&0 \\ 0&1&0&0 \end{pmatrix}$	0.3780	0.6969
$\begin{pmatrix} 1&1&1&1&0&0 \\ 0&1&1&0&0&0 \end{pmatrix}$	0.3860	0.7365
$\begin{pmatrix} 1&1&1&1&1&0&0&0 \\ 0&0&0&1&0&1&0&1 \end{pmatrix}$	0.3066	0.5774

TABLE 1 (to Prop. 3.3; s = 2)

4. INEQUALITIES FOR PRODUCTS OF P-CONCAVE FUNCTIONS

The inequality $\frac{2}{3}\int_0^1 f \int_0^1 g \leq \int_0^1 fg$, valid for positive, concave functions f and g, dates back to 1915 ([2]). A related inequality is due to Tchebycheff: If $f_1,\ldots,f_n$ are positive, increasing, integrable functions on $(0,1)$, then

$$\prod_1^n \int_0^1 f_i \leq \int_0^1 \prod_1^n f_i . \tag{9}$$

There exist many other inequalities of this type, cf. [7], [3], and the literature cited there. In this section, we extend results of [7], Sec. 5, and of [3], Sec. 3, to the case of P-concave functions. Throughout, P denotes a two-sided Pólya matrix.

4.1 THEOREM. <u>For</u> $f,g \in C_P \setminus \{0\}$ <u>and a strictly positive function</u> $h \in L^1(a,b)$, <u>consider</u>

$$R(f,g) = \frac{\int fgh}{\int fh \cdot \int gh} . \tag{10}$$

<u>Furthermore, put</u> $c_{i,j} = R(G_i,G_j)$ $(i,j = 1,\ldots,p)$. <u>Then</u>

$$R(f,g) \geq c_{1,p} \tag{11}$$

<u>holds for all</u> $f,g \in C_P \setminus \{0\}$, <u>and</u>

$$R(f,g) \geq \min(c_{q,q}, c_{q,q+1}, c_{q+1,q+1}) \equiv a \tag{12}$$

<u>holds for all</u> $f,g \in C_P^0 \setminus \{0\}$.

<u>Proof.</u> By the same arguments as in Theorem 3.2, it is clear that R attains its minimum over $C_P \setminus \{0\}$ at some of the functions G_t, $t \in J_q$, and its minimum over $C_P^0 \setminus \{0\}$ at some of the functions G_t, $t \in \{q,q+1\} \cup (a,b)$. Hence for the proof of inequalities (11) and (12) we have to show that $\tilde{R}(s,t) \equiv R(G_s,G_t)$, $s,t \in J_q$, is quasiconcave in both variables. Now for fixed $s \in J_q$, we have $\tilde{R}(s,t) = \tilde{Q}(t)/\int G_s h$ in the notation of Theorem 3.2, where we choose $\nu = h\cdot\lambda$ and $\mu = G_s \cdot h \cdot \lambda$. By [5], Proposition 2.10, the function G_s is quasiconcave and hence μ/ν is quasiconcave in the sense of Section 3. Hence $\tilde{R}(s,t)$ is quasiconcave in the

variable t and, by symmetry, also in the variable s. This implies that the minimum of $R(f,g)$ over $C_P \setminus \{0\}$, resp. $C_P^o \setminus \{0\}$, occurs if $f,g \in \{G_1, G_p\}$, resp. if $f,g \in \{G_q, G_{q+1}\}$. Thus it equals $\min(c_{1,1}, c_{1,p}, c_{p,p})$, resp. $\min(c_{q,q}, c_{q,q+1}, c_{q+1,q+1})$.

To validate (11) we use that $G_1(x)$ is increasing and $G_p(x)$ is decreasing in (a,b), cf. the proof of [5], Lemma 2.11. Consequently, $t \to \tilde{R}(1,t)$ is decreasing and $t \to \tilde{R}(p,t)$ is increasing in J_q. In particular, $c_{1,1} \geq c_{1,p}$ and $c_{1,p} \leq c_{p,p}$, which proves inequality (11). □

4.2 REMARK.

If the first column of P is $\binom{1}{0}$, then $G_1(x) = 1$, thus the constant in (11) is $c_{1,p} = (\int h)^{-1}$. This can also be seen from (a variant of) Tchebycheffs inequality (9) since in this case all $f \in C_P$ are increasing ([5], Proposition 2.10). Similarly, if the first two columns of P are $\begin{pmatrix} 1 & 0 \\ 1 & 0 \end{pmatrix}$ and, say, $(a,b) = (0,1)$, then $G_1(x) = 1 - x$ and $G_p(x) = x$, thus $c_{1,p} = (\int x(1-x)h)/(\int xh \int (1-x)h)$. If $h = 1$, we obtain $c_{1,p} = \frac{2}{3}$, the constant mentioned at the beginning of this section. This is no surprise since the condition on P entails that all $f \in C_P$ are concave. For other examples of specific values of $c_{1,p}$ and a, see the subsequent table.

P	$c_{1,p}$	a
$\begin{pmatrix} 1&1&0 \\ 1&0&0 \end{pmatrix}$	0.6	0.9
$\begin{pmatrix} 1&1&0&1 \\ 1&0&0&0 \end{pmatrix}$	0.6	1.2
$\begin{pmatrix} 1&1&0&0 \\ 1&0&0&1 \end{pmatrix}$	0.6	1.2
$\begin{pmatrix} 1&1&1&0&0 \\ 1&1&0&0&0 \end{pmatrix}$	0.48	1.19
$\begin{pmatrix} 1&1&1&1&0&0 \\ 1&1&0&0&0&0 \end{pmatrix}$	0.45	1.36
$\begin{pmatrix} 1&1&1&1&1&0 \\ 1&0&0&0&0&0 \end{pmatrix}$	0.55	1.64
$\begin{pmatrix} 1&1&1&1&1&0&0&0 \\ 0&0&0&1&0&1&0&1 \end{pmatrix}$	1	2.51

TABLE 2 (to Theorem 4.1; $h \equiv 1$)

We should also like to know the least upper bound for the ratio R. In general, $\tilde{R}$ attains its maximum at the interior of $J_q \times J_q$, so we cannot say much. But there is a simple answer if the first column of P contains only one entry 1. In this case, the constant in (12) can also be specified more precisely:

4.3 THEOREM. *If the first column of* P *is* $\binom{1}{0}$, *then*

(13) $$c_{1,1} \leq R(f,g) \leq c_{p,p}$$

holds for all $f,g \in C_P \setminus \{0\}$, *and*

(14) $$c_{q,q} \leq R(f,g) \leq c_{q+1,q+1}$$

holds for all $f,g \in C_P^0 \setminus \{0\}$. *If the first column is* $\binom{0}{1}$, *then the inequalities are reversed.*

Proof. By [5], Proposition 2.10, the condition on P entails that all $f \in C_P$ are increasing. Therefore the same argument as used in the proof of Theorem 4.1 to show that $\tilde{R}(s,t)$ is quasiconcave in $s,t \in J_q$ now establishes that $\tilde{R}(s,t)$ is increasing in $s,t \in J_q$. From this, both inequalities can be easily deduced. □

Since the above inequalities are sharp, the seemingly different lower bounds in (11) and (13) are equal. Using $G_1(x) = 1$, this can also be seen directly.

We conclude with a multivariate version of the above results. For $f_1,\ldots,f_n \in C_P \setminus \{0\}$, we put

(15) $$R_n(f_1,\ldots,f_n) = \frac{\int(\prod_{i=1}^{n} f_i)h}{\prod_{i=1}^{n} \int f_i h},$$

where again h is a strictly positive function in $L^1(a,b)$.

4.4 THEOREM. *If the first column of* P *is* $\binom{1}{0}$, *then*

(16) $$u_n \leq R_n(f_1,\ldots,f_n) \leq v_n$$

holds $f_i \in C_P$ $(i = 1,\ldots,n)$, *where* $u_n = R_n(1,\ldots,1)$ *and* $v_n = R_n(G_p,\ldots G_p)$. *For* $f_i \in C_P^0$ $(i = 1,\ldots,n)$ *the corresponding inequality is*

(17) $$u_n^0 \leq R_n(f_1,\ldots,f_n) \leq v_n^0,$$

with $u_n^o = R_n(G_q,\dots,G_q)$ and $v_n^o = R_n(G_{q+1},\dots,G_{q+1})$.

<u>Proof.</u> Reasoning as in the proof of Theorem 3.2 we see that the extrema of $\tilde{R}(t_1,\dots,t_n) \equiv R_n(G_{t_1},\dots,G_{t_n})$ $(t_1,\dots,t_n \in J_q)$ have to be determined. If we put $\tilde{R}_j(t_j) \equiv \tilde{R}(t_1,\dots,t_n)$ $(t_j \in J_q)$, and choose $\mu = \prod_{\substack{i=1\\ i\neq j}}^{n} G_{t_i} \cdot h \cdot \lambda$, $\nu = h\cdot\lambda$ in Theorem 3.2, then

$$\tilde{R}_j(t) = \frac{\tilde{Q}(t)}{\prod_{\substack{i=1\\ i\neq j}}^{n} \int G_{t_i} h}$$

in the notation of this theorem. Since the condition on P implies that all $f \in C_P$ are increasing, μ/ν is increasing in the sense of Section 3. Hence $\tilde{R}_j$ is increasing on J_q , that is, $\tilde{R}$ is increasing in each variable. This implies inequalities (16) and (17).

4.5 EXAMPLE. If $P = \left(\begin{smallmatrix}1 & \cdots & 1 & 0\\ 0 & \cdots & 0 & 1\end{smallmatrix}\right)$, $(a,b) = (0,1)$, and $h \equiv 1$, then we obtain

$$1 \leq R_n(f_1,\dots,f_n) \leq \frac{p^n}{(pn - n + 1)}$$

for all $f_i \in C_P \setminus \{0\}$, and

$$\frac{(p-1)^n}{(pn - 2n + 1)} \leq R_n(f_1,\dots,f_n) \leq \frac{p^n}{(pn - n + 1)}$$

for all $f_i \in C_P^o \setminus \{0\}$.

<u>Proof.</u> An easy calculation yields $G_i(x) = \frac{x^{i-1}}{(i-1)!}$ in this case. Using $q = p - 1$, the constants are easily computed.

The results of Theorems 4.3 and 4.4 can be extended to Pólya matrices whose first column is $\left(\begin{smallmatrix}1\\1\end{smallmatrix}\right)$ if one slightly modifies R , resp. R_n . The idea is to use that by the TP_2 property of Γ_P the quotient G_t/G_1 is increasing for all $t \in J_q$. Therefore, if we replace $R_n(f_1,\dots,f_n)$ by

$$S_n(f_1,\ldots,f_n) = \frac{\int(\prod_{i=1}^{n} f_i)G_1^{1-n}h}{\prod_{i=1}^{n}\int f_i h},$$

complete analogues to the inequalities (13), (14), (16), and (17) hold for this functional.

In Table 3, $n = 2$ refers to Theorem 4.3: $v_2 = c_{p,p}$, $u_2^o = c_{q,q}$, $v_2^o = c_{q+1,q+1}$.

P	n	v_n	u_n^o	v_n^o
$\binom{110}{001}$	2	1.8	1.33	1.8
	3	3.86	2	3.86
	6	56.08	9.14	56.08
$\binom{111100}{011000}$	2	2.38	1.57	2.09
	3	6.86	2.70	5.04
	6	240.26	16.71	93.54

TABLE 3 (to Theorems 4.3 and 4.4; $h \equiv 1$)

5. EXAMPLE

It is perhaps instructive to consider one example in detail. We choose, rather arbitrarily, $P = \binom{111100}{011000}$.

Here, $p = 6$ and $q = 4$. The sign formula (4) then yields $\sigma = \sigma_i = 1$ for $i \neq 5$ and $\sigma_5 = -1$. Therefore, if $(a,b) = (0,1)$,

$$C_P = cl\{f \in C^6[0,1] \mid f^{(6)}(x) \geq 0,\ f^{(j)}(0) \geq 0 \quad (j=0,\ldots,3),\ f^{(2)}(1) \leq 0,\ f'(1) \geq 0\},$$

and

$$C_P^o = cl\{f \in C^6[0,1] \mid f^{(6)}(x) \geq 0,\ f^{(j)}(0) = f^{(i)}(1) = 0 \quad (j = 0,\ldots,3\ ;\ i = 1,2)\}.$$

By Corollary 2.3, the polynomials G_4 and G_5 are contained in C_P^o. Hence two of the conditions in the definition of C_P^o could be relaxed: $f^{(3)}(0) \geq 0$, $f^{(2)}(1) \leq 0$. (The same phenomenon is responsible for Theorems 7 and 8 in [7] which, in our terminology, are concerned with C_P^0 for $P = \binom{1100}{1100}$ and $P = \binom{1010}{1010}$, respectively.)

The matrix $D = (\delta_j(x^{i-1}))_{i,j=1,\ldots,p}$ is

$$D = \begin{pmatrix} 1 & 0 & 0 & 0 & 0 & 0 \\ 0 & 1 & 0 & 0 & 0 & 1 \\ 0 & 0 & 2 & 0 & -2 & 2 \\ 0 & 0 & 0 & 6 & -6 & 3 \\ 0 & 0 & 0 & 0 & -12 & 4 \\ 0 & 0 & 0 & 0 & -20 & 5 \end{pmatrix},$$

its inverse being

$$D^{-1} = \begin{pmatrix} 1 & 0 & 0 & 0 & 0 & 0 \\ 0 & 1 & 0 & 0 & -1 & .6 \\ 0 & 0 & .5 & 0 & -.75 & .4 \\ 0 & 0 & 0 & .1\bar{6} & -.25 & .1 \\ 0 & 0 & 0 & 0 & .25 & -.2 \\ 0 & 0 & 0 & 0 & 1 & -.6 \end{pmatrix}.$$

Therefore, the basic polynomials are

$$G_1(x) = 1 \,, \quad G_2(x) = x - x^4 + \tfrac{3}{5}x^5 \,,$$

$$G_3(x) = \tfrac{1}{2}x^2 - \tfrac{3}{4}x^4 + \tfrac{2}{5}x^5 \,, \quad G_4(x) = \tfrac{1}{6}x^3 - \tfrac{1}{4}x^4 + \tfrac{1}{10}x^5 \,,$$

$$G_5(x) = \tfrac{1}{4}x^4 - \tfrac{1}{5}x^5 \,, \quad G_6(x) = x^4 - \tfrac{3}{5}x^5 \,.$$

As an illustration of Proposition 3.3, we compute $c_i = \dfrac{\|G_i\|_2}{\int G_i'}$ $(i = 2,\ldots,6)$ and find the values 0.74, 0.66, 0.63, 0.48, 0.39 in decreasing order as predicted.

REFERENCES

1. E.F. Beckenbach and R. Bellman, Inequalities. Springer-Verlag, Berlin and New York, 2nd Edition, 1965.

2. W. Blaschke and G. Pick, Distanzschätzungen im Funktionenraum II. Math. Ann. 77 (1916), 277 - 300.

3. A. Clausing, Disconjugacy and integral inequalities. Trans. AMS. 260 (1980), 293 - 307.

4. A. Clausing, Pólya operators I: Total positivity. Math. Ann. (to appear).

5. A. Clausing, Pólya operators II: Complete concavity. Math. Ann. (to appear).

6. S. Karlin, Total positivity, Vol. I. Stanford University Press, Stanford, 1968.

7. S. Karlin and Z. Ziegler, Some inequalities for generalized concave functions. J. Approx. Th. 13 (1975), 276 - 293.

8. G. Pólya, Bemerkungen zur Interpolation und zur Näherungstheorie der Balkenbiegung. Z. Angew. Math. Mech. 11 (1931), 445-449.

9. I.J. Schoenberg, On Hermite-Birkhoff interpolation. J. Math. Anal. Appl. 16 (1966), 538 - 543.

Achim Clausing, Institut für Mathematische Statistik,
Westfälische Wilhelms-Universität Münster,
D-4400 Münster, West Germany

International Series of
Numerical Mathematics, Vol. 71

LATTICE OF SOLUTIONS OF A DIFFERENTIAL INEQUALITY

J. Krzyszkowski

Abstract. The family L of solutions of the differential inequality $D^+y \le f(x,y)$ is investigated in terms of lattice theory. Among others, L turns out to be a lattice, and solutions of the differential equation $y' = f(x,y)$ are shown to be $\wedge$-irreducible under proper assumptions regarding f.

1. INTRODUCTION

In this note we shall deal with the family L of solutions of the differential inequality

$$D^+y \le f(x,y) .$$

The family L turns out to be a lattice. Here we shall prove this fact and also some theorems concerning a characterization of the family of solutions of the equation

$$D^+y = f(x,y)$$

in terms of lattice theory. Our results are similar to those obtained by D. Brydak [1] for functional inequalities.

2. NOTATION AND PREREQUISITES

Throughout this article, $I = (a,b)$ is an arbitrary interval in $\mathbb{R}$ $(-\infty \le a < b \le \infty)$, and $S = I \times \mathbb{R}$ is a strip in the plane. For real numbers c, d we write

$$c \wedge d = \min (c,d) \quad \text{and} \quad c \vee d = \max (c,d) .$$

For functions $u,v: I \to \mathbb{R}$, $u \wedge v$ and $u \vee v$ is the pointwise minimum or maximum, and $u \le v$ is defined pointwise, e.g. $(u \vee v)(x) = \max (u(x),v(x))$ for $x \in I$, similarly, $u \le v$ iff $u(x) \le v(x)$ in I. As usual, the four Dini derivatives are denoted by D^+, D^-, D_+, D_-.

The following facts about Dini derivatives will be frequently needed.

LEMMA 1. *Let* $u,v: I \to \mathbb{R}$ *be continuous and* $\xi \in I$.

(a) *If* $u(\xi) < v(\xi)$, *then*

$$D^+(u \wedge v)(\xi) = D^+u(\xi) \quad \text{and} \quad D^+(u \vee v)(\xi) = D^+v(\xi) .$$

(b) *If* $u(\xi) = v(\xi)$, *then*

$$D^+(u \wedge v)(\xi) \le D^+u(\xi) \wedge D^+v(\xi) ,$$

$$D^+(u \vee v)(\xi) = D^+u(\xi) \vee D^+v(\xi) .$$

Similar relations hold for the other Dini derivatives.

Proof. (a) follows immediately from the fact that $u \wedge v = u$ and $u \vee v = v$ in a neighborhood of ξ.

(b) Let $y = u \wedge v$ and $z = u \vee v$. The inequality $y \le u \le z$ implies $D^+y \le D^+u \le D^+z$, and a similar result holds for v. This proves the inequality about $u \wedge v$ and the relation about $u \vee v$ with $\ge$. In order to obtain equality, let (h_n) be a sequence of positive numbers with the property that $h_n \downarrow 0$ and $(z(\xi+h_n)-z(\xi))/h_n \to D^+z$ as $n \to \infty$. Assume that, e.g., $z(\xi+h_n) = u(\xi+h_n)$ for infinitely many n. Then $D^+u(\xi) \ge D^+z(\xi)$, which proves the statement about $u \vee v$ with $\le$. Hence the lemma is proved.

A classical theorem of Scheeffer [3] (see [4; Theorem 2.1]) about Dini derivatives states that if u is continuous and $D^+u \ge 0$ in $I \setminus C$, where C is a countable set, then u is (weakly) increasing, which implies that $Du(x) \ge 0$ for all $x \in I$ and for any Dini derivative D. A simple consequence is the following

LEMMA 2. *Let* $u,h: I \to \mathbb{R}$ *be continuous. If* $D^*u \ge h$ *in* $I \setminus C$, *where* C *is a countable subset of* I *and* D^* *is a fixed Dini derivative, then* $Du \ge h$ *in* I *for any Dini derivative* D.

A similar statement holds with $\ge$ *replaced by* $\le$.

Proof. Let $H(x) = \int h(x)dx$. Then $D^*(u-H) \ge 0$ in $I \setminus C$, hence $D(u-H) \ge 0$ in I by Scheeffer's theorem, which is equivalent to $Du \ge h$ in I.

3. DIFFERENTIAL INEQUALITIES WITH ARBITRARY f

Consider the differential inequality

$$D^+y \le f(x,y) \tag{1}$$

with right-hand side defined in the strip $S = I \times \mathbb{R}$, and denote by L the set of solutions of (1) continuous in the interval I.

THEOREM 1. _The partially ordered set_ $(L,\le)$ _is a lattice_.

Proof. Let $u,v \in L$. We have to show that $u \wedge v \in L$ and $u \vee v \in L$ (see [2; p. 3]). It is obvious that $u \wedge v$ and $u \vee v$ are continuous in I.

Let $\xi \in I$ and assume that $u(\xi) = v(\xi) = \eta$. Then $D^+u(\xi)$, $D^+v(\xi) \le f(\xi,\eta)$, and, according to Lemma 1 (b),

$$D^+(u \wedge v)(\xi) \le f(\xi,\eta) \quad \text{and} \quad D^+(u \vee v)(\xi) \le f(\xi,\eta).$$

Suppose now that $u(\xi) \neq v(\xi)$. Without loss of generality we may assume that $u(\xi) < v(\xi)$. By Lemma 1 (a),

$$D^+(u \wedge v)(\xi) = D^+u(\xi) \le f(\xi,u(\xi)) = f(\xi,(u \wedge v)(\xi)),$$

$$D^+(u \vee v)(\xi) = D^+v(\xi) \le f(\xi,v(\xi)) = f(\xi,(u \vee v)(\xi)).$$

Consequently the functions $u \wedge v$, $u \vee v$ satisfy inequality (1) in I. Therefore $u \wedge v \in L$ and $u \vee v \in L$.

REMARK 1. Theorem 1 is not true for D_+, as the following counterexample shows.

EXAMPLE. Let

$$f(x,y) = \begin{cases} 1 + \frac{1}{|x|} & \text{for } x \neq 0 \\ -1 & \text{for } x = 0 \end{cases}$$

and

$$u(x) = x \sin\frac{1}{x}, \qquad v(x) = x\cos\frac{1}{x} \qquad \text{for } x \neq 0,$$

$u(0) = v(0) = 0$. A simple computation shows that the functions u, v are continuous and satisfy (1), but

$$D_+(u \wedge v)(0) = -1 \quad \text{and} \quad D_+(u \vee v)(0) = -\tfrac{1}{2}\sqrt{2} > -1 .$$

Therefore, the function $u \vee v$ does not satisfy (1) at the point $x = 0$.

REMARK 2. The set of functions continuous in the interval I and satisfying the inequality

(2) $$D_+y \geq f(x,y) \quad \text{for } x \in I$$

is a lattice. This proposition is easily reduced to Theorem 1 by means of the identity $D_+y = -D^+(-y)$. Similar propositions hold for left-sided Dini derivatives, i.e., Theorem 1 remains true with D^+ replaced by D^-, and in (2) we may replace D_+ by D_- and obtain again a lattice.

4. DIFFERENTIAL INEQUALITIES WITH CONTINUOUS f

The function f in the above example is discontinuous, and no such example with continuous f exists. This follows readily from Lemma 2. When f(x,y) and y(x) are continuous, then f(x,y(x)) is also continuous. Hence an inequality $D^*y \leq f(x,y)$ or $D^*y \geq f(x,y)$ for a particular Dini derivative D^* implies $Dy \leq f(x,y)$ or $Dy \geq f(x,y)$ for any other Dini derivative D. It follows from Theorem 1 that the set of functions $y \in C(I)$ satisfying $Dy \leq f(x,y)$ in I, where D is an arbitrary (fixed) Dini derivative, is a lattice (same with $\geq$).

Another simple consequence of Lemma 2 is given by

LEMMA 3. _If the function_ f(x,y) _is continuous in the strip_ $I \times \mathbb{R}$, _then every continuous solution_ y(x) _of_ $Dy = f(x,y)$, _where_ D _is a fixed Dini derivative, belongs to_ $C^1(I)$ _and_

(3) $$y'(x) = f(x,y(x)) \quad \text{for } x \in I .$$

We introduce two hypotheses (G) and (U).

(G) _Global existence._ f is continuous in $S = I \times \mathbb{R}$, and every local solution y of (3) existing, say, in an interval $J \subset I$, can be extended as a solution to the whole interval I.

(U) _Uniqueness to the right._ f is continuous, and any two local solutions u and v of (3) existing in an interval $[\xi,\xi+\varepsilon) \subset I$ ($\varepsilon > 0$) and satisfying $u(\xi) = v(\xi)$ are identical in $[\xi,\xi+\varepsilon)$.

For example, (G) is satisfied for $f \in C^0(S)$ if for every compact interval $J \subset I$ there exists a constant K such that $|f(x,y)| \leq$

$K(1+|y|)$ in $J \times \mathbb{R}$. If $f \in C^{o}(S)$ and if to any $(\xi,\eta) \in S$ there exist two positive constants δ and K such that

$$f(x,y')-f(x,y) \le K(y'-y) \quad \text{for } \xi \le x < \xi+\delta \ , \quad \eta-\delta \le y < y' \le \eta+\delta$$

(local one-sided Lipschitz condition), then (U) is satisfied. In particular, a function $f(x,y)$ which is decreasing in y, satisfies (U).

We denote by L_o the set of functions $y \in C^1(I)$ which are solutions of (3) in I. It follows from Theorem 1 and the corresponding theorem with $\ge$ that L_o is a lattice.

THEOREM 2. *Assume that* $u,v \in L$ *and* $y \in L_o$ *and that* $u \wedge v \le y$. *If* f *satisfies* (U), *then* $u \le y$ *or* $v \le y$.

This result signifies that the element $y \in L_o$ is, in terms of lattice theory, $\wedge$-irreducible in L.

Proof. We may assume that $u(a_n) \le y(a_n)$ for $n = 1,2,\dots$, where $a_n \to a$ (remark that $I = (a,b)$). It follows from (U) that y is in $[a_n,b)$ the right-hand maximal solution of (3) through $(a_n,y(a_n))$. Hence, by a basic theorem on ordinary differential inequalities ([4; Theorem 9.5], [5; Theorem 8.X]), $u \le y$ in $[a_n,b)$. Since this is true for all n, we get $u \le y$ in I.

The following theorem is a partial converse of Theorem 2.

THEOREM 3. *Suppose that the function* $f(x,y)$ *is continuous in* $S = I \times \mathbb{R}$. *Let* $y \in L$ *and assume that the implication*

$$u \wedge v \le y \implies \text{either } u \le y \text{ or } v \le y$$

is true for all $u,v \in L$. *Then* $y \in L_o$.

Proof. Let $h(x) := f(x,y(x))$. We have to show that $D^+y = h$ in I; cf. Lemma 3. Let us assume that this is false. Then, according to Lemma 2, $D^+y(x) < h(x)$ for $x \in M$, where $M \subset I$ is uncountable. Take two points $x_1 < x_2$ from M and denote by K the constant

$$K := \max_H |f(x,y)| \ , \quad \text{where } H = [x_1,x_2] \times [\min y - 1, \max y + 1] \ ,$$

the minimum and maximum taken in the interval $[x_1,x_2]$. Let w be a solution of (3) through $(x_1,y(x_1))$. Choose $\delta > 0$ so small that $x_1+2\delta < x_2$, that $w(x)$ exists in $[x_1,x_1+\delta]$ and that

$$y(x) < w(x) < y(x) + 1 \quad \text{in } (x_1,x_1+\delta]$$

(since $D^+y(x_1) < h(x_1) = w'(x_1)$, we get $y < w$ in a right neighborhood of x_1). Denote by g the linear function $g(x) = w(x_1+\delta) - K_1(x-(x_1+\delta))$, where $K_1 > K$ is so large that $g(x_2) < y(x_2)$. The intermediate value theorem implies that $g(\xi) = y(\xi)$ for some $\xi \in (x_1+\delta,x_2)$, and since $(x,g(x)) \in H$ for $x_1+\delta \le x \le \xi$, we have

$$g'(x) = -K_1 < -K \le f(x,g(x)) \quad \text{for } x_1+\delta \le x \le \xi .$$

Now define the function u as follows:

$$u(x) = \begin{cases} y(x) & \text{for } a < x \le x_1 , \\ w(x) & \text{for } x_1 < x \le x_1+\delta , \\ g(x) & \text{for } x_1+\delta < x \le \xi , \\ y(x) & \text{for } \xi < x < b . \end{cases}$$

It is easily seen that u is continuous in I and that $D^+u \le f(x,u)$ in I, i.e., $u \in L$. In exactly the same way, we construct a function $v \in L$, which is equal to y in $(a,x_2]$, is a solution of (3) through the point $(x_2,y(x_2))$ in an interval $(x_2,x_2+\delta')$,.... These functions u, v have the property that $u \wedge v = y$, but neither $u \le y$ nor $v \le y$. Hence the implication of Theorem 3 is violated and Theorem 3 is proved.

Combining the two preceding theorems, we arrive at the following characterization of solutions to the differential equation (3).

THEOREM 4. _Suppose that_ (U) _is satisfied_. _Then_ $y \in L_0$ _if and only if_ $y \in L$ _and for all_ $u,v \in L$ _satisfying_ $u \wedge v \le y$ _at least one of the relations_ $u \le y$ _or_ $v \le y$ _is true_.

It is natural to ask whether the uniqueness condition (U) is necessary for the validity of Theorem 2 (and Theorem 4). The following theorem gives a positive answer to this question under the additional assumption (G) about global existence (this assumption guarantees that L_o is not empty).

THEOREM 5. <u>Assume that</u> $f \in C^o(S)$, <u>that</u> (G) <u>holds and that</u> (U) <u>is false. Then there exist</u> $y \in L_o$ <u>and</u> $u,v \in L$ <u>such that</u> $u \wedge v \leq y$ <u>but</u> $u \nleq y$ <u>and</u> $v \nleq y$.

<u>Proof.</u> Since (U) does not hold, there exists $(\xi,\eta) \in S$ such that the right-hand maximal solution z through (ξ,η) is different from the corresponding minimal solution y through this point. Because of assumption (G) we may assume that y and z exist in I and that $y(x) = z(x)$ in (a,ξ).

Let w be a solution of

$$w'(x) = f(x,w(x)) - 1 \qquad \text{in } [\xi',\xi] , \qquad w(\xi) = \eta ,$$

where $a < \xi' < \xi$. It is easily seen that $w(x) > y(x)$ in $[\xi',\xi)$. Let w_1 be a solution of (3) in $(a,\xi']$ with initial values $w_1(\xi') = w(\xi')$. Such a solution w_1 exists because of (G), and we may assume that $w_1 \geq y$ in (a,ξ').

Now, the functions u, v are defined as follows: $u = z$ in I,

$$v(x) = \begin{cases} w_1(x) & \text{for } a < x < \xi' , \\ w(x) & \text{for } \xi' \leq x < \xi , \\ y(x) & \text{for } \xi \leq x < b . \end{cases}$$

It is not difficult to see that $u,v \in L$ and that $u \wedge v = y$, but $u \nleq y$, $v \nleq y$.

CONCLUDING REMARK. It has already been pointed out at the beginning of this section that Theorems 2 to 5 remain true, when inequality (1) is replaced by $Dy \leq f(x,y)$ where $D = D_+$, $D = D^-$ or $D = D_-$. All four cases lead to the same family L. But the reader be warned: it is not allowed to define L by the conditions that y is continuous and that at each point x in I the inequality $Dy(x) \leq f(x,y(x))$ holds with some Dini derivative D (depending on x). There are continuous functions with the property that at each point at least one Dini derivative equals $-\infty$.

The differential equation $y' = f(x,y)$ is essentially equivalent to three other differential equations, which are obtained from it by adding a minus sign to x and/or y. The four cases are as follows:

(a) $u(x)$: $u' = f(x,u)$,

(b) $v(x) = -u(x)$: $v' = g(x,v)$ with $g(x,y) = -f(x,-y)$,

(c) $w(x) = u(-x)$: $w' = h(x,w)$ with $h(x,y) = -f(-x,y)$,

(d) $z(x) = -u(-x)$: $z' = k(x,z)$ with $k(x,y) = f(-x,-y)$.

If, instead of (a), the inequality (a') $D^+u \leq f(x,u)$ holds, then (b') $D_+v \geq g(x,v)$, (c') $D^-w \geq h(x,w)$, (d') $D_-w \leq k(x,w)$. Furthermore, the right-hand uniqueness condition (U) transforms into a left-hand uniqueness condition in the cases (c) and (d).

Therefore, all theorems of this section have their pendants, when the differential inequality $Dy \leq f(x,y)$ is replaced by the inverse inequality $Dy \geq f(x,y)$ and/or the uniqueness condition (U) is replaced by a uniqueness condition to the left. It is not our intention to formulate all these theorems.

REFERENCES

1. D. Brydak, On functional inequalities in a single variable. Dissertationes Math. CLX (1979), 1-48.

2. G. Grätzer, Lattice theory. W.H. Freeman and Company, San Francisco, 1971.

3. L. Scheeffer, Zur Theorie der stetigen Funktionen einer reellen Veränderlichen. Acta Math. 5 (1884/5), 183-194 (in particular pp.184-185).

4. J. Szarski, Differential inequalities. PWN, Warszawa, 1967.

5. W. Walter, Differential and Integral Inequalities. Ergebnisse der Mathematik und ihrer Grenzgebiete, Band 55. Springer-Verlag, Berlin, Heidelberg and New York, 1970.

J. Krzyszkowski, Institute of Mathematics, Pedagogical University, 30-011 Kraków, Poland.

International Series of
Numerical Mathematics, Vol. 71

DIFFERENTIAL INEQUALITIES AT RESONANCE†

V. Lakshmikantham

Abstract. It is well known that the theory of differential inequalities for the initial value problems has been very useful in the theory of differential equations [3,8]. Recently, such types of differential inequalities were developed for boundary value problems [1,6] and were used in proving the existence of solutions. It is natural to expect that differential inequalities for problems at resonance will be useful in proving, for example, existence results for periodic boundary value problems. Recently, existence of periodic solutions for first and second order differential equations have been considered by utilizing the method of upper and lower solutions and Lyapunov-Schmidt method [2,4,5]. In this paper following [7] we develop differential inequalities for boundary value problems at resonance for first and second order differential equations. As a simple application, we prove existence of multiple solutions as limits of monotone iterates for first and second order periodic boundary value problems.

1. FIRST ORDER PERIODIC BOUNDARY VALUE PROBLEMS

Consider the periodic boundary value problem (PBVP for short)

$$u' = g(t,u), \; u(0) = u(2\pi), \tag{1.1}$$

where $g \in C[[0,2\pi]\times R,R]$. Let $v \in C[[0,2\pi],R]$. Then a function $G(t,u)$ is said to be a modified function relative to v if $G(t,u) = g(t,p(t,u)) + \frac{p(t,u)-u}{1+u^2}$, where $p(t,u) = \max(v(t),u)$. We shall first prove the following comparison theorem.

THEOREM 1.1. *Assume that*

(i) $m \in C[[0,2\pi],R]$, $m(0) \leq m(2\pi)$ and $D_-m(t) \leq g(t,m(t))$, $0 < t \leq 2\pi$;

(ii) *the PBVP*

$$u' = G(t,u), \; u(0) = u(2\pi) \tag{1.2}$$

admits a solution u *for every lower solution* v *of*

† Research partially supported by U.S. Army Research Grant #DAAG29-80-C-0060.

(1.1);

(iii) _r(t) is the maximal solution for the PBVP (1.1). Then $m(t) \le r(t)$ on $[0,2\pi]$._

Proof. The assertion follows if we prove that $m(t) \le u(t)$, $t \in [0,2\pi]$, where u is a solution of the PBVP (1.2) with $v = m$. Suppose $m(t) \le u(t)$, $t \in [0,2\pi]$ is not true. Then there exists an $\varepsilon > 0$ and a $t_0 \in [0,2\pi]$ such that

(1.3) $m(t_0) = u(t_0) + \varepsilon$ and $m(t) \le u(t) + \varepsilon$, $t \in [0,2\pi]$.

Since $m(t_0) > u(t_0)$, we have $p(t_0,u(t_0)) = m(t_0)$. Hence if $t_0 \in (0,2\pi]$, we obtain from (1.3) and the definition of G that

$$D_-m(t_0) \ge u'(t_0) = g(t_0,p(t_0,u(t_0))) + \frac{p(t_0,u(t_0))-u(t_0)}{1+u^2(t_0)}$$

$$= g(t_0,m(t_0)) + \frac{m(t_0)-u(t_0)}{1+u^2(t_0)} > g(t_0,m(t_0))$$

which contradicts the hypothesis (i). If $t_0 = 0$, we have $m(0) = u(0) + \varepsilon$. Hence we have $m(2\pi) \ge m(0) = u(0) + \varepsilon = u(2\pi) + \varepsilon$. This inequality, together with (1.3), yields $D_-m(2\pi) \ge u'(2\pi)$ which leads to a contradiction as before. The proof of the theorem is complete.

We shall next prove another type of comparison result which deals with upper and lower solutions.

THEOREM 1.2. _Assume that_

(i) $m \in C[[0,2\pi],R]$, $m(0) \le m(2\pi)$ _and_ $D_-m(t) \le g(t,m(t))$, $0 < t \le 2\pi$;

(ii) $\beta \in C[[0,2\pi],R]$, $\beta(0) \ge \beta(2\pi)$ _and_ $D_-\beta(t) \ge g(t,\beta(t))$, $0 < t \le 2\pi$;

(iii) $g(t,u)$ _is strictly decreasing in_ u _for each_ t. _Then_ $m(t) \le \beta(t)$, $t \in [0,2\pi]$.

Proof. If the assertion is not true, then there exists an $\varepsilon > 0$ and a $t_0 \in [0,2\pi]$, such that

(1.4) $m(t_0) = \beta(t_0) + \varepsilon$ and $m(t) \leq \beta(t) + \varepsilon$, $t \in [0,2\pi]$.

If $t_0 \in (0,2\pi]$, then since $m(t_0) > \beta(t_0)$, we have by (iii) and (ii) the inequality

$$D_-m(t_0) \geq D_-\beta(t_0) \geq g(t_0,\beta(t_0)) \geq g(t_0,m(t_0))$$

which contradicts the hypothesis (i). If $t_0 = 0$, we have $m(0) = \beta(0) + \varepsilon$ and hence by (i) and (ii) we get $m(2\pi) \geq m(0) = \beta(0) + \varepsilon \geq \beta(2\pi) + \varepsilon$. Then (1.4) yields $D_-m(2\pi) \geq D_-\beta(2\pi)$ which again leads to a contradiction. This proves the theorem.

The following corollary of Theorem 1.2 is useful.

COROLLARY 1.1 Let $m \in C[[0,2\pi],R]$, $m(0) \leq m(2\pi)$ and $D_-m(t) \leq -Mm(t)$ for $0 < t \leq 2\pi$ for some $M > 0$. Then $m(t) \leq 0$ on $[0,2\pi]$.

As an application of Corollary 1.1, we shall obtain the existence of extremal solutions of PBVP (1.1) as limits of monotone iterates.

THEOREM 1.3. <u>Assume that</u>

(A_0) $\alpha, \beta \in C^1[[0,2\pi],R]$, $\alpha(t) \leq \beta(t)$ <u>on</u> $[0,2\pi]$ <u>and</u> $g(t,u_1) - g(t,u_2) \geq -M(u_1-u_2)$, $t \in [0,2\pi]$, <u>for any</u> u_1,u_2 <u>such that</u> $\alpha(t) \leq u_2 \leq u_1 \leq \beta(t)$ <u>and</u> $M > 0$;

(A_1) $\alpha' \leq g(t,\alpha)$, $t \in (0,2\pi]$ <u>and</u> $\alpha(0) \leq \alpha(2\pi)$;

(A_2) $\beta' \geq g(t,\beta)$, $t \in (0,2\pi]$ <u>and</u> $\beta(0) \geq \beta(2\pi)$.

<u>Then there exist monotone sequences</u> $\{\alpha_n(t)\}$, $\{\beta_n(t)\}$ <u>with</u> $\alpha_0 = \alpha$, $\beta_0 = \beta$ <u>such that</u> $\lim_{n\to\infty} \alpha_n(t) = \rho(t)$, $\lim_{n\to\infty} \beta_n(t) = r(t)$ <u>uniformly and monotonically on</u> $[0,2\pi]$ <u>and that</u> ρ, r <u>are minimal and maximal solutions of PBVP (1.1) respectively</u>.

<u>Proof.</u> For any $\eta \in [\alpha,\beta] = \{\eta \in C[[0,2\pi],R]: \alpha(t) \leq \eta(t) \leq \beta(t), t \in [0,2\pi]\}$, consider the PBVP

(1.5) $$u' = g(t,\eta(t)) - M(u-\eta(t)), \quad u(0) = u(2\pi).$$

Rewriting (1.5) in the form $u' + Mu = \sigma(t)$, $u(0) = u(2\pi)$, where $\sigma(t) \equiv g(t,\eta(t)) + M\eta(t)$, it is easy to see that

$$u(t) = u(0)e^{-Mt} + \int_0^t \sigma(s)e^{-M(t-s)}ds$$

and

$$u(0) = u(2\pi) = \frac{1}{e^{2M\pi}-1}\int_0^{2\pi} \sigma(s)e^{Ms}ds,$$

satisfies the PBVP (1.5). The uniqueness of solutions of (1.5) follows by Corollary 1.1. In fact, if u and v are two distinct solutions of (1.5), then setting $p = u - v$, we get

$$p' = -Mp,\ p(0) = p(2\pi),$$

which implies $p(t) \equiv 0$. Hence, for any $\eta \in [\alpha,\beta]$, we define a mapping A by $A\eta = u$, where u is the unique solution of (1.5). We shall show that

(i) $\alpha \leq A\alpha$, $\beta \geq A\beta$, and

(ii) A is monotone nondecreasing on $[\alpha,\beta]$.

To prove (i), we set $p = \alpha - \alpha_1$ where $\alpha_1 = A\alpha$. We then have

$$p' \leq -Mp,\ p(0) \leq p(2\pi).$$

Hence Corollary 1.1 implies $p(t) \leq 0$ on $[0,2\pi]$ proving $\alpha \leq A\alpha$. A similar proof holds for $\beta \geq A\beta$.

To prove (ii), let $\eta_1, \eta_2 \in [\alpha,\beta]$ such that $\eta_1 \leq \eta_2$. Let $A\eta_1 = u_1$ and $A\eta_2 = u_2$. Setting $p = u_1 - u_2$ and using (A_0), we obtain

$$p' \leq -Mp,\ p(0) = p(2\pi),$$

which implies by Corollary 1.1 that $p(t) \leq 0$. This means that A is monotone on $[\alpha,\beta]$.

It therefore follows that we can define the sequences $\{\alpha_n\}$, $\{\beta_n\}$ with $\alpha_0 = \alpha$, $\beta_0 = \beta$ such that

$$\alpha_n = A\alpha_{n-1},\ \beta_n = A\beta_{n-1}$$

and conclude that on $[0,2\pi]$,

$$\alpha = \alpha_0 \leq \alpha_1 \leq \alpha_2 \leq \dots \leq \alpha_n \leq \beta_n \leq \dots \leq \beta_2 \leq \beta_1 \leq \beta_0 = \beta.$$

The rest of the proof of the theorem is exactly the same as in [4]. Hence the theorem is proved.

2. SECOND ORDER PERIODIC BOUNDARY VALUE PROBLEMS

Consider the second order PBVP

$$-u'' = g(t,u,u'),\ u(0) = u(2\pi),\ u'(0) = u'(2\pi) \tag{2.1}$$

where $g \in C[[0,2\pi]\times R\times R,R]$. Then a function $G(t,u,u')$ is said to be a modified function relative to v if

$$G(t,u,u') = g(t,p(t,u),u') + \frac{p(t,u)-u}{1+u^2}, \tag{2.2}$$

where $p(t,u) = \max(v(t),u)$. Let us prove the following comparison result.

THEOREM 2.1. Assume that

(i) $m \in C^2[[0,2\pi],R]$, $m(0) = m(2\pi)$, $m'(0) \geq m'(2\pi)$ and $-m'' \leq g(t,m,m')$, $0 \leq t \leq 2\pi$;

(ii) the modified PBVP

$$-u'' = G(t,u,u'),\ u(0) = u(2\pi),\ u'(0) = u'(2\pi) \tag{2.3}$$

admits a solution u for every lower solution v of (2.1);

(iii) $r(t)$ is the maximal solution of the PBVP (2.1).

Then $m(t) \leq r(t)$ on $[0,2\pi]$.

Proof. Let u be a solution of PBVP (2.3) where G is the modified function relative to m. If $m(t) \leq u(t)$ on $[0,2\pi]$, then it follows from the definition of G that u is actually a solution of the PBVP (2.1). It is therefore enough to establish $m(t) \leq u(t)$, $t \in [0,2\pi]$. Suppose this is not true. Then there exists an $\varepsilon > 0$ and a $t_0 \in [0,2\pi]$, such that

$$m(t_0) = u(t_0) + \varepsilon \quad\text{and}\quad m(t) \leq u(t) + \varepsilon,\ t \in [0,2\pi]. \tag{2.4}$$

If $t_0 \in (0,2\pi)$, we have $m'(t_0) = u'(t_0)$ and $m''(t_0) \leq u''(t_0)$. Since $m(t_0) > u(t_0)$, we also have $p(t_0,u(t_0)) = m(t_0)$. Consequently, we get

$$- m''(t_0) \geq - u''(t_0) = G(t_0,u(t_0),u'(t_0))$$

$$= g(t_0,m(t_0),m'(t_0)) + \frac{m(t_0)-u(t_0)}{1+u^2(t_0)}$$

$$> g(t_0,m(t_0),m'(t_0))$$

which is a contradiction to (i). If $t_0 = 0$ or 2π, then, by (2.4) we have $m(i) = u(i) + \varepsilon$, $i = 0,2\pi$,

(2.5) $\quad m'(0) \leq u'(0)$ and $m'(2\pi) \geq u'(2\pi)$.

Hence if (i) holds, we get

(2.6) $$m'(2\pi) = u'(2\pi).$$

As a consequence, we are led to a contradiction as before, which proves $m(t) \leq u(t)$ on $[0,2\pi]$. This implies that $u(t)$ is a solution of the PBVP (2.1) and since $u(t) \leq r(t)$ on $[0,2\pi]$, the proof of the theorem is complete.

We shall next prove a comparison result which deals with upper and lower solutions of the PBVP (2.1).

THEOREM 2.2. *Assume that*

(i) $m \in C^2[[0,2\pi],R]$, $m(0) = m(2\pi)$, $m'(0) \geq m'(2\pi)$ *and* $-m'' \leq g(t,m,m')$, $0 \leq t \leq 2\pi$;

(ii) $\beta \in C^2[[0,2\pi],R]$, $\beta(0) = \beta(2\pi)$, $\beta'(0) \leq \beta'(2\pi)$ *and* $-\beta'' \geq g(t,\beta,\beta')$, $0 \leq t \leq 2\pi$;

(iii) $g(t,u,v)$ *is strictly decreasing in* u *for each* (t,v).

Then $m(t) \leq \beta(t)$ *on* $[0,2\pi]$.

Proof. Suppose that the conclusion of the theorem is not true. Then there exists an $\varepsilon > 0$ and a $t_0 \in [0,2\pi]$ such that

(2.7) $\quad m(t_0) = \beta(t_0) + \varepsilon$, $m(t) \leq \beta(t) + \varepsilon$, $0 \leq t \leq 2\pi$.

If $t_0 \in (0,2\pi)$, we have $m'(t_0) = \beta'(t_0)$ and $m''(t_0) \leq \beta''(t_0)$. Noting that $m(t_0) > \beta(t_0)$, we then get

$$g(t_0,m(t_0),m'(t_0)) \geq - m''(t_0) \geq - \beta''(t_0) \geq g(t_0,\beta(t_0),\beta'(t_0)) > g(t_0,m(t_0),m'(t_0))$$

which is a contradiction. If, on the other hand, $t_0 = 0$ or 2π, we obtain from (2.7) the inequalities.

$$m'(0) \leq \beta'(0),\ m'(2\pi) \geq \beta'(2\pi).$$

This, inview of (i) and (ii), yields $m'(0) = \beta'(0)$ and $m'(2\pi) = \beta'(2\pi)$, which leads to a contradiction as before. Hence the proof is complete.

The following corollary of Theorem 2.2 is useful.

COROLLARY 2.1. Let $m \in C^2[[0,2\pi],R]$, $m(0) = m(2\pi)$, $m'(0) \geq m'(2\pi)$ and $-m'' \leq -Mm$, $0 \leq t \leq 2\pi$ for some $M > 0$. Then $m(t) \leq 0$ on $[0,2\pi]$.

As an application of Corollary 2.1, let us consider the existence of extremal solutions of PBVP

$$(2.8)\qquad -u'' = g(t,u),\ u(0) = u(2\pi),\ u'(0) = u'(2\pi),$$

where $g \in C[[0,2\pi]\times R,R]$.

THEOREM 2.3. <u>Assume that</u>

(B_0) $\alpha, \beta \in C^2[[0,2\pi],R]$, $\alpha(t) \leq \beta(t)$ <u>on</u> $[0,2\pi]$ <u>and</u> $g(t,u_1) - g(t,u_2) \geq -M(u_1-u_2)$, $t \in [0,2\pi]$, <u>for any</u> u_1,u_2 <u>such that</u> $\alpha(t) \leq u_2 \leq u_1 \leq \beta(t)$ <u>and</u> $M > 0$;

(B_1) $-\alpha'' \leq g(t,\alpha)$, $t \in [0,2\pi]$, $\alpha(0) = \alpha(2\pi)$, $\alpha'(0) \geq \alpha'(2\pi)$;

(B_2) $-\beta'' \geq g(t,\beta)$, $t \in [0,2\pi]$, $\beta(0) = \beta(2\pi)$, $\beta'(0) \leq \beta'(2\pi)$.

<u>Then there exist monotone sequences</u> $\{\alpha_n(t)\}$, $\{\beta_n(t)\}$ <u>with</u> $\alpha_0 = \alpha$, $\beta_0 = \beta$ <u>such that</u> $\lim_{n\to\infty} \alpha_n(t) = \rho(t)$, $\lim_{n\to\infty} \beta_n(t) = r(t)$ <u>uniformly and monotonically and</u> ρ, r <u>are minimal and maximal solutions of PBVP (2.8) respectively</u>.

<u>Proof.</u> For any η such that $\alpha(t) \leq \eta(t) \leq \beta(t)$, consider the linear PBVP.

$$(2.9)\qquad -u'' = f(t,\eta) - M^2(u-\eta),\ u(0) = u(2\pi),\ u'(0) = u'(2\pi).$$

Writing $g(t,\eta) + M^2\eta \equiv \sigma(t)$, we can find a solution $u(t)$ of (2.9) given by

$$(2.10)\quad u(t) = C_1e^{Mt} + C_2e^{-Mt} - \frac{e^{Mt}}{2M}\int_0^t \sigma(s)e^{-Ms}ds + \frac{e^{-Mt}}{2M}\int_0^t \sigma(s)e^{Ms}ds,$$

where $C_1 = \frac{e^{2M\pi}}{2M(e^{2M\pi}-1)} \int_0^{2\pi} \sigma(s)e^{-Ms}ds,$

and

$$C_2 = \frac{1}{2M(e^{2M\pi}-1)} \int_0^{2\pi} \sigma(s)e^{Ms}ds.$$

We claim that this solution $u(t)$ is unique. If not, let $v(t)$ be another solution of (2.9). Then setting $p(t) = v(t) - u(t)$, we see that

$$-p'' = -M^2p, \ p(0) = p(2\pi), \ p'(0) = p'(2\pi).$$

Hence, by Corollary 2.1, it follows that $p(t) \equiv 0$, which shows $v(t) \equiv u(t)$.

We can now define a mapping A by $A\eta = u$, for any $\eta \in [\alpha,\beta]$, where u is the unique solution of PBVP (2.9). We show that (i) $\alpha \leq A\alpha$, $\beta \geq A\beta$ and (ii) A is monotone nondecreasing on $[\alpha,\beta]$.

To prove (i), let $p = \alpha - \alpha_1$ where $\alpha_1 = A\alpha$. Then, we have

$$-p'' \leq -M^2p, \ p(0) = p(2\pi), \ p'(0) \geq p'(2\pi).$$

Hence, by Corollary 2.1, we get $p(t) \leq 0$, implying $\alpha \leq A\alpha$. Similar arguments hold for $\beta \geq A\beta$.

To prove (ii), let $\eta_1, \eta_2 \in [\alpha,\beta]$ such that $\eta_1 \leq \eta_2$. Let $A\eta_1 = u_1$ and $A\eta_2 = u_2$. Setting $p = u_1 - u_2$ and using (B_0), we get

$$-p'' \leq -M^2p, \ p(0) = p(2\pi), \ p'(0) = p'(2\pi).$$

Hence, by Corollary 2.1, we have $p(t) \leq 0$ and this implies that A is monotone on $[\alpha,\beta]$.

It therefore follows that we can define sequences $\{\alpha_n\}$, $\{\beta_n\}$ such that

$$\alpha_n = A\alpha_{n-1}, \ \beta_n = A\beta_{n-1} \ \text{with} \ \alpha_0 = \alpha, \ \beta_0 = \beta,$$

and on $[0,2\pi]$,

$$\alpha = \alpha_0 \leq \alpha_1 \leq \alpha_2 \leq \cdots \leq \alpha_n \leq \beta_n \leq \cdots \leq \beta_2 \leq \beta_1 \leq \beta_0 = \beta.$$

Furthermore, it is easily seen that there exists a $N > 0$ which depends only on α,β such that $|\alpha_n'|$, $|\beta_n'| \leq N$ on $[0,2\pi]$. It

then follows by employing standard arguments that
$\lim_{n\to\infty} \alpha_n(t) = \rho(t)$, $\lim_{n\to\infty} \beta_n(t) = r(t)$ uniformly and monotonically on $[0,2\pi]$. It is easy to show that $\rho(t)$ and $r(t)$ are solutions of PBVP (2.8) in view of the fact that α_n, β_n satisfy

$$-\alpha_n'' = g(t,\alpha_{n-1}) - M^2(\alpha_n - \alpha_{n-1}),\ \alpha_n(0) = \alpha_n(2\pi),\ \alpha_n'(0) = \alpha_n'(2\pi),$$

$$-\beta_n'' = g(t,\beta_{n-1}) - M^2(\beta_n - \beta_{n-1}),\ \beta_n(0) = \beta_n(2\pi),\ \beta_n'(0) = \beta_n'(2\pi)$$

and possess the integral representation similar to (2.10). It can be proved by induction argument that for any solution u of (2.8) with $\alpha \le u \le \beta$, we have $\alpha \le \alpha_n \le u \le \beta_n \le \beta$ on $[0,2\pi]$ and this proves that ρ, r are extremal solutions of PBVP (1.2). The proof is therefore complete.

REFERENCES

1. J. Chandra, V. Lakshmikantham and S. Leela, Comparison Principle and Theory of Nonlinear Boundary Value Problems. Proceedings of the International Conference on Nonlinear Phenomena in Mathematical Sciences, Academic Press, (1982), 241-248.

2. R. Kannan and V. Lakshmikantham, Periodic solutions of nonlinear boundary value problems. J. Nonlinear Anal. 6 (1982), 1-10.

3. V. Lakshmikantham and S. Leela, Differential and Integral Inequalities. Vol's I and II. Academia Press, New York, 1969.

4. V. Lakshmikantham and S. Leela, Existence and monotone method for periodic solutions of first order differential equations. J. Math. Anal. Appl. Vol. 91, No. 1 (1983), 237-243.

5. S. Leela, Monotone method for second order periodic boundary value problems. J. Nonlinear Anal. Vol. 7, No. 4 (1983), 349-355.

6. J. Schröder, Operator Inequalities. Academic Press, New York, 1980.

7. G. R. Shendge and A. S. Vatsala, Comparison results for first and second order boundary value problems at resonance. Appl. Math. Comput. 12 (1983), 367-380.

8. W. Walter, Differential and Integral Inequalities. Springer-Verlag, New York, 1970.

V. Lakshmikantam, Department of Mathematics, The University of Texas at Arlington, Arlington, Texas 76019, U.S.A.

International Series of
Numerical Mathematics, Vol. 71

ESTIMATES FOR EIGENVALUES OF STURM-LIOUVILLE PROBLEMS

Giorgio Talenti

Abstract. We present sharp estimates (from above and from below) of the first eigenvalue of some Sturm-Liouville problems.

1. Suppose q is real-valued and integrable, and consider the following Sturm-Liouville problem:

$$\begin{cases} -u'' + q(x)u = \lambda u & \text{for } -L < x < L \\ u(-L) = u(L) = 0 \end{cases} \tag{1}$$

We are concerned with some estimates of

$$\lambda(q) = \text{the smallest eigenvalue of (1)}.$$

As usual, we denote by

$$q_+ = \frac{1}{2}(|q| + q) \qquad q_- = \frac{1}{2}(|q| - q) ,$$

the positive and the negative part of q.

THEOREM 1. Let

$$\int_{-L}^{L} q_+(x)\, dx = A . \tag{2}$$

Then

$$\lambda(q) \leq \left(\frac{\pi}{2L}\right)^2 \left[\frac{1}{2} + \sqrt{\frac{1}{4} + \frac{2L}{\pi^2} A}\right]^2 . \tag{3}$$

Equality holds in (3) if q is the step-function defined by

$$q(x) = \begin{cases} \mu^2 & \text{for } |x| < L - (\pi/2\mu) \\ 0 & \text{otherwise} \end{cases} . \tag{4a}$$

Here

$$\mu = \frac{\pi}{2L}\left[\frac{1}{2} + \sqrt{\frac{1}{4} + \frac{2L}{\pi^2}A}\right] , \tag{4b}$$

the positive root of

$$2(L - \frac{\pi}{2\mu})\mu^2 = A .$$

THEOREM 2. *Let*

$$\int_{-L}^{L} q_-(x)\, dx = A . \tag{5}$$

Then

$$\lambda(q) \geq \mu , \tag{6}$$

the smallest eigenvalue of the following problem:

$$\begin{cases} - u'' - A\,\delta(x) u = \mu u & \text{for } -L < x < L , \\ u(-L) = u(L) = 0 . \end{cases} \tag{7}$$

Here δ *is the Dirac mass concentrated at the origin;* μ *is the* (unique) *root of the following transcendental equation:*

$$\frac{\tan(L\sqrt{\mu})}{L\sqrt{\mu}} = \frac{2}{AL} , \quad -\infty < \mu < (\frac{\pi}{2L})^2 . \tag{8}$$

The differential equation in (7) must be understood in the sense of distributions, i.e. $\int_{-L}^{L} u'\phi'\, dx - Au(0)\phi(0) = \mu \int_{-L}^{L} u\phi\, dx$ for all test functions ϕ from $C_o^1(|-L,L|)$. Thus (7) reads

$$\begin{cases} -u'' = \mu u & \text{for } -L < x < 0 \text{ and } 0 < x < L , \\ u'(0+) - u'(0-) + Au(0) = 0 , \\ u(-L) = u(L) = 0 . \end{cases}$$

The left-hand side of (8) involves

$$\frac{\tan\sqrt{z}}{\sqrt{z}} = 1 + \frac{z}{3} + \frac{2z^2}{15} + \frac{17z^3}{315} + \ldots = \sum_{n=0}^{\infty} \frac{2}{\pi^2(n+\frac{1}{2})^2 - z}$$

a meromorphic function of the complex variable z having simple poles at $\pi^2(k+1/2)^2$ $(k=0,1,2,\dots)$. Clearly, the restriction of $\frac{\tan\sqrt{z}}{\sqrt{z}}$ to the real half-line $-\infty < z < (\pi/2)^2$ is an analytic real-valued monotonic convex function, which increases from 0 to $+\infty$ as z increases from $-\infty$ to $(\pi/2)^2$, has the value 1 at $z=0$ and the value $\frac{\tanh\sqrt{-z}}{\sqrt{-z}}$ at any real negative z. Thus (8) reads

$$\frac{\tan(L\sqrt{\mu})}{L\sqrt{\mu}} = \frac{2}{AL} \quad \& \quad 0 < \mu < \left(\frac{\pi}{2L}\right)^2 \quad \text{if} \quad AL < 2, \quad \mu=0 \quad \text{if} \quad AL=2,$$

$$\frac{\tanh(L\sqrt{-\mu})}{L\sqrt{-\mu}} = \frac{2}{AL} \quad \& \quad \mu < 0 \quad \text{if} \quad AL > 2 \;.$$

Let us notice the expansion

$$\mu = \left(\frac{\pi}{2L}\right)^2 - \frac{A}{L} - \frac{A^2}{\pi^2} + \dots \qquad \text{for } A \to 0$$

and the estimates

$$-\left[\frac{A}{2}\tanh\left(\frac{AL}{2}\right)\right]^2 \leq \mu \leq -\left[\frac{A}{2}\tanh\left(\frac{AL}{2}-1\right)\right]^2 .$$

THEOREM 3. *Suppose* q *is constrained by*

(9) $$q(x) \geq 0, \quad \int_{-L}^{L} q(x)\,dx = A, \quad \sup\{q(x) : -L \leq x \leq L\} = B ,$$

where $B > A/(2L)$. *Then* $\lambda(q)$ *achieves its minimum value* μ *on the following* q :

(10) $$q(x) = \begin{cases} 0 & \text{for } |x| < L-(A/2B) \\ B & \text{for } L-(A/2B) < |x| < L \;. \end{cases}$$

Here μ is the smallest positive root of the following equation

$$\frac{1}{\sqrt{\mu}}\cot\left[\sqrt{\mu}\left(L-\frac{A}{2B}\right)\right] = \frac{1}{\sqrt{\mu-B}}\tan\left[\sqrt{\mu-B}\,\frac{A}{2B}\right] .$$

Hence one sees that

$$\mu > \left(\frac{\pi}{2L}\right)^2 ,$$

$$\mu = \left(\frac{\pi}{2L}\right)^2 + \frac{\pi^2 A^3}{48L^3}B^{-2} + \dots \quad \text{for } B \to \infty.$$

Theorem 3 is a special case of [4, theorem 3]. Theorem 1 settles a question by A.G.Ramm [1]. Below we present a proof of theorem 1. A proof of theorem 2 is similar (and simpler), and will be omitted. We refer to G.Chiti [0] for related results.

2. Proof of theorem 1 .

(11) $$\lambda(q) = \min\left\{ \frac{\int_{-L}^{L} \left[(u')^2 + qu^2\right] dx}{\int_{-L}^{L} u^2\, dx} : 0 \not\equiv u \in W_\circ^{1,2}(-L,L) \right\},$$

the variational characterization of the first eigenvalue, tells us that

(12) $$\lambda(q) \leq \min\{ J(u) : 0 \not\equiv u \in W_\circ^{1,2}(L,L)\} ,$$

where

(13) $$J(u) = \frac{\int_{-L}^{L} (u')^2 dx + A(\max |u|)^2}{\int_{-L}^{L} u^2\, dx} .$$

As usual, $W_\circ^{1,2}(-L,L)$ denotes the set of absolutely continuous functions in $[-L,L]$, which vanish at the end points and whose derivative is square integrable.

Step 1. J has a minimum in $W_\circ^{1,2}(-L,L)$. This follows from standard compactness and semicontinuity arguments.

Step 2. A minimizer of J exists in $W_\circ^{1,2}(-L,L)$, which is nonnegative and symmetrically decreasing. In fact, let u be any function from $W_\circ^{1,2}(-L,L)$. The Hardy-Littlewood symmetrically decreasing rearrangement u* of u has the following properties: (i) u* is an even nonnegative function, the restriction of u* to [0,L] decreases monotonically; (ii) u* and |u| are equimeasurable; (iii) u^* is in $W_o^{1,2}(-L,L)$ and $\int_{-L}^{L}(du^*/dx)^2\,dx \leq \int_{-L}^{L}(u')^2dx$. The last property is a special case of Polya-Szegö principle, see [2] or [3] for instance. Property (ii) implies $u^*(0) = \max |u|$ and $\int_{-L}^{L}(u^*)^2dx = \int_{-L}^{L}u^2dx$. Thus we have

$$J(u) \geq J(u^*) \quad ,$$

and the above assertion follows.

Step 3. Any nonnegative minimizer u of J from $W_o^{1,2}(-L,L)$ is concave. Suppose, by contradiction, that u is not concave. Then the restriction of u to some subinterval $[a,b]$ is strictly below the linear function joining $(a,u(a))$ with $(b,u(b))$. Set

$$v(x) = \begin{cases} \frac{1}{b-a}\,[u(b)(x-a)+u(a)(b-x)] & \text{if } a \leq x \leq b, \\ u(x) & \text{otherwise.} \end{cases}$$

Clearly, v is in $W_o^{1,2}(-L,L)$. We have $\max v = \max u$, and $\int_{-L}^{L}v^2dx > \int_{-L}^{L}u^2dx$ since $v(x) > u(x) \geq 0$ for $a < x < b$. Furthermore $\int_{-L}^{L}(u')^2dx > \int_{-L}^{L}(v')^2dx$, since

$$\int_a^b (u')^2dx > [u(b)-u(a)]^2/(b-a) = \int_a^b (v')^2\,dx$$

by Schwarz inequality. Thus we have

$$J(u) > J(v) \quad ,$$

a contradiction.

Step 4. Let u be a nonnegative concave symmetrically decreasing minimizer of J in $W_o^{1,2}(-L,L)$. Then u' is everywhere continuous.

As u is concave, u' decreases monotonically. Thus $u'(\xi-)$ and $u'(\xi+)$ exist and are finite at every point ξ from $]-L,L[$, and

(14a) $$u'(\xi-) \geq u'(\xi+) .$$

Furthermore

(15a) $$u'(0-) = -u'(0+) ,$$

since u is even.

First, we prove the continuity of u' at the origin, i.e.

(15b) $$u'(0+) = 0 .$$

Suppose, by contradiction, that u' is not continuous at the origin, i.e.

$$u'(0+) < 0 .$$

We claim that under the last assumption $J'(u)$, a Gateaux derivative of J at u, exists and has the following values

(16a) $$J'(u)(\phi) = 2\left(\int_{-L}^{L} u^2 dx\right)^{-1}\left(\int_{-L}^{L} u'\phi' dx + Au(0)\phi(0) - J(u)\int_{-L}^{L} u\, dx\right)$$

on test functions ϕ such that

(16b) ϕ is Lipschitz continuous, $\phi(-L) = \phi(L) = 0$.

In fact the concavity of u yields

$$u(x) + \phi(x) \geq u(0)(1 - \frac{|x|}{L}) - |\phi(x)| \geq (L-|x|)(\frac{u(0)}{L} - \sup|\phi'|) ,$$

whenever ϕ is Lipschitz continuous and vanishes at the end points; on the other hand

$$(u(0) + \phi(0)) - (u(x) + \phi(x)) \geq |x|(|u'(0+)| - \sup|\phi'|) .$$

Hence

$$\max |u + t\phi| = u(0) + t\phi(0) ,$$

$$J(u+t\phi) = \left(\int_{-L}^{L}(u+t\phi)^2 dx\right)^{-1}\left(\int_{-L}^{L}(u'+t\phi')^2 dx + A[u(0)+t\phi(0)]^2\right)$$

if $|t|$ is small enough and ϕ is as above. The claim follows, via the formula

$$J'(u)(\phi) = \lim_{t\to 0} \frac{1}{t}\left[J(u+t\phi)-J(u)\right] .$$

Since u minimizes J, we must have

$$J'(u)(\phi) = 0 .$$

Choosing

$$\phi(x) = \begin{cases} 1-\frac{n}{L}|x| & \text{if } |x| \le \frac{L}{n} \\ 0 & \text{otherwise} \end{cases}$$

yields

$$\frac{n}{L}\int_{-L/n}^{0} u'\,dx - \frac{n}{L}\int_{0}^{L/n} u'\,dx + Au(0) = J(u)\int_{-L/n}^{L/n} u\left(1-\frac{n}{L}|x|\right) dx ,$$

whence letting $n\to\infty$ gives

$$u'(0-) - u'(0+) + Au(0) = 0 ,$$

an inconsistent equation with the assumption $u'(0+) < 0$.

Next, we prove

(14b) $$u'(\xi-) \le u'(\xi+)$$

at every point ξ from $]-L,L[$. Formulas (14) show the asserted continuity of u' . Suppose ϕ is a Lipschitz continuous function, which vanishes at the end points and achieves its maximum value at the origin. A slight change in a previous argument shows that in this case the right-hand side of (16a) is the limit of

$$\frac{1}{t}\left[J(u+t\phi) - J(u)\right]$$

as t approaches zero through positive values, i.e. a directional derivative of J at u. Such a directional derivative must be nonnegative, since u minimizes J. Thus the right-hand side of (16a) is nonnegative for all test functions ϕ as above. One of these test functions is

$$\phi(x) = \begin{cases} -1 + \frac{n}{L}|x-\xi| & \text{for } \xi - \frac{L}{n} \leq x \leq \xi + \frac{L}{n} \\ 0 & \text{otherwise} \end{cases},$$

provided $n > L/|\xi|$ and $\xi \neq 0$. In other words, either (15) holds or

$$-\frac{n}{L}\int_{\xi-L/n}^{\xi} u'dx + \frac{n}{L}\int_{\xi}^{\xi-L/n} u'dx \geq -J(u)\int_{\xi-L/n}^{\xi+L/n} udx ,$$

for all $n > L/|\xi|$. Inequality (14b) follows.

<u>Step 5.</u> Let u be as in step 4. Then

(17a) $$u(x) = \begin{cases} u(0) & \text{for } |x| \leq L-\pi/2\mu \\ u(0)\sin\left[\mu(L-|x|)\right] & \text{for } L-\pi/2\mu < |x| \leq L \end{cases}$$

and

(17b) $$J(u) = \mu^2 ,$$

where

(18) $$\mu = (\pi/2L)\left[\frac{1}{2} + \sqrt{\frac{1}{4} + 2\pi^{-2}LA}\right] .$$

Suppose u minimizes J in $W_0^{1,2}(-L,L)$. Then the restriction of u to the following set

(19a) $$\{x \in \left]-L,L\right[\; : \; |u(x)| < \max|u|\}$$

is twice continuously differentiable and the following differential equation

(19b) $$u''(x) + J(u)u(x) = 0$$

holds at every point of that set. This assertion is a straighforward consequence of the fundamental lemma of the calculus of variations and the vanishing of a Gateaux derivative of J at u. Observe in fact that, if u is in $W_o^{1,2}(-L,L)$ and E is an open set such that

(20a) $$\{x \in]-L,L[\; : |u(x)| < \max|u|\} \supseteq E \; ,$$

then J is Gateaux differentiable at u and

(20b) $$J'(u)(\phi) = 2\left(\int_{-L}^{L} u^2 dx\right)^{-1}\left(\int_{-L}^{L} u'\phi' dx - J(u)\int_{-L}^{L} u\phi\, dx\right)$$

for all test functions ϕ such that

(20c) $$\phi \text{ is in } W_o^{1,2}(-L,L) \; , \quad \text{support of } \phi \subset E \; .$$

Suppose u minimizes J in $W_o^{1,2}(-L,L)$ and u is nonnegative concave symmetrically decreasing. Then $\{x \in]-L,L[\; : |u(x)| = \max|u|\}$ is exactly an interval $[-\xi,\xi]$, symmetric about the origin. The set (19a) is just the union of $]-L,-\xi[$ and $]\xi,L[$, and the differential equation (19b) holds in each of these intervals. Integrations yield the following representation

(21a) $$u(x) = \begin{cases} u(0) & \text{for } |x| \le \xi \\ \text{Constant} \times \sin[\sqrt{J(u)}\,(L-|x|)] & \text{for } \xi < |x| \le L \; . \end{cases}$$

Formula (21a) tells us that $\sqrt{J(u)}\,(L-\xi) \le \pi/2$, since u must decrease symmetrically. Actually the constant in (21a) is $u(0)$ and $\sqrt{J(u)}\,(L-\xi) = \pi/2$, or

(21b) $$\xi = L - \frac{\pi}{2} J(u)^{-\frac{1}{2}} \; ,$$

because in step 4 we proved that u' is continuous.

Thus formulas (21) agree with (17) . Putting together (17) and (13) gives

$$\mu^2 = \frac{(\pi/2)\mu + A}{2L - (\pi/2\mu)}$$

a quadratic equation in μ whose positive solution is given by (18).

<u>Step 6.</u> Formulas (17), (18) and inequality (12) show the truth of inequality (3).

One may check that equality holds in (3) if q is given by (4). Indeed, the minimizer u , we dealt with in steps 1 to 5, is nothing but a first eigenfunction of problem (1) in case q is given by (4).

Theorem 1 is fully proved.

REFERENCES

0 G.Chiti, Limitazioni del primo autovalore di $\Delta + a$. Boll. U.M.I. <u>14</u>-A (1977), 136-142.

1 A.G.Ramm, Topics in scattering and spectral theory. Notices of A.M.S. <u>29</u>, no. 4 (June 1982), 327-329.

2 E.Sperner, Symmetrisierung für Funktionen mehrerer reeller Variablen. Manuscripta Math. <u>11</u> (1974), 159-170.

3 G.Talenti, Best constant in Sobolev inequality. Ann.Mat.Pura Appl. <u>110</u> (1976), 353-372.

4 G.Talenti, Elliptic equations and rearrangements. Ann. Scuola Norm.Sup.Pisa <u>3</u> (1976), 697-718.

Giorgio Talenti
Istituto Matematico dell'Università
Viale Morgagni, 67/A
50134 Firenze

International Series of
Numerical Mathematics, Vol. 71

EINSCHLIESSUNG DER LÖSUNGEN VON SYSTEMEN GEWÖHNLICHER DIFFERENTIALGLEICHUNGEN

Peter Volkmann

Abstract. Consider in R^N the initial value problem (+) $u(0) = a$, $u' = f(t,u)$ $(0 \le t \le T)$, where the right hand side of the differential equation satisfies an appropriate uniqueness condition. Then functions v^n, w^n can be constructed, approximating the solution u from below and from above, respectively. As a consequence one gets convergence of successive approximations for (+).

1. EINLEITUNG

Es bezeichne R den Bereich der reellen Zahlen. Für Punkte $x = (x_1,x_2,\dots,x_N)$, $y = (y_1,y_2,\dots,y_N)$ des R^N werde

$$\|x\| = \max\,\{|x_1|,|x_2|,\dots,|x_N|\}$$

gesetzt, und es werde $x \le y$ geschrieben, falls $x_\nu \le y_\nu$ $(\nu = 1,2,\dots,N)$ gilt. Im Weiteren sei $T > 0$ und

$$\omega(t,\sigma):\ [0,T] \times [0,\infty) \to [0,\infty]$$

eine bezüglich der Veränderlichen σ (schwach monoton) wachsende Funktion mit folgender Eigenschaft: Ist $N \in \{1,2,3,\dots\}$, $a \in R^N$ und $f\colon [0,T] \times R^N \to R^N$ eine stetige, beschränkte Funktion, welche der Abschätzung

$$(1) \qquad \|f(t,x)-f(t,\bar{x})\| \le \omega(t,\|x-\bar{x}\|) \qquad (0 \le t \le T;\ x,\bar{x} \in R^N)$$

genügt, so besitzt das Anfangswertproblem

$$(2) \qquad u(0) = a\,, \quad u' = f(t,u) \qquad (0 \le t \le T)$$

genau eine Lösung

$$(3) \qquad u\colon [0,T] \to R^N\,.$$

Unter diesen Voraussetzungen betrachte man eines der Anfangswertprobleme (2) (mit a, f wie oben beschrieben); (3) bezeichne seine Lösung.

In vorliegender Arbeit werden zwei Folgen von Funktionen

(4) $\qquad v^n\colon [0,T] \to \mathbb{R}^N, \quad w^n\colon [0,T] \to \mathbb{R}^N \qquad (n = 0,1,2,\dots)$

so konstruiert, daß die nachstehenden Beziehungen gelten:

(5) $\qquad v^0(t) \le v^1(t) \le v^2(t) \le \dots \le w^2(t) \le w^1(t) \le w^0(t) \quad (0 \le t \le T),$

(6) $\qquad \lim_{n\to\infty} v^n(t) = \lim_{n\to\infty} w^n(t) = u(t) \quad$ (gleichmäßig auf $[0,T]$).

Zur Gewinnung der Folgen (4) werden bekannte Techniken für monotone Operatoren (vgl. z.B. die Bücher von N.S. Kurpel' und B.A. Šuvar [2] und von J. Schröder [4]) mit einer Vorgehensweise aus [5] kombiniert. Als einfache Folgerung erhält man die Konvergenz der sukzessiven Approximationen für das Anfangswertproblem (2); dadurch ist ein neuer Zugang zu Resultaten von E.A. Coddington und N. Levinson [1] sowie von W. Walter [6] und anderen gegeben.

2. EIN LEMMA

Analog zu den Ergebnissen in [5] gilt das nachstehende Lemma.

LEMMA. <u>Es sei</u> Ξ <u>ein metrischer Raum (mit Metrik</u> ρ), $\phi\colon [0,\infty) \to [0,\infty]$ <u>eine (schwach monoton) wachsende und</u> $g\colon \Xi \times \mathbb{R} \to \mathbb{R}$ <u>eine stetige Funktion mit</u>

(7) $\qquad |g(\xi,x) - g(\bar\xi,\bar x)| \le \phi(\max\{\rho(\xi,\bar\xi), |x-\bar x|\}) \quad (\xi,\bar\xi \in \Xi;\ x,\bar x \in \mathbb{R}).$

<u>Definiert man</u>

$$G\colon \Xi \times \mathbb{R} \times \mathbb{R} \to \mathbb{R}$$

<u>durch</u>

(8) $$G(\xi,y,z) = \mathop{\mathrm{M}}_{x=y}^{z} g(\xi,x) \equiv \begin{cases} \min\limits_{y\le x\le z} g(\xi,x) & (y \le z) \\ \max\limits_{z\le x\le y} g(\xi,x) & (z \le y), \end{cases}$$

<u>so gilt</u>

(9) $|G(\xi,y,z)-G(\bar{\xi},\bar{y},\bar{z})| \leq \phi(\max\{\rho(\xi,\bar{\xi}),|y-\bar{y}|,|z-\bar{z}|\})$

$(\xi,\bar{\xi} \in \Xi;\ y,\bar{y},z,\bar{z} \in R)$.

<u>Beweis.</u> Es genügt, für ξ, $\bar{\xi}$ aus Ξ und reelle y, $\bar{y}$, z, $\bar{z}$ die Größe

$$\Delta = G(\xi,y,z) - G(\bar{\xi},\bar{y},\bar{z})$$

durch die rechte Seite von (9) abzuschätzen.

Fall 1: $y \leq z$, $\bar{y} \leq \bar{z}$. Es gilt

$$G(\bar{\xi},\bar{y},\bar{z}) = g(\bar{\xi},\bar{x})$$

mit einem geeigneten $\bar{x} \in [\bar{y},\bar{z}]$. Nun läßt sich leicht ein $x \in [y,z]$ so bestimmen, daß

(10) $$|x-\bar{x}| \leq \max\{|y-\bar{y}|,|z-\bar{z}|\}$$

ausfällt. Damit wird, unter Beachtung von (7),(8),

(11) $$\Delta \leq g(\xi,x)-g(\bar{\xi},\bar{x}) \leq \phi(\max\{\rho(\xi,\bar{\xi}),|y-\bar{y}|,|z-\bar{z}|\}) .$$

Fall 2: $y \leq z$, $\bar{z} \leq \bar{y}$. Hier ist

$$\Delta = g(\xi,x) - g(\bar{\xi},\bar{x})$$

mit geeigneten

(12) $$x \in [y,z] , \qquad \bar{x} \in [\bar{z},\bar{y}] .$$

Aus (12) folgt (10), und damit gilt wieder (11).

Fall 3: $z \leq y$, $\bar{y} \leq \bar{z}$. Dieser Fall kann wie Fall 2 erledigt werden.

Fall 4: $z \leq y$, $\bar{z} \leq \bar{y}$. Hier kann man ähnlich wie bei Fall 1 verfahren.

3. KONSTRUKTION DER FOLGEN (v^n), (w^n)

Wie in der Einleitung gesagt, werde jetzt eines der Anfangswertprobleme (2) betrachtet. Insbesondere wird also

(13) $$||f(t,x)|| \leq L < \infty \quad (0 \leq t \leq T,\ x \in R^N)$$

vorausgesetzt. Bezeichnen $f_\nu: [0,T] \times R^N \rightarrow R$ die Koordinatenfunktionen von f ($\nu = 1,2,\dots,N$), so sei (vgl. (8))

(14) $$F_\nu(t,y_1,y_2,\dots,y_N,z_1,z_2,\dots,z_N) = \underset{x_1=y_1}{\overset{z_1}{M}} \; \underset{x_2=y_2}{\overset{z_2}{M}} \dots \underset{x_N=y_N}{\overset{z_N}{M}} \; f_\nu(t,x_1,x_2,\dots,x_N) .$$

Mit Hilfe des Lemmas ergibt sich aus der Stetigkeit der Funktionen f_ν die Existenz und die Stetigkeit der Funktionen F_ν: $[0,T] \times R^{2N} \to R$ (vgl. auch [5]). Durch die Festsetzung

$$F(t,y,z) = (F_1(t,y,z),F_2(t,y,z),\dots,F_N(t,y,z))$$

für $0 \le t \le T$, $y = (y_1,y_2,\dots,y_N)$, $z = (z_1,z_2,\dots,z_N)$ erhält man demnach eine stetige Funktion

(15) $$F\colon [0,T] \times R^N \times R^N \to R^N .$$

Diese besitzt (nach Konstruktion (14)) zunächst folgende Eigenschaften:

(16) $$F(t,x,x) = f(t,x) \qquad (0 \le t \le T,\ x \in R^N) ,$$

(17) $$F(t,y,z) \le F(t,\bar{y},\bar{z}) \qquad (0 \le t \le T\ ;\ y,\bar{y},z,\bar{z} \in R^N;\ y \le \bar{y},\ \bar{z} \le z).$$

Ferner folgt aus (13) die Beschränktheit von F,

(18) $$\|F(t,y,z)\| \le L \qquad (0 \le t \le T;\ y,z \in R^N) ,$$

und aus (1) ergibt sich mit Hilfe des Lemmas die Abschätzung

(19) $$\|F(t,y,z)-F(t,\bar{y},\bar{z})\| \le \omega(t, \max\{\|y-\bar{y}\|,\|z-\bar{z}\|\}) \qquad (0 \le t \le T;\ y,\bar{y},z,\bar{z} \in R^N).$$

Nach Voraussetzung über ω besitzt also das Anfangswertproblem

(20) $$v(0) = w(0) = a, \quad v' = F(t,v,w), \quad w' = F(t,w,v) \qquad (0 \le t \le T)$$

genau eine Lösung (v,w): $[0,T] \to R^N \times R^N$. Wegen (16) ist daher

(21) $$v(t) = w(t) = u(t) \qquad (0 \le t \le T) ,$$

wobei u die Lösung (3) von (2) bezeichnet.

Nun seien v^o,w^o: $[0,T] \to R^N$ zwei stetige Funktionen, welche den Ungleichungen

(22) $$v^o(t) \le w^o(t) \qquad (0 \le t \le T) ,$$

$$(23)\quad \begin{aligned} v^o(t) \le K(t,v^o,w^o) &\equiv a + \int_0^t F(s,v^o(s),w^o(s))ds , \\ K(t,w^o,v^o) &\le w^o(t) \qquad (0 \le t \le T) \end{aligned}$$

genügen (mit $a = (a_1,a_2,\dots,a_N)$ z.B. $v^o(t) = (a_1-Lt,a_2-Lt,\dots,a_N-Lt)$, $w^o(t) = (a_1+Lt,a_2+Lt,\dots,a_N+Lt)$), und man definiere rekursiv

$$(24)\quad v^{n+1}(t) = K(t,v^n,w^n), \quad w^{n+1}(t) = K(t,w^n,v^n) \qquad (0 \le t \le T; \; n = 0,1,2,\dots).$$

Mit Hilfe von (17) erhält man dann aus (22),(23),(24) ohne weiteres die Ungleichungen (5). Damit existieren (zunächst punktweise) die Grenzwerte

$$(25)\quad v(t) = \lim_{n\to\infty} v^n(t) , \quad w(t) = \lim_{n\to\infty} w^n(t) \qquad (0 \le t \le T) .$$

Nach (18),(24) gilt

$$\|v^n(s)-v^n(t)\| \le L|s-t|, \quad \|w^n(s)-w^n(t)\| \le L|s-t| \qquad (s,t \in [0,T]; \; n = 1,2,3,\dots),$$

also hat man auch

$$\|v(s)-v(t)\| \le L|s-t|, \quad \|w(s)-w(t)\| \le L|s-t| \quad (s,t \in [0,T]),$$

und daher ist die monotone Konvergenz in den Formeln (25) nach einem Satze von Dini gleichmäßig. Grenzübergang $n \to \infty$ in (24) liefert

$$v(t) = K(t,v,w) , \quad w(t) = K(t,w,v) \qquad (0 \le t \le T) ,$$

also gilt (20) und demzufolge auch (21). Somit folgt (6) aus (25).

4. KONVERGENZ VON SUKZESSIVEN APPROXIMATIONEN

Wieder werde eines der Anfangswertprobleme (2) betrachtet. Es sei $u^o\colon [0,T] \to R^N$ eine beliebige stetige Funktion, und es werde

$$(26)\quad u^{n+1}(t) = a + \int_0^t f(s,u^n(s))ds \qquad (0 \le t \le T; \; n = 0,1,2,\dots)$$

gesetzt. Dann kann die Konvergenz

$$(27)\quad \lim_{n\to\infty} u^n(t) = u(t) \qquad \text{(gleichmäßig auf } [0,T])$$

wie folgt gezeigt werden: (26) bedeutet zunächst

(28) $u^{n+1}(t) = K(t,u^n,u^n)$ $(0 \le t \le T;\ n = 0,1,2,\ldots)$

(K gemäß (23)). Wählt man die stetigen Funktionen $v^o, w^o\colon [0,T] \to R^N$ in Nr. 3 derart, daß neben (22),(23) noch

$$v^o(t) \le u^o(t) \le w^o(t) \qquad (0 \le t \le T)$$

gilt, so erhält man aus (24),(28) leicht

$$v^n(t) \le u^n(t) \le w^n(t) \qquad (0 \le t \le T;\ n = 0,1,2,\ldots) .$$

Mit (6) folgt hieraus (27).

5. SCHLUSSBEMERKUNGEN

1. Ein typischer Fall für die hier betrachteten Funktionen ω ist die auf M. Nagumo [3] zurückgehende Funktion $\omega(t,\sigma) = \sigma/t$ (= ∞ für $t = 0$). Die Abschätzung (1) erhält dann für $N = 1$ die Form

(29) $$|f(t,x)-f(t,\bar{x})| \le \frac{|x-\bar{x}|}{t} \qquad (0 < t \le T;\ x,\bar{x} \in R) .$$

Ein einfaches Beispiel zeigt, daß in den Betrachtungen der Nr. 3 die Bedingung (29) nicht durch die schwächere Nagumosche Eindeutigkeitsbedingung

(30) $$f(t,x)-f(t,\bar{x}) \le \frac{x-\bar{x}}{t} \qquad (0 < t \le T;\ x,\bar{x} \in R;\ x > \bar{x})$$

ersetzt werden kann: Dazu sei (mit $T = 1$)

(31) $$f(t,x) = f(x) = -2\,(\operatorname{sgn} x) \min\{1,\sqrt{|x|}\} \qquad (0 \le t \le 1;\ x \in R) .$$

Da f monoton fällt, ist (30) erfüllt, und es gilt $F(t,y,z) = f(z)$. Wählt man noch $a = 0$, so gelten für

(32) $$w^o(t) = -v^o(t) = t^2 \qquad (0 \le t \le 1)$$

die Beziehungen

(33) $$v^o(t) = K(t,v^o,w^o), \quad w^o(t) = K(t,w^o,v^o) \qquad (0 \le t \le 1) ,$$

also können die gemäß (24) konstruierten Folgen $(v^n),(w^n)$ nicht gegen die Lösung $u(t) = 0$ des mit (31) und $a = 0$ gebildeten Anfangswertproblems (2) konvergieren.

2. Die Bedeutung der Abschätzung (19) für das Vorgehen in Nr. 3 kann durch folgendes Beispiel erläutert werden: Es sei

$$q(x) = (\operatorname{sgn} x) \min\{1,\sqrt{|x|}\} \qquad (x \in R) ,$$

und man betrachte (2) in dem trivialen Falle $N = 1$, $T = 1$, $a = 0$,

$$f(t,x) \equiv 0 \qquad (0 \leq t \leq 1,\ x \in R).$$

Setzt man (anders als in Nr. 3)

(34) $$F(t,y,z) = q(y) - q(z) \qquad (0 \leq t \leq 1;\ y,z \in R),$$

so ist

$$F\colon [0,1] \times R \times R \to R$$

eine stetige, beschränkte Funktion, welche (16),(17) erfüllt; die Funktion (34) besitzt damit, abgesehen von (19), alle Eigenschaften der in Nr. 3 konstruierten Funktion (15). Ausgehend von den Funktionen (32) erhält man wieder die Beziehungen (33) (K ist jetzt durch (23) mit $a = 0$ und F gemäß (34) gegeben), also können die durch (24) erklärten Folgen (v^n),(w^n) nicht gegen die Lösung $u(t) = 0$ des Anfangswertproblems

$$u(0) = 0\,, \quad u' = 0 \qquad (0 \leq t \leq 1)$$

konvergieren.

LITERATUR

1. E.A. Coddington und N. Levinson, Uniqueness and the convergence of successive approximations. J. Indian Math. Soc. 16 (1952), 75-81.

2. N.S. Kurpel' und B.A. Šuvar, Dvustoronnie operatornye neravenstva i ih primenenija. Naukova Dumka, Kiev, 1980.

3. M. Nagumo, Eine hinreichende Bedingung für die Unität der Lösung von Differentialgleichungen erster Ordnung. Japan. J. Math. 3 (1926), 107-112.

4. J. Schröder, Operator Inequalities. Academic Press, New York et al., 1980.

5. P. Volkmann (P. Fol'kmann), Zametka ob integral'nyh neravenstvah tipa Vol'terra. Ukrain. Mat. Žurn. 36 (1984), 393-395.

6. W. Walter, Über sukzessive Approximation bei Volterra-Integralgleichungen in mehreren Veränderlichen. Ann. Acad. Sci. Fennicae, Ser. A I, Nr. 345 (1965), 32 Seiten.

Peter Volkmann, Mathematisches Institut I, Universität Karlsruhe, Postfach 6380, 7500 Karlsruhe 1, Westdeutschland

Inequalities in
Economics, Optimization and Applications

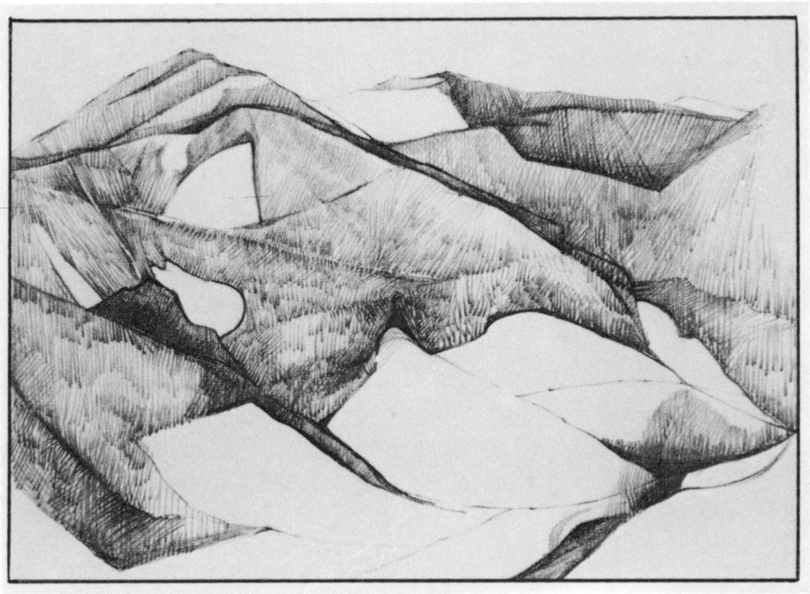

Snow-Slopes

International Series of
Numerical Mathematics, Vol. 71
© 1984 Birkhäuser Verlag Basel

REMARK ON PROGRESSIVE INCOME TAXATION: INEQUALITIES IN THE THEORY OF ECONOMIC INEQUALITY

Wolfgang Eichhorn

Progressive income taxation cuts relative income differentials, thus reducing income inequality. It is interesting to know that this implication can be stated as an equivalence, if one slightly modifies the assumptions made in my REMARK on p. 519 of *General Inequalities 3*.

Let I be a nonempty interval of $\mathbb{R}_{++}$, the set of all positive real numbers. Let p be any function mapping I into $(-\infty,1)$. Interpret $x \in I$ as income before tax and $p(x)$ as average tax rate such that $xp(x)$ is tax liability and $y = (1-p(x))x$ residual income, i.e. income after tax.

This remark focusses on the three principles of taxation being here expressed as properties of $p: I \to (-\infty,1)$.

P.1 *Weakly Progressive Taxation:* p is monotonically increasing in the weak sense;

P.2 *Weakly Incentive Preserving Taxation* (Fei, 1981): Residual income $y(x) = (1-p(x))x$ is weakly monotonically increasing in $x \in I$;

P.3 *Weak Decrease of Income Inequality:* p is such that for all nonidentically distributed $\underline{x} = (x_1, \dots, x_n) \in I^n$ the residual income vector $\underline{y}$ with $y_i := (1-p(x_i))x_i$ weakly dominates $\underline{x}$ in the sense of Lorenz.

Recall that $\underline{y}$ is said to *weakly dominate* $\underline{x}$ *in the sense of Lorenz* if after arranging both in increasing order, $y_{\sigma(1)} \le \dots \le y_{\sigma(n)}$, $x_{\tau(1)} \le \dots \le x_{\tau(n)}$, we obtain

(1) $$\frac{1}{s(y)} \sum_{i=1}^{k} y_{\sigma(i)} \geq \frac{1}{s(x)} \sum_{i=1}^{k} x_{\tau(i)} \quad \text{for all } k = 1,\dots,n-1 ,$$

where

$$s(x) = x_1 + x_2 + \dots + x_n .$$

We speak of *strict Lorenz domination* if strict inequality holds in (1) for $k = 1,\dots,n-1$. [A good reference for the theory of Lorenz domination is Marshall and Olkin, Theory of Majorization and Its Applications, Academic Press, New York 1979. Note that $\underline{y}$ weakly dominates $\underline{x}$ iff $\underline{y}/s(y)$ is majorized by $\underline{x}/s(x)$.]

Properties P.1 - P.3 can equally be stated in strong form P.1*, P.2*, and P.3* by simply substituting "strict(ly)" for "weak(ly)" in P.1 - P.3.

Note that

- neither P.1* nor P.2* implies P.3;
- Properties P.1* and P.2* are logically independent as well as P.1 and P.2;
- there exist functions p that satisfy all properties, for example:

  $$p(x) := \frac{1}{2} \cdot \frac{x}{x+1} \quad ;$$

- Properties P.1 and P.2 imply, for every $\delta > 0$,

  $$0 \leq p(x+\delta) - p(x) \leq \frac{\delta}{x+\delta}(1-p(x)) ,$$

  i.e., continuity of p for all $x > 0$;
- Properties P.1, P.2, and P.3 do *not* imply differentiability of p; see example (2) below at $x = 1$.

THEOREM. (a) Properties P.1 and P.2 hold if and only if Property P.3 holds. (b) Properties P.1* and P.2 hold if and only if Property P.3* holds.

The proof of this theorem can be found in a joint paper with Helmut Funke and Wolfram F. Richter which is submitted to a journal.

Part (b) of the theorem would be wrong if P.2 were replaced by $P.2^*$. Indeed, $P.2^*$ is *not* implied by $P.3^*$, as the following example shows, which is due to Janos Aczél and Wolfgang Walter:

$$p(x) = \begin{cases} \frac{x}{x+1} & \text{for } x \le 1 , \\ \frac{2x-1}{2x} & \text{for } x > 1 . \end{cases} \tag{2}$$

The following example shows that P.3 would no longer imply P.1 if I were allowed to contain $x = 0$. Let $p(0) = 1/2$ and $p(x) = 0$ for $x > 0$. Then $y = (1-p(x))x = x$ for $x \ge 0$. Hence P.3 is satisfied, and p is not increasing.

Wolfgang Eichhorn, Institut für Wirtschaftstheorie und Operations Research, Universität Karlsruhe, D-7500 Karlsruhe, West Germany

International Series of
Numerical Mathematics, Vol. 71

AN APPLICATION OF FARKAS' LEMMA TO A NONCONVEX MINIMIZATION PROBLEM

Raj Jagannathan and Siegfried Schaible

Abstract. A duality theory is developed for a certain type of quasiconvex minimization problem making use of Farkas' Lemma for systems of linear (convex) inequalities.

For the minimization of a convex function on a convex set of R^n (defined by convex inequality constraints) a dual optimization problem can be derived that is intimately related to the given (primal) problem; see for example [6]. If the minimand is only quasiconvex rather than convex [1] similar results can still be obtained as recently shown in [2].

Now consider the following quasiconvex problem that arises in approximation theory, mathematical economics and management science (see, e.g., [3]):

$$\text{(P)} \qquad \operatorname{Inf}\left\{ \max_{1\le i\le p} \frac{a_i^T x + \alpha_i}{b_i^T x + \beta_i} \;\middle|\; Cx \le \gamma,\ x \ge 0 \right\},$$

where $a_i, b_i \in R^n$, $\alpha_i, \beta_i \in R$, $\gamma \in R^m$ and C is a real $m \times n$ matrix. Let A (respectively B) be the $p \times n$ matrix with rows a_i^T (respectively b_i^T) and $\alpha = (\alpha_1, \dots, \alpha_p)^T$, $\beta = (\beta_1, \dots, \beta_p)^T$. For problem (P) a dual optimization problem and duality relations were obtained in [3] by applying the results in [2]. In [5] the authors of this summary have shown that the same duality results can be derived in a much more straightforward way without the rather involved machinery of quasiconvex duality. As shown there, Farkas' Lemma [4] provides the key.

In order to do this, problem (P) is related to the following system of linear inequalities:

$$(A-tB)x \le (-\alpha+t\beta)\ , \quad Cx \le \gamma\ , \quad x \ge 0$$

in the parameter $t \in R$. According to Farkas' Lemma [4] this system is consistent if and only if the related system

$$(A-tB)^T u + C^T v \ge 0\ , \quad u \ge 0\ , \quad v \ge 0\ , \quad (-\alpha+t\beta)^T u + \gamma^T v < 0$$

is inconsistent for given t. The latter system of inequalities can then be used to define a dual problem (D). This is a quasi-concave maximization problem involving again finitely many affine ratios in the objective function. For both (P) and (D) any local optimum is a global optimum. Furthermore, the duality relations obtained resemble and generalize those in linear programming [3].

The duality results for (P) can be extended to convex-concave ratios in the objective function and convex inequality constraints by making use of a suitable generalization of Farkas' Lemma to systems of convex inequalities.

For details of the new duality approach summarized in this paper see [5].

REFERENCES

1. M. Avriel, W.E. Diewert, S. Schaible and W.T. Ziemba, Introduction to concave and generalized concave functions. In: S. Schaible and W.T. Ziemba (eds.), Generalized concavity in optimization and economics, Academic Press, New York, 1981, 21-50.

2. J.P. Crouzeix, A duality framework in quasiconvex programming. In: S. Schaible and W.T. Ziemba (eds.), Generalized concavity in optimization and economics. Academic Press, New York, 1981, 207-225.

3. J.P. Crouzeix, J.A. Ferland and S. Schaible, Duality in generalized linear fractional programming. Math. Programming 27 (1983) (to appear).

4. J. Farkas, Über die Theorie der einfachen Ungleichungen. J. Reine Angew. Math. 124 (1902), 1-24.

5. R. Jagannathan and S. Schaible, Duality in generalized fractional programming via Farkas' Lemma. J. Optim. Theory Appl. 41 (1983), 417-424.

6. R.T. Rockafellar, Convex Analysis. Princeton 1970.

Raj Jagannathan, Department of Management Science, University of Iowa, Iowa City, IA 55240, USA.

Siegfried Schaible, Department of Finance and Management Science, University of Alberta, Edmonton, AB T6G 2G1, Canada.

International Series of
Numerical Mathematics, Vol. 71

RADAR AMBIGUITY FUNCTIONS OF POSITIVE TYPE

Walter Schempp

Abstract. In radar analysis there exists an analogue of the Heisenberg uncertainty principle of quantum mechanics. These uncertainty principles say that it is impossible to determine simultaneously the range (resp. position) and range rate (resp. momentum) of a moving target (resp. non-relativistic quantum-mechanical particle) with an arbitrarily high accuracy at a given instant of time. This analogy suggests that there should be a common mathematical structure behind that explains both of them. The concept of real Heisenberg nilpotent group lies at the foundations of signal theory as well as of quantum mechanics. Adopting the view point of nilpotent harmonic analysis it becomes obvious that the radar auto-ambiguity functions are of pure positive type on the symplectic time-frequency plane which will be identified with the tangent space to the Schrödinger coadjoint orbit of the Heisenberg group. Combining the classical Bochner theorem with Hudson's characterization of the state vectors that give rise to non-negative Wigner quasi-probability densities we identify the Gabor functions as pulse envelopes of those radar auto-ambiguity functions on the time-frequency plane that are of positive type simultaneously with respect to its usual vector group structure as well as to its symplectic structure.

1. RADAR AMBIGUITY FUNCTIONS

The signals received in radar and in most communication systems consist of a monochromatic high-frequency carrier $e^{2\pi i\omega t}$ modulated in amplitude by a complex envelope function f of time t that varies much more slowly than the cycles of the carrier. Radar pulses of finite energy, for instance, are usually transmitted in the form

$$s(t) = f(t)e^{2\pi i\omega t} \qquad (t \in \mathbb{R}).$$

We will assume that the pulse envelope f belongs to the vector space $\mathcal{S}(\mathbb{R})$ of complex-valued functions which are defined and infinitely differentiable on the real line $\mathbb{R}$, and which have the additional property, regulating their decrease at infinity, that all their derivatives tend to zero at infinity, faster than any power of $\frac{1}{|t|}$. It is well known that $\mathcal{S}(\mathbb{R})$ is an everywhere dense vector subspace of the complex Hilbert space $L^2(\mathbb{R})$ and that $\frac{1}{2}||f||_2^2$ is the energy of the signal s under consideration.

To measure the distance of a remote radar target it is necessary to estimate the time x at which the echo from it arrives at the receiver. If time is counted from the transmission of the radar pulse s, the distance from the radar antenna is $\frac{1}{2}cx$, where c is the velocity of electromagnetic radiation. If the radar target is moving toward or away from the antenna, the carrier frequency of the echo differs from that of the transmitted pulse because of the Doppler effect. If we pick the transmitted frequency ω as our basic reference frequency, the echo will have the form

$$s_{echo}(t) = f_{echo}(t)e^{2\pi i\omega t} \qquad (t \in \mathbb{R})$$

where

$$f_{echo}(t) = f(t+x)e^{2\pi iyt}.$$

The Doppler shift y is given by $y = 2\frac{v}{c}\omega$, where v is the radial velocity, i.e., the component of target velocity in the direction of the radar antenna. This range rate furnishes valuable information for tracking the target efficiently.

Whenever it is necessary to distinguish or resolve two narrowband signals in the presence of white Gaussian random noise, the structure of the receiver and its performance actually depend upon mathematical expectation, the scalar product of the envelope functions of the two signals. In the present case the expectation $\langle f_{echo}|f\rangle$ takes the form

$$\int_{\mathbb{R}} f(t+x)\overline{f}(t)e^{2\pi iyt}dt.$$

Disregarding the time independent phase factor $e^{-\pi ixy}$ which is of no importance in the present context, the mathematical expectation can be written in the symmetric form

$$H(f;x,y) = \int_{\mathbb{R}} f(t+\tfrac{1}{2}x)\overline{f}(t-\tfrac{1}{2}x)e^{2\pi iyt}dt.$$

The function $H(f;.,.) \in \mathcal{S}(\mathbb{R} \oplus \mathbb{R})$ defined on the time-frequency plane is called the radar auto-ambiguity function (Woodward [12]). Similarly, if $g \in \mathcal{S}(\mathbb{R})$ denotes the complex envelope of another signal, the radar cross-ambiguity function is defined according to the prescription

$$H(f,g;x,y) = \int_{\mathbb{R}} f(t+\tfrac{1}{2}x)\overline{g}(t-\tfrac{1}{2}x)e^{2\pi iyt}dt.$$

The ranges $H(f;\mathbb{R},\mathbb{R})$ and $H(f,g;\mathbb{R},\mathbb{R})$ are called radar auto- and cross-ambiguity surfaces, respectively.

The following properties (I) and (II) of the radar auto-ambiguity functions are immediate. Property (III) infra is an easy application of the Cauchy-Schwarz inequality.

(I) $H(f;0,0) \geq 0$

(II) $H(f;-x,-y) = \overline{H}(f;x,y)$ for all pairs $(x,y) \in \mathbb{R} \oplus \mathbb{R}$ ("Hermitean central symmetry")

(III) $|H(f;x,y)| \leq H(f;0,0)$ for all pairs $(x,y) \in \mathbb{R} \oplus \mathbb{R}$ ("Peak property")

Thus for every pulse envelope $f \in \mathcal{S}(\mathbb{R})$ the associated radar auto-ambiguity surface is peaked at the origin $(0,0)$ of the time-frequency plane $\mathbb{R} \oplus \mathbb{R}$. The auto-ambiguity function takes on its peak value at the origin the double signal energy $||f||_2^2$.

A second signal arriving with time delay x and Doppler frequency shift y that lie under this central peak will be difficult to distinguish from the first signal. For most types of signals utilized in practice, for instance trains of coherent pulses, the radar auto-ambiguity surface exhibits additional peaks elsewhere over the time-frequency plane $\mathbb{R} \oplus \mathbb{R}$. These "sidelobes" may conceal weak signals with arrival times and carrier frequencies far from those of the first signal. The taller the sidelobes, the greater the probability of gross errors in time delay and Doppler frequency shift and hence in tracking the radar target. It is desirable, therefore, for the central peak of the radar auto-ambiguity surface to be as slender as possible, and for there to be as few and as low adjacent peaks as possible. Thus the geometry of the radar auto-ambiguity surface is of high importance for the practice of radar design.

Every basic treatise on quantum mechanics features Heisenberg's position-momentum indeterminacy principle and its deep implications for the physical measurement process. It is not so well known that in radar analysis there is a similar range-range rate uncertainty principle. The group-theoretic embodiment of both principles is the real Heisenberg nilpotent group [8]. In particular, harmonic analysis on this Lie group determines the geometry of the radar auto-ambiguity surfaces. Therefore we have to summarize some parts of this area in the next section.

2. THE ORBIT PICTURE

In what follows G will be a connected, simply connected nilpotent Lie group. Let $\mathfrak{g}$ denote its Lie algebra and $\mathfrak{g}^*$ its real dual vector space. Then $\exp: \mathfrak{g} \to G$ is a global diffeomorphism which carries Lebesgue measure of $\mathfrak{g}$ to Haar measure of G and G acts on $\mathfrak{g}^*$ by the coadjoint representation Coad, the contragredient of the adjoint representation Ad of G in $\mathfrak{g}$. According to Kirillov theory [4], all the isomorphy classes of irreducible unitary linear representations of G can be constructed by the method of coadjoints orbits. More precisely, if we denote by $U_{\mathcal{O}}$ any continuous irreducible unitary linear representation of G associated with the orbit $\mathcal{O} \in \mathfrak{g}^*/\mathrm{Coad}(G)$, then the assignment $\mathcal{O} \rightsquigarrow U_{\mathcal{O}}$ induces a natural one-to-one correspondence between the space of orbits $\mathfrak{g}^*/\mathrm{Coad}(G)$ and the unitary dual $\hat{G}$ of G. It should be noted that the Kirillov bijection is a homeomorphism when the orbit space $\mathfrak{g}^*/\mathrm{Coad}(G)$ is understood to carry the quotient topology and the dual $\hat{G}$ of G is endowed with the hull-kernel topology (cf. Brown [2]). Finally, recall that any generic orbit $\mathcal{O} \in \mathfrak{g}^*/\mathrm{Coad}(G)$ is a symplectic manifold with respect to the Kirillov form $\omega_{\mathcal{O}}$ and thus it possesses a canonical G-invariant measure given by the volume form $\omega_{\mathcal{O}}^{1/2 \dim \mathcal{O}}$. Furthermore, $\mathcal{O}$ is generically an algebraic variety in $\mathfrak{g}^*$ and more specifically a linear variety if and only if $U_{\mathcal{O}}$ defines a square integrable representation of the group G/K, where K stands for the neutral component of the kernel of $U_{\mathcal{O}}$ (Brezin [1]). In this case it follows that $\mathcal{O}$ is isomorphic to G/Z where Z denotes the center of G and the restriction $U_{\mathcal{O}} | Z$ is a multiple of a central character $\chi_{\mathcal{O}} \in \hat{Z}$ of G (cf. Moore-Wolf [5]).

Suppose, now, that G is a (2n+1)-dimensional, 2-step nilpotent, simply connected analytic group having one dimensional center Z. It follows from linear algebra results that G is isomorphic to the real Heisenberg nilpotent group $\tilde{A}(\mathbb{R}^n)$. Recall that $\tilde{A}(\mathbb{R}^n)$ is the group with underlying space $(\mathbb{R}^n \oplus \mathbb{R}^n) \oplus \mathbb{R}$ and multiplication law

$$(v_1,z_1).(v_2,z_2) = (v_1+v_2, z_1+z_2+\tfrac{1}{2}B_n(v_1,v_1))$$

where we look upon $\mathbb{R}^n \oplus \mathbb{R}^n$ as the vector space of all columns

$$v = \begin{bmatrix} x \\ y \end{bmatrix}, \quad x \in \mathbb{R}^n,\ y \in \mathbb{R}^n$$

and the standard symplectic (= nondegenerate skew symmetric bilinear form) B_n on $\mathbb{R}^n \oplus \mathbb{R}^n$ is defined according to the pre–

scription

$$B_n(v_1,v_2) = B_n(\begin{bmatrix} x_1 \\ y_1 \end{bmatrix}, \begin{bmatrix} x_2 \\ y_2 \end{bmatrix}) = \langle x_1|y_2\rangle - \langle x_2|y_1\rangle.$$

The Lie algebra of $\tilde{A}(\mathbb{R}^n)$ is the (2n+1)-dimensional real Heisenberg Lie algebra

$$\mathfrak{n} = (\mathbb{R}^n \oplus \mathbb{R}^n) \oplus \mathfrak{z}$$

where $\mathfrak{z}$, the Lie algebra of $Z = \{(\begin{bmatrix} 0 \\ 0 \end{bmatrix}, z) \mid z \in \mathbb{R}\}$, denotes the center of $\mathfrak{n}$.

In the case $G = \tilde{A}(\mathbb{R}^n)$ the Kirillov orbit picture yields two families of isomorphy classes of irreducible unitary linear representations:

(1) A "degenerate" family of one dimensional representations which map the center $Z = \mathbb{R}$ of $\tilde{A}(\mathbb{R}^n)$ to $\{0\}$. The corresponding coadjoint orbits are the single points $\kappa = \begin{bmatrix} \xi \\ \eta \end{bmatrix} \in \mathbb{R}^n \oplus \mathbb{R}^n$, and the associated representations are given by

$$U_\kappa(\begin{bmatrix} x \\ y \end{bmatrix}, z) = e^{2\pi i(\langle x|\xi\rangle + \langle y|\eta\rangle)} \mathrm{id}_{\mathbb{C}}.$$

(2) A "generic" family of infinite dimensional representations parametrized by $\lambda \in \mathbb{R}-\{0\}$, acting on the complex Hilbert space $L^2(\mathbb{R}^n)$, which map Z to a central character $\chi_\lambda \in \hat{Z}$. The corresponding coadjoint orbits are the affine hyperplanes $\lambda+\mathfrak{z}^0$, where $\mathfrak{z}^0$ is the annihilator of $\mathfrak{z}$ in $\mathfrak{n}^*$. The associated representations are given by

$$U_\lambda(\begin{bmatrix} x \\ y \end{bmatrix}, z) f(t) = e^{2\pi i\lambda(z + \langle y|t\rangle + \frac{1}{2}\langle x|y\rangle)} f(t+x) \qquad (t \in \mathbb{R}^n)$$

for $f \in \mathcal{S}(\mathbb{R}^n)$ and the Schwartz space $\mathcal{S}(\mathbb{R}^n)$ forms precisely the vector space of smooth vectors for U_λ $(\lambda \in \mathbb{R}-\{0\})$ acting on $L^2(\mathbb{R}^n)$.

It follows that all the infinite dimensional irreducible unitary linear representations of $\tilde{A}(\mathbb{R}^n)$ are square integrable mod Z, i.e., they belong to the discrete series of $\tilde{A}(\mathbb{R}^n)$. Among these the prototype U_1 which is unique up to dilation and duality is called the linear Schrödinger representation of $\tilde{A}(\mathbb{R}^n)$. In the case n=1 the action

$$U_1\left(\begin{bmatrix} x \\ y \end{bmatrix}, 0\right) f(t) = e^{2\pi i (ty + \frac{1}{2}xy)} f(t+x) \qquad (t \in \mathbb{R})$$

shows that when the linear Schrödinger representation U_1 of $\tilde{A}(\mathbb{R})$ is restricted to the polarized cross-section $\{(\begin{bmatrix} x \\ y \end{bmatrix}, 0) \mid x \in \mathbb{R}, y \in \mathbb{R}\}$ to Z in $\tilde{A}(\mathbb{R})$ and then applied to the complex pulse envelope $f \in \mathcal{S}(\mathbb{R})$ it produces a time delay and a frequency shift similar to the reflection of the signal pulse s by a moving target.

3. FUNCTIONS OF POSITIVE TYPE

The tangent space to the coadjoint orbit $1+\mathfrak{z}^{\circ}$ in $\mathfrak{n}^*$ at the point 1 carries the symplectic form

$$\omega_1 : \left(\begin{bmatrix} \xi_1 \\ \eta_1 \end{bmatrix}, \begin{bmatrix} \xi_2 \\ \eta_2 \end{bmatrix}\right) \rightsquigarrow 1.\left[\begin{bmatrix} \xi_2 \\ \eta_2 \end{bmatrix}, \begin{bmatrix} \xi_1 \\ \eta_1 \end{bmatrix}\right] = -\langle \xi_2 | \eta_1 \rangle + \langle \xi_1 | \eta_2 \rangle$$
$$= B_n\left(\begin{bmatrix} \xi_1 \\ \eta_1 \end{bmatrix}, \begin{bmatrix} \xi_2 \\ \eta_2 \end{bmatrix}\right)$$

where $[.,.]$ denotes the bracket operation in $\mathfrak{n}$. The tangent space $(\mathfrak{z}^{\circ}; \omega_1)$ and the standard space $(\mathbb{R}^n \oplus \mathbb{R}^n; B_n)$ are isomorphic symplectic vector spaces and will henceforth be identified.

DEFINITION 1. A continuous complex-valued function F on $\mathbb{R}^n \oplus \mathbb{R}^n$ is said to be of positive type on the standard symplectic vector space $(\mathbb{R}^n \oplus \mathbb{R}^n; B_n)$ if, for any vectors $(v_j)_{1 \leq j \leq m}$ of $\mathbb{R}^n \oplus \mathbb{R}^n$, the matrix

$$\left(e^{-\pi i B_n(v_j, v_k)} F(v_j - v_k)\right)_{1 \leq j, k \leq m}$$

is positive hermitean.

In other words, for any vectors $(v_j)_{1\leq j\leq m}$ of $\mathbb{R}^n \oplus \mathbb{R}^n$ and complex numbers $(\alpha_j)_{1\leq j\leq m}$, we have

$$\sum_{1\leq j,k\leq m} \alpha_j \bar{\alpha}_k e^{-\pi i B_n(v_j,v_k)} F(v_j - v_k) \geq 0.$$

DEFINITION 2. A continuous complex-valued function F of positive type on the standard symplectic vector space $(\mathbb{R}^n \oplus \mathbb{R}^n; B_n)$ is said to be pure if in every decomposition

$$F = F_1 + F_2$$

of F into a sum of two continuous functions F_1 and F_2 of positive type on $(\mathbb{R}^n \oplus \mathbb{R}^n; B_n)$ the functions F_1 and F_2 are proportional to F.

In other words, the continuous functions of pure positive type on $(\mathbb{R}^n \oplus \mathbb{R}^n; B_n)$ can be identified with the extremal points of the convex vaguely compact set of normalized continuous functions of positive type on $(\mathbb{R}^n \oplus \mathbb{R}^n; B_n)$. A standard construction due to Gelfand-Raikov and Godement (cf. [10]) shows that any given complex-valued function $F \in \mathcal{S}(\mathbb{R}^n \oplus \mathbb{R}^n)$ is of pure positive type if and only if there exists a Schwartz function $f \in \mathcal{S}(\mathbb{R}^n)$ such that

$$F(x,y) = \langle U_1(\begin{bmatrix} x \\ y \end{bmatrix}, 0) f \mid f \rangle$$

holds for all pairs $\begin{bmatrix} x \\ y \end{bmatrix} \in \mathbb{R}^n \oplus \mathbb{R}^n$. In the case n=1 the identity

$$H(f,g;x,y) = \langle U_1(\begin{bmatrix} x \\ y \end{bmatrix}, 0) f \mid g \rangle$$

shows that the radar cross-ambiguity function of the envelopes $f,g \in \mathcal{S}(\mathbb{R})$ is the matrix coefficient of the linear Schrödinger representation restricted to the polarized cross-section to Z in $\tilde{A}(\mathbb{R})$. Thus we conclude

THEOREM 1. <u>The radar auto-ambiguity function</u> $H(f;.,.) \in \mathcal{S}(\mathbb{R} \oplus \mathbb{R})$ <u>is a function of pure positive type on the symplectic time-frequency plane</u> $(\mathbb{R} \oplus \mathbb{R}; B_1)$ <u>for every choice of the pulse envelope</u> $f \in \mathcal{S}(\mathbb{R})$.

The preceding result shows that harmonic analysis on the three dimensional real Heisenberg nilpotent group $\tilde{A}(\mathbb{R})$ furnishes a suitable framework for describing simultaneously signals in terms of both time and frequency by considering the symplectic time-frequency plane $(\mathbb{R} \oplus \mathbb{R}; B_1)$ as the tangent space to the Schrödinger coadjoint orbit $(1+\mathfrak{z}^{o}; \omega_1) \in \mathfrak{n}^{*}/\mathrm{Coad}(\tilde{A}(\mathbb{R}))$ at the point 1.

The properties (I) and (III) supra of the radar auto-ambiguity functions $H(f;.,.)$ suggest that these are of positive type also on the additive group $\mathbb{R} \oplus \mathbb{R}$. However this is true only for a special class of pulse envelopes $f \in \mathcal{S}(\mathbb{R})$. This assertion stands in stark contrast to Theorem 1 supra.

4. GABOR FUNCTIONS

DEFINITION 3. A function $f: \mathbb{R} \to \mathbb{C}$ is said to be a Gabor function if f is the exponential e^q of a quadratic polynomial q with complex coefficients, the leading coefficient having negative real part.

Obviously the Gabor functions belong to the vector space $\mathcal{S}(\mathbb{R})$. Fix a pulse envelope $f \in \mathcal{S}(\mathbb{R})$ and consider the Fourier transform

$$P(f;.,.) = \mathcal{F}_{\mathbb{R}^2} H(f;.,.)$$

of the radar auto-ambiguity function $H(f;.,.)$. Then

$$P(f;y',x') = \int_{\mathbb{R}} f(x'+\tfrac{1}{2}t)\overline{f}(x'-\tfrac{1}{2}t)e^{-2\pi i y' t}dt$$

denotes the Wigner quasiprobability distribution function (cf. Wigner [11]). In view of Bochner's theorem, $H(f;.,.)$ is a function of positive type on the additive group $\mathbb{R} \oplus \mathbb{R}$ if and only if $P(f;.,.) \geq 0$. By Hudson's theorem (cf. Hudson [3]) which on its part is a consequence of the theorem of Hadamard in complex analysis, this holds if and only if the wavefunction $f \in \mathcal{S}(\mathbb{R})$ is a Gabor function. Summarizing we established the following result:

THEOREM 2. The radar auto-ambiguity function $H(f;.,.) \in \mathcal{S}(\mathbb{R} \oplus \mathbb{R})$ is of positive type on the additive group $\mathbb{R} \oplus \mathbb{R}$ if and only if the complex pulse envelope $f \in \mathcal{S}(\mathbb{R})$ is a Gabor function.

Radar pulses with Gabor functions as envelopes play a rôle, for instance, in seismic signal processing. For recent developments in seismic processing, see Morlet [6].

5. GROUPS ATTACHED TO RADAR AUTO-AMBIGUITY SURFACES

Of course, signals with the same energy but different pulse envelopes may generate the same radar auto-ambiguity surface. In order to determine the complex envelopes $f \in \mathcal{S}(\mathbb{R})$ that give rise to the same ambiguity surface, let $Mp(1,\mathbb{R})$ denote the metaplectic group which forms a twofold covering group of the symplectic group $Sp(1,\mathbb{R}) = SL(2,\mathbb{R})$. Let $\varepsilon: \tilde{\sigma} \rightsquigarrow \sigma$ be the covering homomorphism so that

$$\{0\} \to \mathbb{Z}/2\mathbb{Z} \to Mp(1,\mathbb{R}) \xrightarrow{\varepsilon} Sp(1,\mathbb{R}) \to \{1\}$$

forms a short exact sequence. Notice that $Mp(1,\mathbb{R})$ has a well-known infinite dimensional unitary linear representation $\tilde{\sigma} \rightsquigarrow T_{\tilde{\sigma}}$ acting on $L^2(\mathbb{R})$, the so-called unitary oscillator representation which has certain formal resemblances to the spin representation of the double covering of $SO(1)$. An application of Segal's formula then yields the following result:

THEOREM 3. *Let* S *denote a unitary automorphism of the complex prehilbert space* $\mathcal{S}(\mathbb{R})$. *Suppose that for all* $f \in \mathcal{S}(\mathbb{R})$ *the radar auto-ambiguity surfaces generated by the pulse envelopes* f *and* Sf *are the same, i.e., for any pair* $(x,y) \in \mathbb{R} \oplus \mathbb{R}$ *there exists a pair* $(x',y') \in \mathbb{R} \oplus \mathbb{R}$ *such that the identity*

$$H(f;x,y) = H(Sf;x',y')$$

holds. Then there exists a unique mapping $\sigma \in Sp(1,\mathbb{R})$ *and a complex number* $\zeta, |\zeta|=1$, *such that*

$$\sigma(x,y) = (x',y'), \qquad S = \zeta T_{\tilde{\sigma}}$$

holds.

The unitary automorphisms $T_{\tilde{\sigma}}$ $(\sigma \in Sp(1,\mathbb{R}))$ of $\mathcal{S}(\mathbb{R})$ can be computed explicitly in terms of polarizations of $\mathfrak{n}$ and appropriate partial Fourier transforms. In particular, we obtain

THEOREM 4. *Let the complex pulse envelope* $f \in \mathcal{S}(\mathbb{R})$ *have* L^2-*norm* $||f||_2=1$. *The radar auto-ambiguity function* $H(f;.,.)$ *is radially symmetric and of positive type on the vector group* $\mathbb{R} \oplus \mathbb{R}$ *if and only if* f *is up to a phase factor* $\zeta \in \mathbb{C}, |\zeta|=1$, *the standardized Gaussian.*

For further details, the reader is referred to the papers [7] and [9].

REFERENCES

1. Brezin, J.: Geometry and the method of Kirillov. In: Non-Commutative Harmonic Analysis, pp. 13-25. J. Carmona, J. Dixmier, and M. Vergne, editors. Lecture Notes in Mathematics, Vol. 466. Berlin-Heidelberg-New York: Springer 1975

2. Brown, I.D.: Dual topology of a nilpotent Lie group. Ann. scient. Éc. Norm. Sup. 6, 407-411 (1973)

3. Hudson, R.L.: When is the Wigner quasi-probability density non-negative? Rep. Math. Phys. 6, 249-252 (1974)

4. Kirillov, A.A.: Unitary representations of nilpotent Lie groups. Uspehi Mat. Nauk 17, 57-110 (1962). Also in: Russian Math. Surveys 17, 53-104 (1962)

5. Moore, C.C., Wolf, J.A.: Square integrable representations of nilpotent groups. Trans. Amer. Math. Soc. 185, 445-462 (1973)

6. Morlet, J.: Sampling theory and wave propagation. In: Issues in Acoustic Signal/Image Processing and Recognition, pp. 233-261. C.H. Chen, editor. NATO ASI Series, Vol. F1. Berlin-Heidelberg-New York-Tokyo: Springer 1983

7. Schempp, W.: Radar ambiguity functions, nilpotent harmonic analysis, and holomorphic theta series. In: Special Functions: Group Theoretical Aspects and Applications. R.A. Askey, T.H. Koornwinder, and W. Schempp, editors. MIA Series. Dordrecht-Boston-London: D. Reidel (to appear)

8. Schempp, W.: Radar ambiguity functions and the linear Schrödinger representation. In: Functional Analysis and Approximation. P.L. Butzer, B.Sz.-Nagy, and R.L. Stens, editors. ISNM Series. Basel-Boston-Stuttgart: Birkhäuser (to appear)

9. Schempp, W.: Radar ambiguity functions, the Heisenberg group, and holomorphic theta series (to appear)

10. Schempp, W.: Positive-definite functions on nilpotent Lie groups with linear coadjoint orbits (to appear)

11. Wigner, E.P.: Quantum-mechanical distribution functions revisited. In: Perspectives in Quantum Theory, pp. 25-36. W. Yourgrau, and A. van der Merwe, editors. New York: Dover Publications 1979

12. Woodward, P.M: Probability and Information Theory with Applications to Radar. 2nd edition. New York: McGraw-Hill 1965

Walter Schempp, Lehrstuhl für Mathematik I, Universität Siegen, Hölderlinstraße 3, D-5900 Siegen, West Germany

International Series of
Numerical Mathematics, Vol. 71

INEQUALITIES AND MATHEMATICAL PROGRAMMING, II

Chung-lie Wang

Abstract. This paper is a continuation of the paper "Inequalities and Mathematical Programming" which was presented at the "Third International Conference on General Inequalities" in 1981. Several problems including inventory models in economics are studied. Examples are also given to demonstrate the concept of a transition constraint.

1. INTRODUCTION

As has been done in [8], inequalities are again used to establish mathematical programming problems in a unified and simple manner. In this connection, we treat several problems including inventory models in economics mainly by means of the usual A-G (arithmetic-geometric mean) inequality. Further, after the introduction of the concept of a transition constraint, we demonstrate its novelty by three simple examples. For the classical inequalities cited here without mentioning any source, one should refer to Beckenbach and Bellman [1], Hardy, Littlewood and Pólya [3], or Mitrinović [5].

2. SIMPLE OPTIMAL PROBLEMS

We use the A-G inequality (see, e.g., [1,3,5]) to solve several nonlinear optimal problems. These problems are not only models for many problems in calculus but also models for certain problems dealing with costs and profits in economics [2]. In the following, we use opt to designate max or min, and $\overline{\text{opt}} = \min$ if opt $= \max$, etc. The letter x denotes a variable while the other letters denote parameters (or arbitrary constants).

PROBLEM 1. Let

(1) $$f(x) = (ax^2+bx+c)/(px+q)$$

be a rational function. Consider the problem

$$\text{opt } f(x)$$

subject to

(2) $$px+q,\ p,\ aA > 0,\ (a \neq 0),$$

where

$$A = c + \frac{aq^2}{p^2} - \frac{bq}{p}.$$

Since

(3) $$f(x) = \frac{b}{p} - 2\frac{aq}{p^2} + \frac{a}{p^2}(px+q) + \frac{A}{px+q},$$

a use of the A-G inequality on the last two terms of (3) yields that

$$f(x) \geq B + C, \text{ if } a,\ A > 0$$

or

$$f(x) \leq B - C, \text{ if } a,\ A < 0;$$

where

$$B = \frac{b}{p} - 2\frac{aq}{p^2}$$

and

$$C = \frac{2a}{p}\left[\frac{c}{a} + \left(\frac{q}{p}\right)^2 - \frac{bq}{ap}\right]^{\frac{1}{2}}.$$

Hence

$$\min f(x) = B + C, \text{ if } a,\ A > 0$$

or

$$\max f(x) = B - C, \text{ if } a, A < 0.$$

Either the minimum value or the maximum value is attained at

$$\frac{a}{p^2}(px+q) = A/(px+q)$$

or

(4) $$x = -\frac{q}{p} + [\frac{c}{a} + (\frac{q}{p})^2 - \frac{bq}{ap}]^{\frac{1}{2}}.$$

PROBLEM 2. Let $g(x) = 1/f(x)$, where $f(x)$ is given in (1). Then consider the problem opt $g(x)$ subject to (2).

PROBLEM 3. Let $h(x) = (x^2+ux+v)/(ax^2+bx+c)$ with $u = p+(b/a)$, $v = q+(c/a)$. Then consider the problem opt $h(x)$ subject to (2).

The solutions of problems 2 and 3 are easy consequences of that of problem 1:

$$\text{opt } g(x) = 1/\overline{\text{opt}}\ f(x)$$

and

$$\text{opt } h(x) = (1/a) + [1/\overline{\text{opt}}\ f(x)],$$

where the optimal value is attained at x given in (4).

PROBLEM 4. Let

(5) $$k(x) = (ax^2+bx+c)^{\frac{1}{2}} - dx.$$

Then consider the problem

$$\text{opt } k(x)$$

subject to

$$ax^2+bx+c \geq 0, \ PR > 0 \ (a \neq 0),$$

where

$$P = a - d^2 \text{ and } R = 4ac - b^2.$$

In solving problem 4, we present two different methods to establish the following:

(6) $$\min k(x) = [bd + (PR)^{\frac{1}{2}}]/2a$$

is attained at $x = [-b + d(R/P)^{\frac{1}{2}}]/2a$, if $P, R > 0$ or

(7) $$\max k(x) = [bd - (PR)^{\frac{1}{2}}]/2a$$

is attained at $x = [-b - d(R/P)^{\frac{1}{2}}]/2a$, if $P, R < 0$.

I. Assuming that there exists an $\hat{x}$ such that

(8) $$\text{opt } k(x) = k(\hat{x}) = k,$$

(5) and (8) yield

(9) $$P\hat{x}^2 + (b-2dk)\hat{x} + c - k^2 = 0.$$

The nonnegativity of the discriminant of the quadratic equation (9) is equivalent to

(10) $$4ak^2 - 4bdk + b^2 - 4cP \geq 0.$$

The solutions of (10) are

(11) $$k \leq [bd - (PR)^{\frac{1}{2}}]/2a$$

or

(12) $$k \geq [bd + (PR)^{\frac{1}{2}}]/2a.$$

From (9), (11) and (12), it follows

(13) $$\hat{x} = (-b + 2d(k)/2P$$

$$= [-b + d(R/P)^{\frac{1}{2}}]/2a, \text{ if } P, R > 0,$$

or

$$= [-b - d(R/P)^{\frac{1}{2}}]/2a, \text{ if } P, R > 0.$$

It is now clear that (6) or (7) follows from (8) and (13).

II. To establish (6) ($P, R > 0$), we rewrite (5) as follows

$$k(x) = a^{\frac{1}{2}}[(x + \frac{b}{2a})^2 + (\frac{R^{\frac{1}{2}}}{2a})^2]^{\frac{1}{2}} - dx. \tag{14}$$

Substituting

$$x = (R^{\frac{1}{2}}\tan y - b)/2a \tag{15}$$

into (14), we have

$$k(x) = \frac{bd}{2a} + \frac{R^{\frac{1}{2}}}{2a}[(a^{\frac{1}{2}} - d \sin y)^2/\cos^2 y]^{\frac{1}{2}}$$

$$= \frac{bd}{2a} + \frac{R^{\frac{1}{2}}}{2a}[P + (a^{\frac{1}{2}}\sin y - d)^2/\cos^2 y]^{\frac{1}{2}}$$

$$\geq [bd + (PR)^{\frac{1}{2}}]/2a.$$

$\min k(x) = [bd + (PR)^{\frac{1}{2}}]/2a$ is attained at

$$\sin y = d/a^{\frac{1}{2}}. \tag{16}$$

Noting that (15) and (16) yield (13), we establish (6).

To establish (7) ($P, R < 0$), we substitute

$$x = [(-R)^{\frac{1}{2}}\sec y - b]/2a$$

into (14) and have

$$k(x) = \frac{bd}{2a} - \frac{(-R)^{\frac{1}{2}}}{2a}[(d - a^{\frac{1}{2}}\sin y)^2/\cos^2 y]^{\frac{1}{2}}$$

$$= \frac{bd}{2a} - \frac{(-R)^{\frac{1}{2}}}{2a}[-P + (d \sin y - a^{\frac{1}{2}})^2/\cos^2 y]^{\frac{1}{2}}$$

$$\leq [bd - (PR)^{\frac{1}{2}}]/2a.$$

$\max k(x) = [bd - (PR)^{\frac{1}{2}}]/2a$ is attained at $\sin y = a^{\frac{1}{2}}/d$. We now establish (7) immediately.

3. INVENTORY MODELS

We apply further the A-G inequality (or/and the Jensen inequality) to treat two problems concerning the economic order quantity in inventory models as follows.

PROBLEM 5. (see [6,p.613; 7,p.327]) Consider the cost function

$$C = C(Q,S) = a(d/Q) + h(S^2/2Q) + b[(Q-S)^2/2Q]$$

where Q and S are variables; the other letters are constants. Determine min C.

Since (by the Jensen and A-G inequalities)

$$C = \frac{ad}{Q} + \frac{1}{2}\{\frac{b}{h+b}([\frac{h(h+b)}{b}]^{\frac{1}{2}}SQ^{-\frac{1}{2}})^2 + \frac{h}{h+b}([\frac{b(h+b)}{h}]^{\frac{1}{2}}(Q^{\frac{1}{2}}-SQ^{-\frac{1}{2}}))^2\}$$

$$\geq \frac{ad}{Q} + \frac{1}{2}\{\frac{b}{h+b}[\frac{h(h+b)}{b}]^{\frac{1}{2}}SQ^{-\frac{1}{2}} + \frac{h}{h+b}[\frac{b(h+b)}{h}]^{\frac{1}{2}}(Q^{\frac{1}{2}}-SQ^{-\frac{1}{2}})\}^2$$

$$= \frac{ad}{Q} + \frac{hbQ}{2(h+b)} \geq [2adhb/(h+b)]^{\frac{1}{2}}$$

min $C = [2adhb/(h+b)]^{\frac{1}{2}}$ is attained at

$$[\frac{h(h+b)}{b}]^{\frac{1}{2}}SQ^{-\frac{1}{2}} = [\frac{b(h+b)}{h}]^{\frac{1}{2}}(Q^{\frac{1}{2}}-SQ^{-\frac{1}{2}}) \text{ and } \frac{2ad}{Q} = \frac{hbQ}{h+b}$$

or

$$Q = [\frac{2ad}{h}(\frac{h+b}{b})]^{\frac{1}{2}} \text{ and } S = [\frac{2ad}{h}(\frac{b}{h+b})]^{\frac{1}{2}}.$$

PROBLEM 6. (see [4,p.44]) Consider the cost function

$$K = K(Q,S) = \frac{\lambda}{Q}A + \frac{1}{2Q}IC(Q-S)^2 + \frac{1}{Q}(\pi\lambda S + \frac{1}{2}\hat{\pi}S^2)$$

where Q and S are variables and the other letters are constants. Determine min K.

Since

$$K(Q,S) = \frac{1}{2Q}[(\hat{\pi}+IC)S^2 + 2\pi\lambda S + 2\lambda A] + \frac{IC}{2}Q - ICS. \tag{17}$$

a use of the A-G inequality on the first two terms of (17) yields

$$K(Q,S) \geq \hat{K}(S) \tag{18}$$

where

(19) $$\hat{K}(S) = [IC(\hat{\pi}+IC)S^2 + 2IC\pi\lambda S + 2\lambda AIC]^{\frac{1}{2}} - ICS$$

the sign of equality holding in (18) if and only if

(20) $$Q^2 = [(\hat{\pi}+IC)S^2 + 2\pi\lambda S + 2\lambda A]/IC.$$

Furthermore, determining min $\hat{K}(S)$, we compare (19) with (5) and see that

$$P = \hat{\pi}IC, \quad R = 4(IC)^2[2\lambda A(\hat{\pi}+IC) - (\pi\lambda)^2];$$

$$a = IC(\hat{\pi}+IC), \quad b = 2IC\pi\lambda, \quad d = IC.$$

Substituting the above into (6), we obtain that

(21) $$\min \hat{K}(S) = (\hat{\pi}+IC)^{-1}(IC\pi\lambda + [\hat{\pi}IC\{2\lambda A(\hat{\pi}+IC) - (\pi\lambda)^2\}]^{\frac{1}{2}})$$

is attained by

(22) $$S = (\hat{\pi}+IC)^{-1}\{-\pi\lambda + [2\lambda AIC(1 + \frac{IC}{\hat{\pi}}) - \frac{IC}{\hat{\pi}}(\pi\lambda)^2]^{\frac{1}{2}}\}, \quad \hat{\pi} \neq 0$$

From (18), (20), (21) and (22), we conclude that min K = min $\hat{K}(S)$ is attained at S given in (22) and

$$Q = (\hat{\pi}IC)^{-\frac{1}{2}}[2\lambda\Lambda(\hat{\pi}+IC) - (\pi\lambda)^2]^{\frac{1}{2}}, \quad \hat{\pi} \neq 0$$

which is derived from (20) and (22).

REMARK. We can also solve problem 6 by eliminating first S and then Q as was done in problem 5. This indicates that our method is more flexible then the usual one adopted in [4,6,7].

4. TRANSITION CONSTRAINT

The concept of a transition constraint for mathematical programming problems is a natural one. In fact, it plays a role similar to that of an auxiliary line in solving problems in geometry or an auxiliary function in proving mean value theorems in analysis. Let us give a formal definition (e.g., see [10,11]) as follows.

DEFINITION. A transition constraint of a mathematical programming problem is an additional constraint consistent with the original constraint (or constraints), designed to facilitate solving the problem.

As the definition indicated, we may introduce at liberty a constraint (or constraints) for a mathematical programming problem without constraints so as to suit the problem. In order to demonstrate its general usage, three simple examples are in order.

EXAMPLE 1. Consider the problem

$$\max \min\left(x, \frac{y}{1+y}\right)$$

subject to

$$x + y = a, \ a \geq 0,$$

$$x, \ y \geq 0.$$

Let

(23) $$x + \frac{y}{1+y} = b, \ b \geq 0$$

be a transition constraint. Since

$$\min\left(x, \frac{y}{1+y}\right) \leq \frac{1}{2}\left(x + \frac{y}{1+y}\right) = \frac{b}{2},$$

$\max \min\left(x, \frac{y}{1+y}\right) = \frac{b}{2}$ is attained at

$$x = y/(1+y) = b/2$$

or

(24) $$x = b/2 \text{ and } y = b/(2-b).$$

Substituting (24) into (23), a simple manipulation yields

(25) $$b^2 - 2(a+2)b + 4a = 0.$$

Solving (25) we obtain

$$b = a + 2 - (a^2+4)^{\frac{1}{2}}, \quad (0 \leq b < 2).$$

Hence

$$\max \min(x, \frac{x}{1+y}) = [a+2-(a^2+4)^{\frac{1}{2}}]/2$$

is attained at

$$x = [a+2-(a^2+4)^{\frac{1}{2}}]/2 \text{ and } y = [a-2+(a^2+4)^{\frac{1}{2}}]/2.$$

EXAMPLE 2. Consider the problem

$$\max[x_1 + \min(x_2, x_3 + x_4)]$$

subject to

$$\max(x_1, x_2 + \max(x_3, x_4)) = a, \quad a \geq 0.$$

$$x_1, x_2, x_3, x_4 \geq 0.$$

Let

$$\max(x_3, x_4) = b, \quad b \geq 0$$

and

$$x_2 + b = c, \quad c \geq 0$$

be two transition constraints. Then

$$x_3 + x_4 = 2\max(x_3, x_4) = 2b$$

is attained at

(26) $$x_3 = x_4 = b,$$

while

$$\min(x_2, x_3+x_4) = \min(x_2, 2b)$$

$$= \frac{2}{3}x_2 + \frac{1}{3}(2b) = \frac{2}{3}c$$

is attained at

(27) $$x_2 = 2b = \frac{2}{3}c.$$

Now, the original problem becomes a simpler problem as follows

$$\max(x_1 + \frac{2}{3}c)$$

subject to

$$\max(x_1, c) = a, \ a \geq 0,$$

$$x_1, \ c \geq 0.$$

Since

$$x_1 + \frac{2}{3}c = \frac{5}{3}(\frac{3}{5}x_1 + \frac{2}{5}c) \leq \frac{5}{3}\max(x_1, c) = \frac{5}{3}a,$$

$\max(x_1 + \frac{2}{3}c) = \frac{5}{3}a$ is attained at

(28) $$x_1 = c = a.$$

Finally, from (26), (27) and (28), we obtain that

$$\max[x_1 + \min(x_2, x_3+x_4)] = \frac{5}{3}a$$

is attained at

$$x_1 = a, \ x_2 = \frac{2}{3}a, \ x_3 = x_4 = \frac{1}{3}a.$$

NOTE. S. Iwamoto established examples 1 and 2 by using the dynamic programming approach (in the preprint of his work in Japanese).

EXAMPLE 3. Determine max F(x), where
$F(x) = (px+q)^{1/3} - rx$ with $pr > 0$, $(px+q \geq 0)$.

Let $px+q = c$ be a transition constraint $(c \geq 0)$. Since (by the A-G inequality)

$$F(x) = \hat{F}(c) = \frac{qr}{p} + c^{1/3}(1-\frac{r}{p}c^{2/3})$$

$$= \frac{qr}{p} + (\frac{p}{2r})^{1/2}[(\frac{2r}{p}c^{2/3})^{1/3}(1-\frac{r}{p}c^{2/3})^{2/3}]^{3/2}$$

$$\leq \frac{qr}{p} + (\frac{p}{2r})^{1/2}[\frac{1}{3}(\frac{2r}{p}c^{2/3}) + \frac{2}{3}(1-\frac{r}{p}c^{2/3})]^{3/2}$$

$$= \frac{qr}{p} + \frac{2}{3}(\frac{r}{3p})^{1/2},$$

$\max F(x) = \max \hat{F}(c) = \frac{qr}{p} + \frac{2}{3}(\frac{r}{3p})^{1/2}$ is attained at

$$\frac{2r}{p}c^{2/3} = 1-\frac{r}{p}c^{2/3}$$

or

$$c = \frac{p}{3r}(\frac{p}{3r})^{1/2} \text{ with } x = (c-q)/p.$$

Hence

$$\max F(x) = \frac{qr}{p} + \frac{2}{3}(\frac{r}{3p})^{1/2}$$

is attained at

$$x = \frac{1}{3r}(\frac{p}{3r})^{1/2} - \frac{q}{p}.$$

5. CONCLUSION

The objective functions in problems 1-4 and examples 3 are differentiable functions in one variable, while the objective functions in problems 5 and 6 are differentiable functions in two variables. In all the cases, the calculus method can also be used to solve them as was done in [4,6,7]. However, the objective functions in examples 1 and 2 are non-differentiable functions (as are most of the practical problems in the mathematical programming setting (e.g. see,

[6,7]). For the latter case, the usual calculus method cannot be directly applied to establish solutions.

In view of the results given here and in [8-11] we have demonstrated that basic inequalities can be used to establish mathematical programming problems in a systematic and simple manner. Moreover, using the concept of a transition constraint, a further development of the idea for solving mathematical programming problems by suitable inequalities appears to be very promising.

The research of this paper has been supported in part by the NSERC of Canada (Grant No. A4091) and the President's NSERC Funds of the University of Regina.

REFERENCES

1. E.F. Beckenbach and R. Bellman, Inequalities. Springer-Verlag, Berlin, 2nd rev. ed., 1965.

2. W.R. Derrick and J.L. Derrick, Finite Mathematics with Calculus for the Management, Life, and Social Sciences. Addison-Wesley, Reading, Massachusetts and London, 1979.

3. G.H. Hardy, J.E. Littlewood, and G. Pólya, Inequalities. Cambridge Univ. Press, Cambridge, 2nd ed., 1952.

4. G. Hadley and T.M. Whitin, Analysis of Inventory System. Prentice-Hall, Englewood Cliffs, N.J., 1963.

5. D.S. Mitrinović, Analytic Inequalities. Springer-Verlag, Berlin, 1970.

6. H. Moskowitz and G.P. Wright, Operations Research Techniques for Management. Prentice-Hall, Englewood Cliffs, N.J., 1979.

7. D.T. Phillips, A. Ravindran and J.J. Solberg, Operations Research: Principles and Practice. John Wiley, New York and London, 1976.

8. C.-L. Wang, Inequalities and mathematical programming. In: E.F. Beckenbach and W. Walter (eds.), General Inequalities 3 (Proceedings of the Third International Conference on General Inequalities, Oberwolfach), 149-164. Birkhäuser Verlag, Basel and Stuttgart, 1983.

9. C.-L. Wang, Introduction to inequalities (in Chinese). Unpublished (138 pages).

10. C.-L. Wang, Functional equation approach to inequalities, V. J. Math. Anal. Appl. Submitted for publication 1980.

11. C.-L. Wang, The van der Waerden inequality and mathematical programming. Submitted for publication.

Chung-lie Wang, Dept. of Math & Stats, University of Regina, Regina, Saskatchewan, Canada S4S 0A2

Problems and Remarks

Young Birches

International Series of
Numerical Mathematics, Vol. 71
© 1984 Birkhäuser Verlag Basel

ON SOME METRICS INDUCED BY COPULAS

Claudi Alsina

Let T be a t-norm and also a copula, i.e., T is a binary operation on $[0,1]$, such that $([0,1],T,\leq,0,1)$ is a topologically ordered Abelian semigroup with 1 as a unit and 0 as a null element which in addition satisfies $T(c,a)-T(b,a) \leq c-b$ for all $a,b,c \in [0,1]$, $b \leq c$. Let T^* be the t-conorm associated to T, i.e., $T^*(x,y)$. We have

THEOREM. *If* T *is a* t-*norm and a copula, then the function* d_T: $[0,1]\times[0,1] \to [0,1]$, *defined by* $d_T(x,y) = 0$ *if* $x = y$ *and* $d_T(x,y) = T^*(x,y)-T(x,y)$ *if* $x \neq y$, *is a metric on* $[0,1]$.

Proof. For any $x,y,z \in [0,1]$, $x \neq y \neq z \neq x$, we have

$$T(1-x,1-z) \leq \mathrm{Min}(1-x,1-z) , \tag{1}$$

$$T(x,z) \leq \mathrm{Min}(x,z) , \tag{2}$$

$$T(1-z,1-y) - T(1-x,1-y) \leq \mathrm{Max}(x-z,0) , \tag{3}$$

$$T(z,y)-T(x,y) \leq \mathrm{Max}(z-x,0) . \tag{4}$$

Adding (1), (2), (3) and (4) we immediately obtain

$$T(1-x,1-z) + T(x,z) + T(1-z,1-y) - T(1-x,1-y) + T(z,y) - T(x,y) \leq 1 ,$$

which in turn implies

$$d_T(x,y) \leq d_T(x,z) + d_T(z,y) .$$

The other metric properties are immediate.

Claudi Alsina, Departament de Matemàtiques i Estadîstica (ETSAB), Universitat Politècnica de Barcelona, Diagonal 649, Barcelona 28, Spain

International Series of
Numerical Mathematics, Vol. 71

FROM MID-POINT TO FULL CONVEXITY

Claudi Alsina

Let f be a continuous real function defined on an interval. After applying appropriate transformations, we may assume without loss of generality that 0 is in I and that $f(0) = 0$. If f satisfies the Jensen inequality

(1) $$f\left(\frac{x+y}{2}\right) \le \frac{f(x) + f(y)}{2} \quad \text{for all } x,y \in I ,$$

then, taking $y = 0$ in (1), we obtain $f(x/2) \le f(x)/2$. It follows by induction that, for all natural numbers n and $x_1, x_2, \dots, x_n$ in dom f,

(2) $$f\left(\sum_{i=1}^{n} \frac{x_i}{2^i}\right) \le \sum_{i=1}^{n} \frac{f(x_i)}{2^i} .$$

Given any λ in $[0,1]$, let $\lambda = \sum_{i=1}^{\infty} \lambda_i/2^i$ ($\lambda_i \in \{0,1\}$ for all i) be a binary expansion of λ. In view of (2) and the continuity of f, we have, whenever x and y are in I,

$$f(\lambda x+(1-\lambda)y) = f\left(\sum_{i=1}^{\infty} \frac{\lambda_i x+(1-\lambda_i)y}{2^i}\right) \le \sum_{i=1}^{\infty} \frac{f(\lambda_i x+(1-\lambda_i)y)}{2^i}$$

$$= \sum_{i=1}^{\infty} \frac{\lambda_i f(x)+(1-\lambda_i)f(y)}{2^i} = \lambda f(x) + (1-\lambda)f(y) ,$$

whence f is convex.

REFERENCES

1. R.P. Boas, A Primer of real functions. The Carus Math. Monographs No.13, The Mathematical Association of America (1972).

2. G.H. Hardy, J.E. Littlewood and G. Pólya, Inequalities. Cambridge University Press, Cambridge, 2nd Edition, 1952.

Claudi Alsina, Departament de Matemàtiques i Estadística (ETSAB), Universitat Politècnica, Diagonal 649, Barcelona 28, Spain

International Series of
Numerical Mathematics, Vol. 71

A PROBLEM CONCERNING LATTICES OF SOLUTIONS

Dobiesław Brydak

Consider the iterative functional inequality

$$\Psi(f(x)) \le g(x,\Psi(x)) ,$$

where Ψ is an unknown function, g is an increasing function with respect to the second variable, and the differential inequaltiy

$$D^{+(-)}\Psi(x) \le f(x,\Psi(x)) ,$$

where $D^{+(-)}$ denotes upper Dini's derivatives. The family L of solutions of each of these inequalities is a lattice with respect to the natural operations

$$\Psi_1 \vee \Psi_2(x) = \max(\Psi_1(x),\Psi_2(x)) ,$$

$$\Psi_1 \wedge \Psi_2(x) = \min(\Psi_1(x),\Psi_2(x)) .$$

This is not true for the lower Dini's derivatives.

Consider an isotonic operator $A: X \to X$, where X is a linear lattice.

The family L of solutions of the inequality $Ax \le x$, where $\le$ is the lattice order, is not a sublattice of X even if A is a linear operator.

PROBLEM: How to characterize such operators for which L is a lattice?

Dobiesław Brydak, Institute of Mathematics, Pedagogical University, 30-001 Kraków, Poland

International Series of
Numerical Mathematics, Vol. 71

REMARK ON A THEOREM OF P. VOLKMANN

Bogdan Choczewski

The following version of a theorem due to P. Volkmann [1] has been recently proved by Z. Powązka (Kraków) and then used, among others, for finding conditions under which any solution of the functional inequality

$$\Psi(G(x,y)) \le F(\Psi(x),\Psi(y)) ,$$

defined and continuous in an interval of reals, is a solution of the corresponding functional equation.

THEOREM. *Let* $g: [0,b) \to [0,b)$ *be a continuous, strictly increasing function, satisfying further the conditions* $g(0) = 0$, $g(a) = a$, *for an* $a \in (0,b)$. *If* $f: R \to R$ *is a function with the properties*

(*) $f(x+y) \ge f(x) + f(y)$ *and* $f(g(|x|)) \ge g(|f(x)|)$

for $x,y,x+y \in [-a,a]$ *and* $f(a) = a$,

then necessarily

$$f(x) = x \quad \textit{for} \quad x \in R .$$

Volkmann's theorem corresponds to the case where $b = \infty$, $a = 1$ and inequalities (*) are assumed to hold in all of R. The proof of the theorem, however, does not essentially differ from that supplied by P. Volkmann.

REFERENCE

1. P. Volkmann, Sur une systême d'inéquations fonctionnelles. C.R.Math.Rep.Acad.Sci.Canada, Vol.IV, 3 (1982), 155-158.

Bogdan Choczewski, Institute of Mathematics, University of Mining and Metallurgy, 30-059 Kraków, Poland

International Series of
Numerical Mathematics, Vol. 71
© 1984 Birkhäuser Verlag Basel

A PROBLEM CONCERNING MAJORIZATION

Achim Clausing

For $x > 0$, $t \in \mathbb{R}$ put

$$g_t(x) := \begin{cases} -x^t, & \text{if } t \in (0,1), \\ x^t & \text{otherwise}. \end{cases}$$

The functions g_t are convex, hence the following partial order $\preccurlyeq$ on $(0,\infty)^n$ is weaker than the usual majorization order $\prec$:

$$x \preccurlyeq y : \iff \sum_{i=1}^{n} g_t(x_i) \leq \sum_{i=1}^{n} g_t(y_i) \quad (t \in \mathbb{R}).$$

Give an example where $x \preccurlyeq y$, but not $x \prec y$.

What is the smallest $n \in \mathbb{N}$ for which such an example exists?

In which subset of $(0,\infty)^n$ do the two orders coincide?

(The problem arose in connection with inequalities for L^p-means, c.f. A. Clausing, On Quotients of L^p-means. In: E.F. Beckenbach and W. Walter, General Inequalities 3, pp.43-68. Birkhäuser Verlag, Basel - Boston - Stuttgart, 1983.)

Achim Clausing, Institut für Mathematische Statistik, Westfälische Wilhelms-Universität Münster, D-4400 Münster, West Germany

International Series of
Numerical Mathematics, Vol. 71
© 1984 Birkhäuser Verlag Basel

A PROBLEM CONCERNING THE GAMMA FUNCTION

William N. Everitt

It is known that (Γ is the gamma function)

$$\operatorname{im}\left[\frac{\lambda^2\Gamma(-\lambda)}{\Gamma(\frac{1}{2}-\lambda)}\right] < 0 \qquad \text{for } \lambda = re^{i\pi/3} \text{ and all } r \in (0,\infty) .$$

Is it true that

$$\operatorname{im}\left[\frac{\lambda^2\Gamma(-\frac{1}{2}-\lambda)}{\Gamma(-\lambda)}\right] < 0 \qquad \text{for } \lambda = re^{i\pi/3} \text{ and all } r \in (0,\infty) \text{ ?}$$

These inequalities are connected with Hardy-Littlewood type integral inequalities which are reported on elsewhere in these Proceedings.

W.N. Everitt, Department of Mathematics, University of Birmingham, Birmingham B15 2TT, England

International Series of
Numerical Mathematics, Vol. 71

AN INEQUALITY CONCERNING SUBHARMONIC FUNCTIONS

Werner Haußmann

The following problem arises from the best L^1 approximation of subharmonic functions by harmonic functions on the unit disk.

Let

$$D := \{(x,y) \in \mathbb{R}^2 : \sqrt{x^2+y^2} < 1\}$$

and

$$D_o := \{(x,y) \in \mathbb{R}^2 : \sqrt{x^2+y^2} < \tfrac{1}{2}\sqrt{2}\} .$$

Denote the set of all subharmonic functions on D by

$$S(D) := \{h \in C^2(D) \cap C(\bar{D}) : \Delta h \geq 0 \text{ in } D\} ,$$

and the set of harmonic functions on D by

$$H(D) := \{u \in C^2(D) \cap C(\bar{D}) : \Delta u = 0 \text{ in } D\} ,$$

where Δ is the Laplacian differential operator. Given $h \in S(D) \setminus H(D)$ and $u \in H(D)$ such that

(*) $$(h-u)\big|_{\partial D_o} = 0 ,$$

then the maximum principle implies (if $h \not\equiv u$ on D_o) that

$$h - u < 0 \quad \text{in } D_o .$$

For which $h \in S(D) \setminus H(D)$ and $u \in H(D)$ satisfying (*) does the following inequality hold:

$$h - u > 0 \qquad \text{a.e. in } D \setminus \bar{D}_o \ ?$$

REFERENCE

1. M. Goldstein, W. Haußmann and K. Jetter: Best harmonic L^1 approximation to subharmonic functions. J. London Math. Soc. (to appear).

Werner Haußmann, Department of Mathematics, University of Duisburg, D-4100 Duisburg, West Germany

International Series of
Numerical Mathematics, Vol. 71

A CYCLIC INEQUALITY

Alexander Kovačec

Determine the best constant C in the cyclic inequality

$$C \sum_{i=1}^{n} \sqrt{x_i} \leq \sum_{i=1}^{n} \frac{x_i}{\sqrt{x_i + x_{i+1}}}, \qquad x_i > 0, \quad x_{n+1} = x_1 .$$

In [1] it is shown that the inequality holds with $C = \left(1 - \frac{1}{2\sqrt{2}}\right)$.

During the conference partial results on the inequality were obtained by G. Talenti, Hermann König, C.H. FitzGerald, D.K. Ross and Z. Páles.

REFERENCE

[1] A. Kovačec, Two contributions to inequalities. In: W. Walter (Ed.), General Inequalities 4, pp. 37-46. Birkhäuser-Verlag, Basel.

Alexander Kovačec, Institut für Mathematik, Universität Wien, Strudlhofgasse 4, A-1090 Wien, Austria.

International Series of
Numerical Mathematics, Vol. 71

THE FOMIN CLASSES $\mathcal{F}_p$

E.R. Love

For $p > 1$, G.A. Fomin introduced a class $\mathcal{F}_p$ of null-sequences $\underset{\sim}{a}$ defined by the property that $F_p(\underset{\sim}{a}) < \infty$, where $\Delta a_n = a_n - a_{n+1}$ and

$$F_p(\underset{\sim}{a}) := \sum_{n=1}^{\infty} \left(\frac{1}{n} \sum_{r=n}^{\infty} |\Delta a_r|^p \right)^{1/p}$$

[G.A. Fomin, A class of trigonometric series. Mat. Zametki 23 (1978), 213-222]. Another writer claimed that "the class $\mathcal{F}_p$ is wider when p is closer to 1", which is equivalent to claiming that $F_p(\underset{\sim}{a})$ is an increasing function of p for each null-sequence $\underset{\sim}{a}$ [C.V. Stanojević, Classes of L^1-convergence of Fourier and Fourier-Stieltjes series. Proc. Amer. Math. Soc. 82 (1981), 209-215].

Actually there are null-sequences $\underset{\sim}{a}$ for which $F_p(\underset{\sim}{a})$ is not monotonic on $p > 1$. Let $a_1 = 71$, $a_2 = 11$, $a_3 = a_4 = a_5 = \dots = 0$. Then

$$F_p(\underset{\sim}{a}) = (60^p + 11^p)^{1/p} + \left(\tfrac{1}{2}\right)^{1/p} \cdot 11 , \quad \text{continuous in } p > 0 .$$

$$F_1(\underset{\sim}{a}) = 71 + \frac{1}{2} \cdot 11 = 76\frac{1}{2} ,$$

$$F_2(\underset{\sim}{a}) = 61 + \sqrt{1/2} \cdot 11 < 69 ,$$

$$F_\infty(\underset{\sim}{a}) = 60 + 1 \cdot 11 = 71 .$$

E.R. Love, Department of Mathematics, The University of Melbourne, Parkville, Victoria 3052, Australia

International Series of
Numerical Mathematics, Vol. 71

A MINIMUM PROPERTY OF THE SQUARE

Jürg Rätz

Let n be a positive integer, $n \geq 3$, C the unit circle line in R^2, and z the center of C. Let $a_1, \dots, a_n$ be n pairwise distinct points of C, lying on C in this cyclic order, such that z is an interior point of $A := \text{conv}\,\{a_1, \dots, a_n\}$; put $a_{n+1} = a_1$. Finally, let B be the n-gon circumscribed to C which is obtained as the intersection of the closed supporting half-planes of C in $a_1, \dots, a_n$. When is the area sum $F(A) + F(B)$ minimal as n and $a_1, \dots, a_n$ vary? J. Aczél and L. Fuchs [1] showed that the square and only it realizes the minimum. J. Aczél [2] asked whether the calculus proof in [1] could be replaced by a truly elementary proof not using calculus.

If $x_1, \dots, x_n$ denote the halves of the central angles $\measuredangle\, a_i z a_{i+1}$ we have $0 < x_i < \frac{\pi}{2}$ $(i = 1, \dots, n)$, and $F(A) + F(B) = f(x_1) + \dots + f(x_n)$ where $f(x) := \frac{1}{2} \sin 2x + \tan x$, $0 \leq x < \frac{\pi}{2}$.

The following three steps of the proof given in [1] can be confirmed on the basis of the functional equations for cos and sin, of some elementary limit results and of elementary inequality techniques, but without reference to the mean value theorem:

(I) $f\left(\frac{\pi}{4} + x\right) > f\left(\frac{\pi}{4}\right) + f(x)$ for $0 < x < \frac{\pi}{4}$.

(II) For $\theta := \arccos 2^{-1/4}$, f is strictly concave on $[0, \theta]$ and strictly convex on $[\theta, \frac{\pi}{2}[$.

(III) $5 \cdot f(x) + f(\pi - 5x) > 5 \cdot f\left(\frac{\pi}{5}\right)$ for $\theta \leq x < \frac{\pi}{5}$.

A similarly elementary proof of

(IV) $4 \cdot f(x) + f(\pi - 4x) > 4 \cdot f\left(\frac{\pi}{4}\right)$ for $\frac{\pi}{5} \leq x < \frac{\pi}{4}$

is not known to the author; possibly, the ideas of [3] might be of help. In the case of success, we would have an elementary proof of the minimum problem. An alternative procedure using (I) as well as

(V) $f(x+y) < f(x) + f(y) \quad \left(0 < x < \frac{\pi}{4},\ 0 < y < \frac{\pi}{4},\ x + y \le \frac{\pi}{4}\right)$

"(VI)" $f\left(\frac{\pi}{4}\right) + f\left(x + y - \frac{\pi}{4}\right) < f(x) + f(y) \quad \left(0 < x < \frac{\pi}{4},\ 0 < y < \frac{\pi}{4} < x + y\right)$

would have been nice. But (VI) refuses its service by being false, for the minimum of $f(x) + f(y) - f\left(x + y - \frac{\pi}{4}\right)$ on the domain given in (VI) is approximately 1.49 while $f\left(\frac{\pi}{4}\right) = \frac{3}{2}$.

REFERENCES

1. J. Aczél and L. Fuchs, A minimum problem on areas of inscribed and circumscribed polygons of a circle. Compos. Math. 8 (1950), 61-67.

2. J. Aczél, A minimum property of the square. In: E.F. Beckenbach (ed.), General Inequalities 2 (Proceedings of the Second International Conference on General Inequalities, Oberwolfach), p.467, Birkhäuser Verlag, Basel - Boston - Stuttgart, 1980.

3. A. Clausing, Type t entropy and majorization. SIAM J. Math. Anal. 14 (1983), 203-208.

Jürg Rätz, Universität Bern, Mathematisches Institut, Sidlerstr. 5, CH-3012 Bern, Switzerland

International Series of
Numerical Mathematics, Vol. 71

A CHARACTERIZATION OF DIMENSION THREE

In memoriam Professor Hugo Hadwiger (1908-1981)

Jürg Rätz

For Euclidean n-space $(R^n, \|\cdot\|)$, $n \geq 2$, an arbitrary unit vector e belonging to R^n, and angles α, β satisfying $0 \leq \alpha \leq \beta \leq \pi$, we consider the following convex sets: The unit ball $U_n := \{x \in R^n: \|x\| \leq 1\}$, the spherical segment (Kugelschicht) $H_n[\alpha,\beta] := \{x \in U_n: \cos\beta \leq \langle x,e\rangle \leq \cos\alpha\}$, the Archimedean cylinder $A_n := \{x'+\lambda e: x' \in H_n[\frac{\pi}{2},\frac{\pi}{2}], |\lambda| \leq 1\}$, and the cylindrical segment (Zylinderschicht) $Z_n[\alpha,\beta] := \{x \in A_n: \cos\beta \leq \langle x,e\rangle \leq \cos\alpha\}$. Furthermore, F_n, V_n, M_n denote the n-dimensional surface area, Peano-Jordan content, lateral area (Mantelfläche) functionals, respectively (cf. [1; p. 206 ff]). By evaluation of the Minkowski "Quermassintegral" (cf. [1; p 209 ff]), the following statements can be proved (cf. [2]):

a) The sequences with the terms $V_n(U_n)$, $F_n(U_n)$, $V_n(A_n)$, $F_n(A_n)$ strictly increase to their respective maximum values $V_5(U_5)$, $F_7(U_7)$, $V_6(A_6)$, $F_8(A_8)$ and then strictly decrease to zero.

b) The isoperimetric quotients $F_n^n(U_n)/V_n^{n-1}(U_n)$ and $F_n^n(A_n)/V_n^{n-1}(A_n)$ strictly increase to $+\infty$ as n tends to infinity.

c) For all $0 \leq \alpha < \beta \leq \pi$, the Archimedean quotients $a_n[\alpha,\beta] := M_n(H_n[\alpha,\beta])/M_n(Z_n[\alpha,\beta])$ strictly decrease to zero, and $a_3[\alpha,\beta] = 1$.

As a corollary of c) we get $M_n(H_n[\alpha,\beta]) = M_n(Z_n[\alpha,\beta])$ if and only if $n = 3$. For a related result cf. [3].

REFERENCES

1. H. Hadwiger, Vorlesungen über Inhalt, Oberfläche und Isoperimetrie. Springer-Verlag, Berlin - Göttingen - Heidelberg, 1957.

2. J. Rätz, Über Inhalt und Oberfläche von Kugel und Zylinder. Math.-phys. Semesterber. 15 (1968), 88-93.

3. W. Rudin, A generalization of a theorem of Archimedes. Amer. Math. Monthly 80 (1973), 794-796.

H. Hadwiger's text [1] beautifully exhibits the power and strength of inequalities in many respects.

Jürg Rätz, Universität Bern, Mathematisches Institut, Sidlerstr.5, CH-3012 Bern, Switzerland

International Series of
Numerical Mathematics, Vol. 71

THE UNIT BALL OF NORMED VECTOR SPACES AS AN INTERSECTION OF HALF-SPACES

Jürg Rätz

Let $(X, \|\cdot\|)$ be a real normed space, $X \neq \{0\}$, $U := \{x \in X: \|x\| \leq 1\}$ the unit ball and $S := \{x \in X: \|x\| = 1\}$ the unit sphere of X. Let $h(\|\cdot\|)$ denote the minimum cardinal number of families M of closed half-spaces of X such that $U = \bigcap M$. If T stands for the norm topology on X, we denote the subspace topology of T on S by $T|S$. For the density character $d(T) := \min \{\text{card}\, A: A \subset X, \text{cl}\, A = X\}$ we have $d(T|S) = 2$ for $\dim_{\mathbf{R}} X = 1$ and $d(T|S) = d(T)$ for $\dim_{\mathbf{R}} X \geq 2$. [Notice that in topological spaces, even in separated topological vector spaces, a subspace may have a density character larger than that of the whole space (cf. [2]); but in pseudometrizable spaces this phenomenon cannot occur.] It was shown in [1] that

$$h(\|\cdot\|) \leq d(T|S),$$

with equality whenever X is (i) locally uniformly convex, or (ii) separable and strictly convex. For the special case $X = \mathbf{R}^n$ cf. [3]. In an unpublished note in 1980, R. Phelps proved that the norm $\|\cdot\|$ on l^∞ given by $\|x\| := \sup |x_n| + (\sum 2^{-n} x_n^2)^{1/2}$ is strictly convex and satisfies $h(\|\cdot\|) < d(T|S)$ which shows that separability in (ii) cannot be dropped without being replaced by some other condition.

REFERENCES

1. H. Briggen and J. Rätz, Die Einheitskugel normierter Vektorräume als Durchschnitt von Stützhalbräumen. Arch. Math. (Basel) 27 (1976), 636-639.

2. G. Henriques, Ein nicht d-separabler linearer Unterraum eines d-separablen tonnelierten Raumes. Arch. Math. (Basel) 15 (1964), 448-449.

3. A. Kirsch, Konvexe Figuren als Durchschnitte abzählbar vieler Halbräume. Arch. Math. (Basel) 18 (1967), 313-319.

Jürg Rätz, Universität Bern, Mathematisches Institut, Sidlerstr. 5, CH-3012 Bern, Switzerland

International Series of
Numerical Mathematics, Vol. 71

A NUMBER-THEORETIC PROBLEM

Dieter K. Ross

The four numbers 1, 3, 8, 120 are such that the product of any two, increased by 1, is a perfect square.

At Oberwolfach in March 1968, Professor J.H. van Lint asked the question: "Is there any other number N that can replace the 120 ?". Clearly any such N may be written in the form x^2-1 where x is a positive integer, and the only nontrivial case corresponds to a value $x > 11$. The problem may be stated in terms of the simultaneous Diophantine equations

$$3x^2 - 2 = y^2,$$

$$8x^2 - 1 = z^2.$$

In 1969 A. Baker and H. Davenport showed that no solution with $x > 11$ existed. Is it possible to find an elementary proof of this result?

As a first approach to this problem the following equivalent theorem appears to be useful.

THEOREM. _If_ $\{A_n\}$ _and_ $\{B_n\}$ _are two integer sequences defined by the recurrence relations_

$$A_{n+1} = 4A_n - A_{n-1} \quad \text{with } A_o = 1, \quad A_1 = 6,$$

$$B_{n+1} = 6B_n - B_{n-1} \quad \text{with } B_o = 1, \quad B_1 = 4$$

for $n = 1,2,3,\ldots$, _then there exists an integer_ $N > 120$ _such that four numbers_ 1, 3, 8, N _have the property that the product of any two increased by_ 1 _is a perfect square iff_ $A_i = B_j$ _for some integer_ $i,j \geq 3$.

Prove that $A_i \neq B_j$ unless $i = j = 0$ or 2.

CONJECTURE. If $\{A_n\}$ and $\{B_n\}$ are two integer sequences defined by the recurrence relations

$$A_{n+1} = a_1 A_n - a_2 A_{n-1} \quad \text{with } A_o = 1 , \; A_1 > 0$$

and

$$B_{n+1} = b_1 B_n - b_2 B_{n-1} \quad \text{with } B_o = 1 , \; B_1 > 0$$

for $n = 1, 2, 3, \ldots$ and where a_1, a_2, b_1, b_2 are nonnegative integers and if, in addition, three distinct points of the sequences $\{A_n\}$ and $\{B_n\}$ coincide, then infinitely many coincide.

This result is a discrete analogy of a Sturm type interlacing of zeroes theorem that arises in the study of comparison theorems for ordinary second order linear differential equations.

REFERENCE

1. A. Baker and H. Davenport, The equations $3x^2 - 2 = y^2$ and $8x^2 - 7 = z^2$. Quart. J. Math. Oxford Ser. (2) 20 (1969), 129-137.

Dieter K. Ross, Applied Mathematics Department, La Trobe University, Victoria, 3083, Australia

International Series of
Numerical Mathematics, Vol. 71

A PROBLEM CONCERNING THE GAMMA FUNCTION

Dennis C. Russell

In a recent paper by A. Jakimowski and D.C. Russell [1], we required the following

LEMMA 1. Let $G(z) := z\left(\frac{d}{dz}\right)^2 \log \Gamma(z)$, $z \neq 0,-1,-2,-3,\ldots$. If $z = u + it$, $u \geq 0$, $t \geq 0$, $z \neq 0$, then $\operatorname{Im} G(z) \leq 0$.

This was proved by making use of two different representations of $G(z)$ (one a series valid for $z \neq 0,-1,-2,\ldots$, the other an integral valid for $\operatorname{Re} z > 0$), showing that $\operatorname{Im} G(z) \leq 0$ on the boundary of a closed region $\Delta = \Delta(r,R)$ in the positive quadrant (which tended to the domain in question as $r \to 0+$, $R \to +\infty$) together with the fact that $\operatorname{Im} G(z)$ is harmonic in Δ and hence $\operatorname{Im} G(z) \leq 0$ throughout Δ. Although not required for the work in question, it was clear by continuity, since

$$\operatorname{Im} G(it) = -\frac{1}{2t}\left[1 + \left(\frac{\pi t}{\sinh \pi t}\right)^2\right] < 0 \quad \text{for } t > 0,$$

that the inequality $\operatorname{Im} G(z) < 0$ persists for some values of z in the second quadrant. A tentative conjecture was that $\operatorname{Im} z > 0 \Longrightarrow \operatorname{Im} G(z) < 0$, but W.N. Everitt, C. FitzGerald and F. Huckemann all pointed out that this was untenable due to the double poles of $G(t)$ at $z = -1,-2,-3,\ldots$.

Nevertheless, the question still arises as to the exact form of the inverse image of the set $T_1 := \{\operatorname{Im} G(z) < 0\}$ under the mapping $w = G(z)$.

A related question arises from

LEMMA 2. *If* $g(z) = \left(\frac{d}{dz}\right)^2 \log \Gamma(z)$, *then* $\mathrm{Re}\, z \geq 1 \implies \mathrm{Re}\, g(z) > 0$.

Here we would like to know the inverse image of $T_2 := \{\mathrm{Re}\, g(z) > 0\}$ under the mapping $w = g(z)$.

REFERENCE

1. A. Jakimowski and D.C. Russell, Mercerian theorems involving Cesàro means of positive order, Mohnatsh. Math. 96 (1983), 119-131.

Dennis C. Russell, York University, Department of Mathematics, 4700 Keele Street, Downsview, Ontario M3J 1P3, Canada

International Series of
Numerical Mathematics, Vol. 71

A PROBLEM CONCERNING SPECIAL FUNCTIONS

Abe Sklar

In 1963 Schweizer and Sklar [J. Math. Phys. 42] published the following Minkowski-type inequalities for the confluent hypergeometric function ${}_1F_1$: Let s_1, s_2, s_3 be nonnegative numbers and σ_1, σ_2, σ_3 positive numbers such that $2\max(s_1,s_2,s_3) \le s_1 + s_2 + s_3$ and $2\max(\sigma_1^2,\sigma_2^2,\sigma_3^2) \le \sigma_1^2 + \sigma_2^2 + \sigma_3^2$. Let n be a positive integer. Then for $\beta \ge 1$

$$\sigma_3\left[{}_1F_1\left(-\frac{\beta}{2};\frac{n}{2};-\frac{s_3^2}{2\sigma_3^2}\right)\right]^{1/\beta} \le \sigma_1\left[{}_1F_1\left(-\frac{\beta}{2};\frac{n}{2};-\frac{s_1^2}{2\sigma_1^2}\right)\right]^{1/\beta} + \sigma_2\left[{}_1F_1\left(-\frac{\beta}{2};\frac{n}{2};-\frac{s_2^2}{2\sigma_2^2}\right)\right]^{1/\beta} ,$$

while for $0 < \beta \le 1$

$$\sigma_3^\beta\,{}_1F_1\left(-\frac{\beta}{2};\frac{n}{2};-\frac{s_3^2}{2\sigma_3^2}\right) \le \sigma_1^\beta\,{}_1F_1\left(-\frac{\beta}{2};\frac{n}{2};-\frac{s_1^2}{2\sigma_1^2}\right) + \sigma_2^\beta\,{}_1F_1\left(-\frac{\beta}{2};\frac{n}{2};-\frac{s_2^2}{2\sigma_2^2}\right) .$$

These inequalities were derived from some fairly deep results in the theory of probabilistic metric spaces; see § 10.7 of the book [1]. It would be worthwhile to have simpler and, if possible, purely analytic proofs, but to date such exist only for the cases $n = \beta = 1$ and $\beta \in [1,2]$, $n \ge 4-\beta$.

It is an open problem to find reasonably simple analytic proofs for all real nonnegative β and n, and another to find similar inequalities involving other special functions.

REFERENCE

1. B. Schweizer and A. Sklar, Probabilistic Metric Spaces. Elsevier North Holland, New York, 1982.

Abe Sklar, Department of Mathematics, Illinois Institute of Technology, Chicago, Illinois 60616, USA

INDEX